Modern API Development with Spring 6 and Spring Boot 3

Design scalable, viable, and reactive APIs with REST, gRPC, and GraphQL using Java 17 and Spring Boot 3

Sourabh Sharma

BIRMINGHAM—MUMBAI

Modern API Development with Spring 6 and Spring Boot 3

Group Product Manager: Rohit Rajkumar
Publishing Product Manager: Bhavya Rao
Senior Editor: Divya Anne Selvaraj
Technical Editor: Simran Haresh Udasi
Copy Editor: Safis Editing
Book Project Manager: Sonam Pandey
Proofreader: Safis Editing
Indexer: Subalakshmi Govindhan
Production Designer: Aparna Bhagat
Marketing Coordinator: Nivedita Pandey and Namita Velgekar

First published: June 2021

Second edition: September 2023

Production reference: 1250823

Published by Packt Publishing Ltd.
Grosvenor House
11 St Paul's Square
Birmingham
B3 1RB, UK

ISBN 978-1-80461-327-6

www.packtpub.com

To my adored wife, Vanaja, and son, Sanmaya, for their unquestioning faith, support, and love.

To my parents, Mrs. Asha and Mr. Ramswaroop, for their blessings.

– Sourabh Sharma

Contributors

About the author

Sourabh Sharma works at Oracle as a lead technical member, where he is responsible for developing and designing the key components of the blueprint solutions. He was a key member of the team that designed the architecture being used by various Oracle products. He has over 20 years of experience in delivering enterprise products and applications for leading companies. His expertise lies in conceptualizing, modeling, designing, and developing N-tier and cloud-based applications, as well as leading teams. He has vast experience in developing microservice-based solutions and implementing various types of workflow and orchestration engines. He also believes in continuous learning and sharing knowledge through his books and training.

I would like to thank Divya Anne Selvaraj for her hard work and critical review feedback, Sonam Pandey, Chayan Majumdar, Mark D'Souza, Manthan Patel, and the entire Packt team for their support. I would also like to thank Bhavya Rao and Packt Publishing for providing me with the opportunity to publish this book. I am also thankful for the in-depth technical insight provided by Eric Pirard, who was one of the technical reviewers of this book.

About the reviewers

Eric Pirard has been a Java developer for a few years now. Eric is interested in new technologies that help developers in their day-to-day jobs to satisfy customers' requirements as quickly as possible. Eric is always interested in working on projects that find a way to use technology to solve customer problems quickly. He also likes to help his friends and colleagues solve problems or progress in their projects whenever possible.

Eric believes there are many things one can do in life in addition to having an exciting job. With this philosophy, he has more time to spend with family and friends, play sports, travel, and, in short, enjoy life.

Harsh Mishra is a software developer with nine years of experience in the industry. His expertise spans different technical stacks, with a specialization in financial web-based applications. Harsh is proficient in utilizing Spring Boot, a popular Java framework, to develop robust and scalable web applications for the finance industry. He also possesses extensive knowledge of technologies such as frontend, Google Cloud Platform (GCP), CockroachDB, and APIs, and is skilled in handling large-scale datasets using Kafka Streaming, BigQuery, and Apache Beam technology.

Throughout his career, Harsh has worked for several giants in the industry, including Publicis Sapient, TCS, GE Capital, Ericsson, Goldman Sachs, and Lloyds Bank.

This is the second book that Harsh has been involved in, having previously worked on *Learning Spring Boot 3.0*.

Ravi Kant Soni is a full stack engineer with years of experience in software development. Additionally, he is an AWS Certified Solutions Architect and an author, storyteller, and entrepreneur. Ravi is a published author of three books and one video course on software development: *Spring Boot with React and AWS, Build Microservices with Spring Cloud and Spring Boot, Full Stack AngularJS for Java Developers*, and *Learning Spring Application Development*

Ravi has worn many hats throughout his tenure, ranging from software development to multi-tenant application design and writing books.

4:00 AM @ Bangalore International Airport was his first fictional story, published in May 2023.

Contact this self-driven, competent, and amazing team player on LinkedIn.

Table of Contents

2

Spring Concepts and REST APIs   23

3

API Specifications and Implementation   49

4

Writing Business Logic for APIs   75

5

Asynchronous API Design   97

Part 2 – Security, UI, Testing, and Deployment

6

7

Part 3 – gRPC, Logging, and Monitoring

10

Getting Started with gRPC   287

11

gRPC API Development and Testing   305

12

Adding Logging and Tracing to Services   351

Part 4 – GraphQL

13

Getting Started with GraphQL   383

14

GraphQL API Development and Testing   405

Preface

This book is an in-depth guide to using Spring 6 and Spring Boot 3 for web development. Spring is a powerful and widely used framework for building scalable and reliable web applications in Java. Spring Boot is a popular extension to the framework that simplifies the setup and configuration of Spring-based applications. This book will teach you how to use these technologies to build modern and robust web APIs and services.

The book covers a wide range of topics that are essential for API development, such as fundamentals of REST/GraphQL/gRPC, Spring concepts, and API specifications and implementation. Additionally, the book covers topics such as asynchronous API design, security, designing user interfaces, testing APIs, and the deployment of web services. The book provides a highly contextual real-world sample app that readers can use as a reference for building different types of APIs for real-world apps, including the persistent database layer. The approach taken in the book is to guide readers through the entire development cycle of API development, including design and specification, implementation, testing, and deployment.

By the end of this book, you will have learned how to design, develop, test, and deploy scalable and maintainable modern APIs using Spring 6 and Spring Boot 3, along with best practices to ensure the security and reliability of your applications, and practical ideas to improve your application's functionalities.

Who this book is for

This book is aimed at novice Java programmers, recent computer science graduates, coding boot camp alumni, and professionals who are new to creating real-world web APIs and services. It is also a valuable resource for Java developers interested in transitioning to web development and seeking a comprehensive introduction to web service development. The ideal reader possesses knowledge of fundamental programming constructs, data structures, and algorithms in Java, but lacks the practical web development experience necessary to begin working as a web developer.

What this book covers

Chapter 1, *RESTful Web Service Fundamentals*, drives you through the fundamentals of RESTful APIs, or, for short, REST APIs, and their design paradigm. These basics will provide you with a solid platform for developing a RESTful web service. You will also learn about the best practices while designing APIs. This chapter will also introduce the example e-commerce app that will be used across the book while learning about the different aspects of API development.

Chapter 2, Spring Concepts and REST APIs, explores the Spring fundamentals and features that are required to implement REST, gRPC, and GraphQL APIs using the Spring Framework. This will provide the technical perspective required for developing an example e-commerce app.

Chapter 3, API Specifications and Implementation, makes use of OpenAPI and Spring to implement the REST APIs. We have chosen a design-first approach to implementation. You will make use of the OpenAPI specification to first design the APIs and later implement them. You will also learn how to handle the errors that occur during the serving of the request. Here, the APIs of the example e-commerce app will be designed and implemented for reference.

Chapter 4, Writing Business Logic for APIs, helps you implement the API's code in terms of business logic, along with data persistence in the H2 database. You will write services and repositories for implementation. You will also add hypermedia and ETag headers to API responses for optimal performance and caching.

Chapter 5, Asynchronous API Design, covers asynchronous or reactive API design, where calls will be asynchronous and non-blocking. We'll develop these APIs using Spring WebFlux, which is itself based on Project Reactor (`https://projectreactor.io`). First, we'll walk through the reactive programming fundamentals and then migrate the existing e-commerce REST APIs (the previous chapter's code) to asynchronous (reactive) APIs to make things easier by correlating and comparing the existing (imperative) way and reactive way of programming.

Chapter 6, Securing REST Endpoints Using Authorization and Authentication, explains how you can secure these REST endpoints using Spring Security. You'll implement token-based authentication and authorization for REST endpoints. Successful authentication will provide two types of tokens – a **JSON Web Token (JWT)** as an access token, and a refresh token in response. The JWT-based access token then will be used to access secured URLs. A refresh token will be used to request a new JWT if the existing JWT has expired. A valid request token can provide a new JWT to use. You'll associate users with roles such as `Admin`, `User`, and so on. These roles will be used as authorization to make sure that REST endpoints can only be accessed if the user holds a certain role. We'll also briefly discuss **cross-site request forgery (CSRF)** and **cross-origin resource sharing (CORS)**.

Chapter 7, Designing a User Interface, concludes the end-to-end development and communication between different layers of the online shopping app. This UI app will be a **single-page application (SPA)** that consists of interactive components such as `Login`, `Product Listing`, `Product Detail`, `Cart`, and `Order Listing`. By the end of the chapter, you will have learned about SPA and UI component development using React and consuming REST APIs using the browser's in-built Fetch API.

Chapter 8, Testing APIs, introduces manual and automated testing of APIs. You will learn about unit and integration test automation. After learning about automation in this chapter, you will be able to make both types of testing an integral part of the build. You will also set up the Java code coverage tool to calculate the different code coverage metrics.

Chapter 9, Deployment of Web Services, explains the fundamentals of containerization, Docker, and Kubernetes. You will then use this concept to containerize the example e-commerce app using Docker. This container will then be deployed in a Kubernetes cluster. You are going to use minikube for Kubernetes, which makes learning and Kubernetes-based development easier.

Chapter 10, Getting Started with gRPC, introduces the gRPC fundamentals.

Chapter 11, gRPC API Development and Testing, implements gRPC-based APIs. You will learn how to write a gRPC server and client, along with writing APIs based on gRPC. In the latter part of the chapter, you will be introduced to microservices and how they will help you to design modern, scalable architecture. Here, you will go through the implementation of two services – a gRPC server and a gRPC client.

Chapter 12, Adding Logging and Tracing to Services, explores the logging and monitoring tool called the **Elasticsearch, Logstash, Kibana** (**ELK**) stack, and Zipkin. These tools will then be used to implement the distributed logging and tracing of the request/response of the API calls. You will learn how to publish and analyze the logging and tracing of different requests and logs related to responses. You will also use Zipkin to monitor the performance of API calls.

Chapter 13, Getting Started with GraphQL, talks about the fundamentals of GraphQL – the **schema definition language** (**SDL**), queries, mutations, and subscriptions. This knowledge will help you in the next chapter, where you will implement an API based on GraphQL. Over the course of this chapter, you will learn about the basics of the GraphQL schema and solving the N+1 problem.

Chapter 14, GraphQL API Development and Testing, explains GraphQL-based API development and its testing. You will implement GraphQL-based APIs for an example application in this chapter. A GraphQL server implementation will be developed based on the design-first approach.

To get the most out of this book

Ensure you have the following hardware and software:

Software/hardware covered in the book	Operating system requirements
Java 17	Windows, macOS, or Linux (any)
Any Java IDE such as Netbeans, IntelliJ, or Eclipse	An internet connection to clone the code from GitHub and download the dependencies and libraries
Docker	
Kubernetes (minikube)	
cURL or any API client such as Insomnia	
Node 18.x	
VS Code	
The ELK stack and Zipkin	

Each chapter will contain special instructions to install the required tools if applicable.

If you are using the digital version of this book, we advise you to type the code yourself or access the code from the book's GitHub repository (a link is available in the next section). Doing so will help you avoid any potential errors related to the copying and pasting of code.

Download the example code files

You can download the example code files for this book from GitHub at https://github.com/ PacktPublishing/Modern-API-Development-with-Spring-6-and-Spring-Boot-3. If there's an update to the code, it will be updated in the GitHub repository.

We also have other code bundles from our rich catalog of books and videos available at https:// github.com/PacktPublishing/. Check them out!

Conventions used

There are a number of text conventions used throughout this book.

Code in text: Indicates code words in text, database table names, folder names, filenames, file extensions, pathnames, dummy URLs, user input, and Twitter handles. Here is an example: "If we are using Link in any of the models, then the generated models would use the mapped org. springframework.hateoas.Link class instead of the model defined in the YAML file."

A block of code is set as follows:

```
const Footer = () => {
  return (
    <div>
      <footer
        className="text-center p-2 border-t-2 bggray-
          200 border-gray-300 text-sm">
        No &copy; by Ecommerce App.{" "}
        <a href=https://github.com/PacktPublishing/Modern-
          API-Development-with-Spring-and-Spring-Boot>
          Modern API development with Spring and Spring Boot
        </a>
      </footer>
    </div>
  );
};
export default Footer;
```

When we wish to draw your attention to a particular part of a code block, the relevant lines or items are set in bold:

```
<Error>
    <errorCode>PACKT-0001</errorCode>
    <message>The system is unable to complete the request.
        Contact system support.</message>
    <status>500</status>
    <url>http://localhost:8080/api/v1/carts/1</url>
    <reqMethod>GET</reqMethod>
</Error>
```

Any command-line input or output is written as follows:

```
$ curl --request POST 'http://localhost:8080/api/v1/carts/1/items' \
  --header 'Content-Type: application/json' \
  --header 'Accept: application/json' \
  --data-raw '{
  "id": "1",
  "quantity": 1,
  "unitPrice": 2.5
  }'
[]
```

Bold: Indicates a new term, an important word, or words that you see onscreen. For instance, words in menus or dialog boxes appear in **bold**. Here is an example: "Select **System info** from the **Administration** panel."

> Tips or important notes
> Appear like this.

Get in touch

Feedback from our readers is always welcome.

General feedback: If you have questions about any aspect of this book, email us at customercare@ packtpub.com and mention the book title in the subject of your message.

Errata: Although we have taken every care to ensure the accuracy of our content, mistakes do happen. If you have found a mistake in this book, we would be grateful if you would report this to us. Please visit www.packtpub.com/support/errata and fill in the form.

Piracy: If you come across any illegal copies of our works in any form on the internet, we would be grateful if you would provide us with the location address or website name. Please contact us at copyright@packt.com with a link to the material.

If you are interested in becoming an author: If there is a topic that you have expertise in and you are interested in either writing or contributing to a book, please visit authors.packtpub.com.

Share Your Thoughts

Once you've read *Modern API Development with Spring 6 and Spring Boot 3*, we'd love to hear your thoughts! Scan the QR code below to go straight to the Amazon review page for this book and share your feedback.

https://packt.link/r/1-804-61327-4

Your review is important to us and the tech community and will help us make sure we're delivering excellent quality content.

Download a free PDF copy of this book

Thanks for purchasing this book!

Do you like to read on the go but are unable to carry your print books everywhere? Is your eBook purchase not compatible with the device of your choice?

Don't worry, now with every Packt book you get a DRM-free PDF version of that book at no cost.

Read anywhere, any place, on any device. Search, copy, and paste code from your favorite technical books directly into your application.

The perks don't stop there, you can get exclusive access to discounts, newsletters, and great free content in your inbox daily

Follow these simple steps to get the benefits:

1. Scan the QR code or visit the link below

https://packt.link/free-ebook/9781804613276

2. Submit your proof of purchase
3. That's it! We'll send your free PDF and other benefits to your email directly

Part 1 –
RESTful Web Services

In this part, you will develop and test production-ready and evolving REST-based APIs supported by HATEOAS and ETags. API specifications will be written using OpenAPI specifications (Swagger). You will learn the fundamentals of reactive API development using Spring WebFlux. By the end of this part, you will know the fundamentals of REST APIs, their best practices, and how to write evolving APIs. After completing this part, you will be able to develop sync and async (reactive) non-blocking APIs.

This part contains the following chapters:

- *Chapter 1, RESTful Web Service Fundamentals*
- *Chapter 2, Spring Concepts and REST APIs*
- *Chapter 3, API Specifications and Implementation*
- *Chapter 4, Writing Business Logic for APIs*
- *Chapter 5, Asynchronous API Design*

1

RESTful Web Service Fundamentals

In this chapter, we will go through the fundamentals of RESTful APIs (or REST APIs for short) and their design paradigms. We will take a brief look at the history of REST, learn how resources are formed, and understand methods and status codes before we move on to exploring **Hypermedia As The Engine Of Application State** (**HATEOAS**). These basics should provide a solid platform to enable you to develop a RESTful web service. You will also learn the best practices for designing **application programming interfaces** (**APIs**).

This chapter will also introduce a sample e-commerce app, which will be used throughout the book as you learn about the different aspects of API development.

In this chapter, we will cover the following topics:

- Introducing REST APIs
- Handling resources and **Uniform Resource Identifiers** (**URIs**)
- Exploring **Hypertext Transfer Protocol** (**HTTP**) methods and status codes
- What is HATEOAS?
- Best practices for designing REST APIs
- Overview of an e-commerce app (our sample app)

Technical requirements

This chapter does not require any specific software. However, knowledge of HTTP is necessary.

Introducing REST APIs

An API is how a piece of code communicates with another piece of code. You might have already written and consumed APIs for your programs; for example, Java provides APIs through classes wrapped in different modules, such as collection, input/output, and streams.

Java's SDK APIs allow one part of a program to communicate with another part of a program. You can write a function and then expose it with public access modifiers so that other classes can use it. That function signature is an API for that class. However, APIs that are exposed using these classes or libraries only allow internal communication inside a single application or an individual service. So, what happens when two or more applications (or services) want to communicate with each other, or, in other words, you would like to integrate two or more services? This is where system-wide APIs help us.

Historically, there were different ways to integrate one application with another – RPC, **Simple Object Access Protocol** (**SOAP**)-based services, and more. The integration of apps has become an integral part of software architectures, especially after the boom of the cloud and mobile phones. You now have social logins, such as Facebook, Google, and GitHub, which means you can develop your application even without writing an independent login module and get around security issues such as storing passwords securely.

These social logins provide APIs using REST and GraphQL. Currently, REST is the most popular, and it has become the standard for writing APIs for integration and web app consumption. We'll also discuss GraphQL in detail in the final chapters of this book (in *Chapter 13, Getting Started with GraphQL*, and *Chapter 14, GraphQL API Development and Testing*).

REST stands for **REpresentational State Transfer**, which is a style of software architecture. Web services that adhere to the REST style are called RESTful web services. In the following sections, we will take a quick look at the history of REST to understand its fundamentals.

The history of REST

Before the adoption of REST, when the internet was just starting to become widely known and Yahoo and Hotmail were the popular mail and social messaging apps, there was no standard software architecture that offered a homogenous way to integrate with web applications. People were using SOAP-based web services, which, ironically, were not simple at all.

Then came the light. Roy Fielding, in his doctoral research, *Architectural Styles and the Design of Network-Based Software Architectures* (`https://www.ics.uci.edu/~fielding/pubs/dissertation/top.htm`), came up with REST in 2000. REST's architecture style allowed any server to communicate with any other server over the network. It simplified communication and made integration easier. REST was made to work on top of HTTP, which enables it to be used all over the web and in internal networks.

eBay was the first to exploit REST-based APIs. It introduced the REST API with selected partners in November 2000. Later, Amazon, Delicious (a site-bookmarking web app), and Flickr (the photo-sharing app) started providing REST-based APIs. Then, **Amazon Web Services** (**AWS**) took advantage of Web 2.0 (with the invention of REST) and provided REST APIs to developers for AWS cloud consumption in 2006.

Later, Facebook, Twitter, Google, and other companies started using it. Nowadays (in 2023), you will hardly find any web applications that have been developed without a REST API. However, the GraphQL-based API for mobile apps is getting close in terms of popularity.

REST fundamentals

REST works on top of the HTTP protocol. Each URI works as an API resource. Therefore, we should use nouns as endpoints instead of verbs. RPC-style endpoints use verbs, for example, `/api/v1/getPersons`. In comparison, in REST, this endpoint could be simply written as `/api/v1/persons`. You must be wondering, then, how we can differentiate between the different actions performed on a REST resource. This is where HTTP methods help us. We can make our HTTP methods act as verbs, for example, `GET`, `DELETE`, `POST` (for creating), `PUT` (for modifying), and `PATCH` (for partial updating). We'll discuss this in more detail later. For now, the `getPerson` RPC-style endpoint is translated into `GET /api/v1/persons` in REST.

> **Note**
>
> The REST endpoint is a unique URI that represents a REST resource. For example, `https://demo.app/api/v1/persons` is a REST endpoint. Additionally, `/api/v1/persons` is the endpoint path and `persons` is the REST resource.

Here, there is client and server communication. Therefore, REST is based on the *client-server* concept. The client calls the REST API and the server responds. REST allows a client (that is, a program, web service, or UI app) to talk to a remotely (or locally) running server (or web service) using HTTP requests and responses. The client sends an API command wrapped in an HTTP request to the web service. This HTTP request may contain a payload (or input) in the form of query parameters, headers, or request bodies. The called web service responds with a success/failure indicator and the response data wrapped inside the HTTP response. The HTTP status code normally denotes the status, and the response body contains the response data. For example, an HTTP status code of `200 OK` normally represents success.

From a REST perspective, an HTTP request is self-descriptive and has enough context for the server to process it. Therefore, REST calls are *stateless*. States are either managed on the client side or on the server side. A REST API does not maintain its state. It only transfers states from the server to the client or vice versa. This is why it is called *REpresentational State Transfer*, or REST for short.

REST also makes use of HTTP cache control, which makes REST APIs *cacheable*. Therefore, the client can also cache the representation (that is, the HTTP response) because every representation is self-descriptive.

REST operates using three key components:

- Resources and URIs
- HTTP methods
- HATEOAS

A sample REST call in plain text looks like the following:

```
GET /licenses HTTP/2
Host: api.github.com
```

Here, the /licenses path denotes the licenses resource. GET is an HTTP method. 2 at the end of the first line denotes the HTTP protocol version. The second line shares the host to call.

GitHub responds with a JSON object. The status is 200 OK and the JSON object is wrapped in a response body, as follows:

```
HTTP/2 200 OK
date: Mon, 10 Jul 2023 17:44:04 GMT
content-type: application/json; charset=utf-8
server: GitHub.com
status: 200 OK
cache-control: public, max-age=60, s-maxage=60
vary: Accept, Accept-Encoding, Accept, X-Requested-With,
      Accept-Encoding  etag:W/"3cbb5a2e38ac6fc92b3d798667e
          828c7e3584af278aa314f6eb1857bbf2593ba"
… <bunch of other headers>
Accept-Ranges: bytes
Content-Length: 2037
X-GitHub-Request-Id: 1C03:5C22:640347:81F9C5:5F70D372
[
  {
    "key": "agpl-3.0",
    "name": "GNU Affero General Public License v3.0",
    "spdx_id": "AGPL-3.0",
    "url": "https://api.github.com/licenses/agpl-3.0",
    "node_id": "MDc6TGljZW5zZTE="
  },
  {
    "key": "apache-2.0",
```

```
    "name": "Apache License 2.0",
    "spdx_id": "Apache-2.0",
    "url": "https://api.github.com/licenses/apache-2.0",
    "node_id": "MDc6TGljZW5zZTI="
  },
  …
]
```

If you note the third line in this response, it tells you the value of the content type. It is good practice to have JSON as the content type for both the request and the response.

Now that we have familiarized ourselves with the fundamentals of REST, we are going to dive a bit deeper into REST's first concept, resources and URIs, and learn what they are and how they are generally used.

Handling resources and URIs

Every document on the **World Wide Web** (**WWW**) is represented as a resource in terms of HTTP. This resource is represented as a URI, which is an endpoint that represents a unique resource on a server.

Roy Fielding in his doctoral research states that a URI is known by many names – a WWW address, a **Universal Document Identifier** (**UDI**), a URI, a **Uniform Resource Locator** (**URL**), and a **Uniform Resource Name** (**URN**).

So, what is a URI? A URI is a string (that is, a sequence of characters) that identifies a resource by its location, name, or both (in the WWW world). There are two types of URIs: URLs and URNs.

URLs are widely used and even known to non-developer users. URLs are not only restricted to HTTP but are also used for many other protocols, such as FTP, JDBC, and MAILTO. A URL is an identifier that identifies the network location of a resource. We will go into more detail in the later sections.

The URI syntax

The URI syntax is as follows:

```
scheme:[//authority]path[?query][#fragment]
```

As per the syntax, the following is the list of components of a URI:

- **Scheme**: This refers to a non-empty sequence of characters followed by a colon (:). scheme starts with a letter and is followed by any combination of digits, letters, periods (.), hyphens (-), or plus characters (+).

  Scheme examples include HTTP, HTTPS, MAILTO, FILE, and FTP. URI schemes must be registered with the **Internet Assigned Numbers Authority** (**IANA**).

- **Authority**: This is an optional field and is preceded by `//`. It consists of the following optional subfields:

  - **Userinfo**: This is a subcomponent that might contain a username and a password, which are both optional.

  - **Host**: This is a subcomponent containing either an IP address or a registered host or domain name.

  - **Port**: This is an optional subcomponent that is followed by a colon (`:`).

- **Path**: A path contains a sequence of segments separated by slash characters (`/`). In the preceding GitHub REST API example, `/licenses` is the path.

- **Query**: This is an optional component and is preceded by a question mark (`?`). The query component contains a query string of non-hierarchical data. Each parameter is separated by an ampersand (`&`) in the query component and parameter values are assigned using an equals (`=`) operator.

- **Fragment**: This is an optional field and is preceded by a hash (`#`). The fragment component includes a fragment identifier that gives direction to a secondary resource.

The following list contains examples of URIs:

- `www.packt.com`: This doesn't contain the scheme. It just contains the domain name. There is no port either, which means it points to the default port.

- `index.html`: This contains no scheme nor authority. It only contains the path.

- `https://www.packt.com/index.html`: This contains the scheme, authority, and path.

  Here are some examples of different scheme URIs:

  - `mailto:support@packt.com`

  - `telnet://192.168.0.1:23/`

  - `ldap://[2020:ab9::9]/c=AB?objectClass?obj`

> **Note**
>
> From a REST perspective, the path component of a URI is very important because it represents the resource path and your API endpoint paths are formed based on it. For example, take a look at the following:
>
> `GET https://www.domain.com/api/v1/order/1`
>
> Here, `/api/v1/order/1` represents the path and `GET` represents the HTTP method.

What is a URL?

If you look closely, most of the URI examples mentioned earlier can also be called URLs. A URI is an identifier; on the other hand, a URL is not only an identifier, but it also tells you how to get to it.

Request for Comments (RFC)

As per RFC-3986 on URIs (`https://datatracker.ietf.org/doc/html/rfc3986`), the term URL refers to the subset of URIs that, in addition to identifying a resource, provide a means of locating the resource by describing its primary access mechanism (for example, its network *location*).

A URL represents the full web address of a resource, including the protocol name (the scheme), the hostname port (in case the HTTP port is not `80`; for HTTPS, the default port is `443`), part of the authority component, the path, and optional query and fragment subcomponents.

What is a URN?

URNs are not commonly used. They are also a type of URI that starts with a scheme – `urn`. The following URN example is directly taken from RFC-3986 for URIs (`https://www.ietf.org/rfc/rfc3986.txt`):

```
urn:oasis:names:specification:docbook:dtd:xml:4.1.2
```

This example follows the `"urn:"` `<NID>` `":"` `<NSS>` syntax, where `<NID>` is the namespace identifier, and `<NSS>` is the namespace-specific string. We are not going to use URNs in our REST implementation. However, you can read more about them at RFC-2141 (`https://tools.ietf.org/html/rfc2141`).

Note

As per RFC-3986 on URIs (`https://datatracker.ietf.org/doc/html/rfc3986`), the term URN has been used historically to refer to both URIs under the *"urn"* scheme RFC-2141, which are required to remain globally unique and persistent even when the resource ceases to exist or becomes unavailable, and to any other URI with the properties of a name.

Now that you understand the difference between a URI and a URN and how they make up URIs, let's learn about the second concept that makes up REST: HTTP methods and status codes.

Exploring HTTP methods and status codes

HTTP provides various HTTP methods. However, you are primarily going to use only five of them. To begin with, you want to have **Create**, **Read**, **Update**, and **Delete** (**CRUD**) operations associated with HTTP methods:

- POST: Create or search
- GET: Read
- PUT: Update
- DELETE: Delete
- PATCH: Partial update

Some organizations also provide the HEAD method for scenarios where you just want to retrieve the header responses from the REST endpoints. You can hit any GitHub API with the HEAD operation to retrieve only headers; for example, `curl --head https://api.github.com/users`.

> **Note**
>
> REST has no requirement that specifies which method should be used for which operation. However, widely used industry guidelines and practices suggest following certain rules.

Let's discuss each method in detail.

POST

The HTTP POST method is normally what you want to associate with creating resource operations. However, there are certain exceptions when you might want to use the POST method for read operations. However, it should be put into practice after a well-thought-out process. One such exception is a search operation where the filter criteria have too many parameters, which might cross the GET call's length limit.

A GET query string has a limit of 256 characters. Additionally, the HTTP GET method is limited to a maximum of 2,048 characters minus the number of characters in the actual path. On the other hand, the POST method is not limited by the size of the URL for submitting name and value pairs.

You may also want to use the POST method with HTTPS for a read call if the submitted input parameters contain any private or secure information.

For successful create operations, you can respond with the 201 Created status, and for successful search or read operations, you should use the 200 OK or 204 No Content status codes, although the call is made using the HTTP POST method.

For failed operations, REST responses may have different error status codes based on the error type, which we will look at later in this section.

GET

The HTTP GET method is what you usually want to associate with read resource operations. Similarly, you must have observed the GitHub GET /licenses call that returns the available licenses in the GitHub system. Additionally, successful GET operations should be associated with the 200 OK status code if the response contains data, or 204 No Content if the response contains no data.

PUT

The HTTP PUT method is what you usually want to associate with update resource operations. Additionally, successful update operations should be associated with a 200 OK status code if the response contains data, or 204 No Content if the response contains no data. Some developers use the PUT HTTP method to replace existing resources. For example, GitHub API v3 uses PUT to replace the existing resource.

DELETE

The HTTP DELETE method is what you want to associate with resource deletion operations. GitHub does not provide the DELETE operation on the licenses resource. However, if you assume it exists, it will certainly look very similar to DELETE / licenses/agpl-3.0. A successful DELETE call should delete the resource associated with the agpl-3.0 key. Additionally, successful DELETE operations should be associated with the 204 No Content status code.

PATCH

The HTTP PATCH method is what you want to associate with partial update resource operations. Additionally, successful PATCH operations should be associated with a 200 OK status code. PATCH is relatively new as compared to other HTTP operations. In fact, a few years ago, Spring did not have state-of-the-art support for this method for REST implementation due to the old Java HTTP library. However, currently, Spring provides built-in support for the PATCH method in REST implementation.

HTTP status codes

There are five categories of HTTP status codes, as follows:

- Informational responses (100–199)
- Successful responses (200–299)
- Redirects (300–399)
- Client errors (400–499)
- Server errors (500–599)

You can view a complete list of status codes at MDN Web Docs (`https://developer.mozilla.org/en-US/docs/Web/HTTP/Status`) or RFC-7231 (`https://tools.ietf.org/html/rfc7231`). However, you can find the most commonly used REST response status codes in the following table:

HTTP Status Code	Description
`200 OK`	For successful requests other than those already created.
`201 Created`	For successful creation requests.
`202 Accepted`	The request has been received but not yet acted upon. This is used when the server accepts the request, but the response cannot be sent immediately, for example, in batch processing.
`204 No Content`	For successful operations that contain no data in the response.
`304 Not Modified`	This is used for caching. The server responds to the client that the resource is not modified; therefore, the same cache resource can be used.
`400 Bad Request`	This is for failed operations when input parameters either are incorrect or missing or the request itself is incomplete.
`401 Unauthorized`	This is for operations that have failed due to unauthenticated requests. The specification says it's unauthorized, but semantically, it means unauthenticated.
`403 Forbidden`	This is for failed operations that the invoker is not authorized to perform.
`404 Not Found`	This is for failed operations when the requested resource doesn't exist.
`405 Method Not Allowed`	This is for failed operations when the method is not allowed for the requested resource.
`409 Conflict`	This is for failed operations when an attempt is made for a duplicate create operation.
`429 Too Many Requests`	This is for failed operations when a user sends too many requests in a given amount of time (rate limiting).
`500 Internal Server Error`	This is for failed operations due to server errors. It's a generic error.
`502 Bad Gateway`	This is for failed operations when upstream server calls fail, for example, when an app calls a third-party payment service, but the call fails.
`503 Service Unavailable`	This is for failed operations when something unexpected has happened at the server, for example, an overload or a service fails.

We have discussed the key components of REST, such as endpoints in the form of URIs, methods, and status codes. Let's explore HATEOAS, the backbone of REST concepts that differentiates it from RPC style.

What is HATEOAS?

With HATEOAS, RESTful web services provide information dynamically through hypermedia. Hypermedia is a part of the content that you receive from a REST call response. This hypermedia content contains links to different types of media, such as text, images, and videos.

Hypermedia links can be contained either in HTTP headers or the response body. If you look at GitHub APIs, you will find that GitHub APIs provide hypermedia links in both headers and the response body. GitHub uses the header named `Link` to contain the paging-related links. Additionally, if you look at the responses of GitHub APIs, you'll also find other resource-related links with keys that have a postfix of `url`. Let's look at an example. We'll hit the `GET /users` resource and analyze the response:

```
$ curl -v https://api.github.com/users
```

This command execution provides an output similar to the following:

```
*    Trying 20.207.73.85:443...* Connected to api.github.com
(20.207.73.85) port 443 (#0)
… < more info>

…
> GET /users HTTP/2
> Host: api.github.com
> user-agent: curl/7.78.0
… < more info >
< HTTP/2 200
< server: GitHub.com
< date: Sun, 28 Aug 2022 04:31:50 GMT status: 200 OK
< content-type: application/json; charset=utf-8

…
< link: <https://api.github.com/users?since=46>; rel="next", <https://
api.github.com/users{?since}>; rel="first"

…
[
  {
    "login": "mojombo",
    "id": 1,
    "node_id": "MDQ6VXNlcjE=",
    "avatar_url":
        "https://avatars.githubusercontent.com/u/1?v=4",
    "gravatar_id": "",
```

```
    "url": "https://api.github.com/users/mojombo",
    "html_url": "https://github.com/mojombo",
    "followers_url":
        "https://api.github.com/users/mojombo/followers",
    "following_url":

"https://api.github.com/users
/mojombo/following{/other_user}",
    "gists_url": "https://api.github.com/users/mojombo/gists{/gist_
        id}",
    "starred_url":
"https://api.github.com/users/mojombo/starred{/owner}{/repo}",
    "subscriptions_url":
        "https://api.github.com/users/mojombo/subscriptions",
    "organizations_url":
        "https://api.github.com/users/mojombo/orgs",
    "repos_url":
        "https://api.github.com/users/mojombo/repos",
    "events_url":    "https://api.github.com/users/mojombo/events{/
        privacy}",
    "received_events_url":
        "https://api.github.com/users/mojombo/received_events",
    "type": "User",
    "site_admin": false
},
…
… < more data >
]
```

In the preceding output, you'll find that the Link header contains the pagination information. Links to the next page and the first page are given as a part of the response. Additionally, you can find many URLs in the response body, such as avatar_url or followers_url, which provide links to other hypermedia.

REST clients should possess a generic understanding of hypermedia so they can interact with RESTful web services without having any specific knowledge of how to interact with the server. You just call any static REST API endpoint, and you will receive the dynamic links as a part of the response to interact further. REST allows clients to dynamically navigate to the appropriate resource by traversing the links. It empowers machines, as REST clients can navigate to different resources in a similar way to how humans look at a web page and click on any link. Put simply, the REST client uses these links to navigate.

HATEOAS is a very important concept of REST. It is one of the concepts that differentiate REST from RPC. Even Roy Fielding was so concerned with certain REST API implementations that he published the following blog on his website in 2008: *REST APIs must be hypertext-driven* (`https://roy.gbiv.com/untangled/2008/rest-apis-must-be-hypertext-driven`).

You must be wondering what the difference between hypertext and hypermedia is. Essentially, hypermedia is just an extended version of hypertext.

> **What is the difference between hypermedia and hypertext?**
>
> As Roy Fielding states: *"When I say hypertext, I mean the simultaneous presentation of information and controls such that the information becomes the affordance through which the user (or automaton) obtains choices and selects actions. Hypermedia is just an expansion on what text means to include temporal anchors within a media stream; most researchers have dropped the distinction. Hypertext does not need to be HTML on a browser. Machines can follow links when they understand the data format and relationship types."*

Now that you are familiar with REST, let's explore REST best practices in the next section.

Best practices for designing REST APIs

It is too early to talk about the best practices for implementing APIs. APIs are designed first and implemented later. Therefore, you'll find design-related best practices mentioned in the next sections. You'll also find best practices for going forward during REST API implementation.

Using nouns and not verbs when naming a resource in the endpoint path

We previously discussed HTTP methods. HTTP methods use verbs. Therefore, it would be redundant to use verbs yourself, and it would make your call look like an RPC endpoint, for example, `GET /getlicenses`. In REST, we should always use the resource name because, according to REST, you transfer the states and not the instructions. For example, let's take another look at the GitHub license API, which retrieves licenses. It is `GET /licenses`. That is perfect. Let's assume that if you use verbs for this endpoint, then it will be `GET /getlicenses`. It will still work, but semantically, it doesn't follow REST because it conveys the processing instruction rather than state transfer. Therefore, only use resource names.

However, GitHub's public API only offers `read` operations on the `licenses` resource, out of all the CRUD operations. If we need to design the rest of the operations, their paths should look like the following:

- `POST /licenses`: This is for creating a new license.

- `PATCH /licenses/{license_key}`: This is for partial updates. Here, the path has a parameter (that is, an identifier) that makes the path dynamic. Here, the license key is a unique value in the license collection and is being used as an identifier. Each license will have a unique key. This call should make the update in the given license. Please remember that GitHub uses `PUT` for the replacement of the resource.

- `DELETE /licenses/{license_key}`: This is for retrieving license information. You can try this with any licenses that you receive in the response to the `GET /licenses` call. One example is `GET /licenses/agpl-3.0`.

You can see how having a noun in the resource path with the HTTP methods sorts out any ambiguity.

Using the plural form for naming the collection resource in the endpoint path

If you observe the GitHub license API, you might find that a resource name is given in the plural form. It is a good practice to use the plural form if the resource represents a collection. Therefore, we can use `/licenses` instead of `/license`. A `GET` call returns the collection of licenses. GitHub doesn't allow create, update, or delete public operations on a licensed resource. Hypothetically, if it allowed this, then a `POST` call would create a new license in the existing license collection. Similarly, for `DELETE` and `PATCH` calls, a license key is used to identify the specific license for performing delete and minor update operations respectively.

Using hypermedia (HATEOAS)

Hypermedia (that is, links to other resources) makes the REST client's job easier. There are two advantages if you provide explicit URL links in a response. First, the REST client is not required to construct the REST URLs on their own. Second, any upgrade in the endpoint path will be taken care of automatically and this, therefore, makes upgrades easier for clients and developers

Versioning your APIs

The versioning of APIs is key for future upgrades. Over time, APIs keep changing, and you may have customers who are still using an older version. Therefore, you need to support multiple versions of APIs.

There are different ways you can version your APIs:

- *Using headers*: The GitHub API uses this approach. You can add an `Accept` header that tells you which API version should serve the request; for example, consider the following:

  ```
  Accept: application/vnd.github.v3+json
  ```

  This approach gives you the advantage of setting the default version. If there is no `Accept` header, it should lead to the default version. However, if a REST client that uses a versioning header is not changed after a recent upgrade of APIs, it may break the functionality. Therefore, it is recommended that you use a versioning header.

- *Using an endpoint path*: In this approach, you add a version in the endpoint path itself; for example, `https://demo.app/api/v1/persons`. Here, `v1` denotes that version 1 is being added to the path itself.

  You cannot set default versioning out of the box. However, you can overcome this limitation by using other methods, such as request forwarding. Clients always use the intended versions of the APIs in this approach.

Based on your preferences and views, you can choose either of the preceding approaches for versioning. However, the important point is that you should always use versioning.

Nesting resources

Consider this very interesting question: how are you going to construct the endpoint for resources that are nested or have a certain relationship? Let's take a look at some examples of customer resources from an e-commerce perspective:

- `GET /customers/1/addresses`: This returns the collection of addresses for customer 1
- `GET /customers/1/addresses/2`: This returns the second address of customer 1
- `POST /customers/1/addresses`: This adds a new address to customer 1's addresses
- `PUT /customers/1/addresses/2`: This replaces the second address of customer 1
- `PATCH /customers/1/addresses/2`: This partially updates the second address of customer 1
- `DELETE /customers/1/addresses/2`: This deletes the second address of customer 1

So far so good. Now, can we have an altogether separate address resource endpoint (`GET /addresses/2`)? It makes sense, and you can do that if there is a relationship that requires it; for example, orders and payments. Instead of `/orders/1/payments/1`, you might prefer a separate `/payments/1` endpoint. In the microservice world, this makes more sense; for instance, you would have two separate RESTful web services for both orders and payments.

Now, if you combine this approach with hypermedia, it makes things easier. When you make a REST API request to customer 1, it will provide the customer 1 data and address links as hypermedia (that is, links). The same applies to orders. For orders, the payment link will be available as hypermedia.

However, in some cases, you might wish to have a complete response in a single request rather than using the hypermedia-provided URLs to fetch the related resource. This reduces your web hits. However, there is no rule of thumb. For a flag operation, it makes sense to use the nested endpoint approach; for example, `PUT /gist/2/star` (which adds a star) and `DELETE /gist/2/star` (which undoes the star) in the case of the GitHub API.

Additionally, in some scenarios, you might not find a suitable resource name when multiple resources are involved, for example, in a search operation. In that case, you should use a `direct/search` endpoint. This is an exception.

Securing APIs

Securing your API is another expectation that requires diligent attention. Here are some recommendations:

- Always use HTTPS for encrypted communication.
- Always look for OWASP's top API security threats and vulnerabilities. These can be found on their website (`https://owasp.org/www-project-api-security/`) or their GitHub repository (`https://github.com/OWASP/API-Security`).
- Secure REST APIs should have authentication in place. REST APIs are stateless; therefore, REST APIs should not use cookies or sessions. Instead, they should be secured using JWT or OAuth 2.0-based tokens.

Maintaining documentation

Documentation should be easily accessible and up to date with the latest implementation with their respective versioning. It is always good to provide sample code and examples. It makes the developer's integration job easier.

A change log or a release log should list all the affected libraries, and if some APIs are deprecated, then replacement APIs or workarounds should be elaborated upon inside the documentation.

Complying with recommended status codes

We have already learned about status codes in the *Exploring HTTP methods and status codes* section. Please follow the same guidelines discussed there.

Ensuring caching

HTTP already provides a caching mechanism. You just have to provide additional headers in the REST API response. Then, the REST client makes use of the validation to make sure whether to make a call or use the cached response. There are two ways to do this:

- **ETag**: ETag is a special header value that contains the hash or checksum value of the resource representation (that is, the response object). This value must change with respect to the response representation. It will remain the same if the resource response doesn't change. Now, the client can send a request with another header field, called `If-None-Match`, which contains the `ETag` value. When the server receives this request and finds that the hash or checksum value of the resource representation value is different from `If-None-Match`, only then should it return the response with a new representation and this hash value in the `ETag` header. If it finds them to be equal, then the server should simply respond with a `304 (Not Modified)` status code.

- **Last-Modified**: This approach is identical to the `ETag` way. However, instead of using the hash or checksum, it uses the timestamp value in RFC-1123 (`http://www.ietf.org/rfc/rfc1123.txt`) in the format: `Last-Modified: Wed, 21 Oct 2015 07:28:00 GMT`. It is less accurate than `ETag` and should only be used as a fallback.

  In the `Last-Modified` approach, the client sends the `If-Modified-Since` header with the value received in the `Last-Modified` response header. The server compares the resource-modified timestamp value with the `If-Modified-Since` header value and sends a `304` status if there is a match; otherwise, it sends the response with a new `Last-Modified` header.

Maintaining the rate limit

Maintaining the rate limit is important if you want to prevent the overuse of APIs. The HTTP status code `429 Too Many Requests` is used when the rate limit is infringed. Currently, there is no standard to communicate any warning to the client before the rate limit goes over. However, there is a popular way to communicate about it using response headers. These response headers are as follows:

- `X-Ratelimit-Limit`: The number of allowed requests in the current period, for example, `X-Ratelimit-Limit: 60`.

- `X-Ratelimit-Remaining`: The number of remaining requests in the current period, for example, `X-Ratelimit-Remaining: 55`.

- `X-Ratelimit-Reset`: The number of seconds left in the current period, for example, `X-Ratelimit-Reset: 1601299930`.

- `X-Ratelimit-Used`: The number of requests used in the current period, for example, `X-Ratelimit-Used: 5`. This information then might be used by the client to keep track of the total number of available API calls for the given period.

So far, we have discussed various concepts related to REST. Next, let me introduce you to the app we will be building in this book using these concepts.

Introducing our e-commerce app

The e-commerce app we will be building will be a simple online shopping application with the following features for users:

- Browsing through the products

- Adding/removing/updating the products in the cart

- Placing an order

- Modifying the shipping address

- Support for a single currency

E-commerce is a very popular domain. If we look at the features, we can divide the application into the following subdomains using bounded contexts:

- **Users**: This subdomain is related to users. We'll add the `users` RESTful web service, which provides REST APIs for user management.

- **Carts**: This subdomain is related to the cart. We'll add the `carts` RESTful web service, which provides REST APIs for cart management. Users can perform CRUD operations on cart items.

- **Products**: This subdomain is related to the products catalog. We'll add the `products` RESTful web service, which provides REST APIs to search and retrieve the products.

- **Orders**: This subdomain is related to orders. We'll add the `orders` RESTful web service, which provides REST APIs for users to place orders.

- **Payments**: This subdomain is related to payments. We'll add the `payments` RESTful web service, which provides REST APIs for payment processing.

- **Shippings**: This subdomain is related to shipping. We'll add the `shippings` RESTful web service, which provides REST APIs for order tracking and shipping.

Here's a visual representation of our app's architecture:

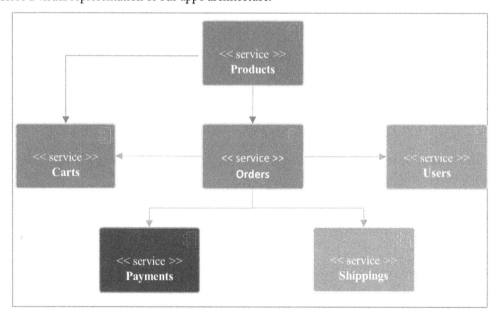

Figure 1.1 – The e-commerce app architecture

We'll implement a RESTful web service for each of the subdomains. We'll keep the implementation simple, and we will focus on learning these concepts throughout this book.

Summary

In this chapter, you learned about the basic concepts of the REST architecture style and its key concepts – resources, URI, HTTP methods, and HATEOAS. Now, you know how REST, which is based on HTTP, simplifies and makes the integration of different applications and services easier.

We also explored the different HTTP concepts that allow you to write REST APIs in a meaningful way. We also learned why HATEOAS is an integral part of REST implementation. Additionally, we learned the best practices for designing REST APIs. We also went through an overview of our e-commerce app. This sample app will be used throughout the book.

The REST concepts you learned in this chapter will provide the foundation for REST implementation. Now, you can make use of the best practices you learned in this chapter to design and implement state-of-the-art REST APIs.

In the next chapter, you'll learn about the fundamentals of the Spring Framework.

Questions

1. Why have RESTful web services become so popular and, arguably, the industry standard?

2. What is the difference between RPC and REST?

3. How would you explain HATEOAS?

4. What error codes should be used for server-related issues?

5. Should verbs be used to form REST endpoints, and why?

Answers

1. RESTful services became popular because they work on top of HTTP, which is the backbone of the internet. You don't need separate protocol implementations such as SOAP. You can use existing web technologies to implement the REST APIs with simple application integration compared to other technologies available. REST APIs make application integration simpler than other technologies available at the time.

    RESTful services work on REST, which works on web resources. Resources represent domain models. Actions are defined using HTTP methods, which are performed on web resources. REST also allows clients to perform actions based on links available through HATEOAS implementation, like a human who can navigate in the browser.

2. RPC is more like functions that perform actions. RPC endpoints are directly formed based on verbs that lead to separate URLs for each action. Whereas REST URLs represent nouns and could be the same for different operations, for example:

    ```
    RPC: GET localhost/orders/getAllOrdersREST: GET localhost/orders
    RPC: POST localhost/orders/createOrderREST: POST localhost/
    orders
    ```

3. With HATEOAS, RESTful web services provide information dynamically through hypermedia. Hypermedia is the part of the content you receive from a REST call response. This hypermedia content contains links to different types of media such as text, images, and videos. Machines, aka REST clients/browsers, can follow links when they understand the data format and relationship types.

4. Status code 500 should be used for generic server errors. The 502 status code should be used when an upstream server fails. Status code 503 is for unexpected server events such as an overload.

5. Verbs should not be used to form REST endpoints. Instead, you should use the noun that represents the domain model as a resource. HTTP methods are used to define the actions performed on resources, such as POST for creating and GET for retrieving.

Further reading

- *Architectural Styles and the Design of Network-based Software Architectures*: https://www.ics.uci.edu/~fielding/pubs/dissertation/top.htm

- The URI Generic Syntax (*RFC-3986*): https://tools.ietf.org/html/rfc3986

- The URN Syntax (*RFC-2141*): https://tools.ietf.org/html/rfc2141

- HTTP Response Status Codes – *RFC 7231*: https://tools.ietf.org/html/rfc7231

- HTTP Response Status Codes – Mozilla Developer Network: https://developer.mozilla.org/en-US/docs/Web/HTTP/Status

- *REST APIs must be hypertext-driven*: https://roy.gbiv.com/untangled/2008/rest-apis-must-be-hypertext-driven

- The RFC for the URI template: https://tools.ietf.org/html/rfc6570

- The OWASP API security project: https://owasp.org/www-project-api-security/ and https://github.com/OWASP/API-Security

2

Spring Concepts and REST APIs

In the previous chapter, we learned about the REST architecture style. Before we go and implement RESTful web services using Spring and Spring Boot, we need to have a proper understanding of the basic Spring concepts. In this chapter, you will learn about the Spring fundamentals and features that are required to implement RESTful web services using the Spring Framework. This will provide the technical perspective required for developing the example e-commerce app. If you are already aware of the Spring fundamentals required for implementing RESTful APIs, you can move on to the next chapter.

We'll cover the following topics as part of this chapter:

- Introduction to Spring
- Understanding the basic concepts of the Spring Framework
- Working with the servlet dispatcher

Technical requirements

This chapter covers concepts and does not involve writing actual code. However, you'll need basic Java knowledge.

Please visit the following link to download the code files: `https://github.com/PacktPublishing/Modern-API-Development-with-Spring-6-and-Spring-Boot-3/tree/main/Chapter02`.

Understanding the patterns and paradigms of Spring

Spring is a framework written in the Java language. It provides lots of modules, such as Spring Data, Spring Security, Spring Cloud, Spring Web, and so on. It is popular for building enterprise applications. Initially, it was looked at as a **Java Enterprise Edition** (**JEE**) alternative. However, over the years, it has become preferred over JEE. Spring supports **dependency injection** (**DI**), also known as **inversion of control** (**IoC**), and **aspect-oriented programming** (**AOP**) out of the box at its core. Apart from Java, Spring also supports other JVM languages such as Groovy and Kotlin.

With the introduction of Spring Boot, the turnaround time for the development of web services was reduced. We can hit the ground running. This is huge and one of the reasons why Spring has become so popular lately.

Covering Spring fundamentals itself requires a dedicated book. I'll try to be concise and cover all the features required for you to go ahead and grasp the required knowledge of REST implementation in a granular way.

However, before we proceed, we should understand the principles and design patterns that form Spring's foundations, particularly IoC, DI, and AOP.

What is IoC?

Traditional CLI programs are the typical method for implementations of procedural programming, where the flow is determined by the programmer and the code runs sequentially, meaning one piece after another. However, UI-based OS applications determine the flow of programs based on user inputs and events, which are dynamic.

Long ago, when mostly procedural ways of programming were dominant, you would have to look for a way to move the control of flow from the traditional procedural way (where the programmer dictates the flow) to external sources such as a framework or components that determined the control flow of the program. This movement is what is called IoC. It is a very generic principle and part of most frameworks.

With the arrival of the **object-oriented programming (OOP)** approach, frameworks soon began to offer the IoC container pattern implementation, which supports DI.

What is DI?

Let's say you are writing a program that needs to get some data from a database. The program therefore requires a database connection. You could use the JDBC database connection object, instantiating and assigning the database connection object instantly in the program.

Or you could simply take the connection object as a constructor or a setter/factory method parameter. Then, the framework will create the connection object per the configuration and assign that object to your program at runtime. Here, the framework essentially injects the connection object at runtime. This is called DI. Spring supports DI for class compositions.

> **Note**
>
> The Spring Framework throws an error at runtime if any dependency is unavailable, or if the proper object name is not marked when more than one type of object is available. In contrast, there are some frameworks that also check these dependencies at compile time, such as Dagger, for example.

DI is a type of IoC. IoC containers construct and maintain implementation objects. These types of objects (objects required by other objects – a form of dependency) are injected into objects that need them in a constructor, setter, or interface. This decouples the instantiation and allows DI at runtime. DI can also be achieved using the Service Locator pattern. However, we'll stick to the IoC pattern approach.

We'll look at IoC more closely with a code example in the following main section of this chapter.

What is AOP?

AOP is a programming paradigm that works in tandem with OOP. It's a good practice in OOP to handle only a single responsibility in a particular class – this principle is called the **single-responsibility principle** (**SRP**), which is applicable to modules/classes/methods. For example, if you are writing a Gear class in an application for the automotive domain, then the Gear class should only allow functions related to the gear object, or it should not be allowed to perform other functions such as braking. However, in programming models, you often need a feature/function that extends across more than one class. In fact, in some applications, most classes use features such as logging or metrics.

Features such as logging, security, transaction management, and metrics are required across multiple classes/modules. The code of these features is also scattered across multiple classes. In OOP, there is no way to abstract and encapsulate such features. This is where AOP comes to the rescue. These features (read: aspects) are cross-cutting concerns that range across multiple points in the object model. AOP provides a way to let you handle these aspects across multiple classes/modules.

AOP allows you to do the following:

- Abstract and encapsulate cross-cutting concerns.
- Add aspects' behavior around your code.
- Make the code for cross-cutting concerns modular to allow us to easily maintain and extend it.
- Focus on your business logic inside the code. This makes code clean. Cross-cutting concerns are encapsulated and maintained separately.

Without AOP, it is very difficult and complex to achieve all the preceding points.

This section should have helped you to understand IoC, DI, and AOP conceptually. Throughout the rest of this chapter, we'll take a deep dive into the code implementation of these patterns and paradigms.

Now, we will go through the fundamentals of the Spring Framework and its basic building blocks.

Understanding the application of IoC containers

The Spring Framework's backbone is the IoC container that is responsible for a bean's life cycle. In the Spring world, a Java object can be a **bean** if it is instantiated, assembled, and managed by the IoC container. You create a great number of beans, or objects, for your application. A bean may have dependencies that require other objects to work. The IoC container is responsible for injecting the object's dependencies when it creates that bean. In the Spring context, IoC is also known as DI.

> **Note**
>
> You can refer to the Spring documentation (`https://docs.spring.io/spring-framework/docs/current/reference/html/`) for more information about the Spring Framework.

The Spring Framework's IoC container core is defined in two packages: `org.springframework.beans` and `org.springframework.context`.

BeanFactory (`org.springframework.beans.factory.BeanFactory`) and **ApplicationContext** (`org.springframework.context.ApplicationContext`) are two important interfaces that provide the basis for IoC containers. *BeanFactory* provides the configuration framework and basic functionality and takes care of bean instantiation and wiring.

ApplicationContext can also take care of bean instantiation and wiring. However, it primarily provides more enterprise-specific functionalities, as follows:

- Integrated life cycle management

- Automatic registration of `BeanPostProcessor` and `BeanFactoryPostProcessor`

- Internationalization (message resource handling) with easy access to `MessageSource`

- Event publication using a built-in `ApplicationEvent`

- Provision of `WebApplicationContext`, an application-layer-specific context for web applications

`ApplicationContext` is a sub-interface of `BeanFactory`. Let's look at its class signature:

```
public interface ApplicationContext extends EnvironmentCapable,
ListableBeanFactory, HierarchicalBeanFactory,MessageSource,
ApplicationEventPublisher, ResourcePatternResolver {…}
```

Here, `ListableBeanFactory` and `HierarchicalBeanFactory` are sub-interfaces of `BeanFactory`.

Spring recommends the use of `ApplicationContext` due to these additional features as well as its state-of-the-art bean management functionalities.

Now you know that the `ApplicationContext` interface represents the IoC container and manages the beans, but you must be wondering how it determines which beans to instantiate, assemble, and configure. Where does it get its instructions? The answer is configuration metadata. Configuration metadata allows you to express your application objects and the interdependencies among those objects. Configuration metadata can be represented in three ways: through XML configuration, Java annotations, and Java code. You write the business objects and provide the configuration metadata, and the Spring container generates a fully configured, ready-to-use system as shown in *Figure 2.1*.

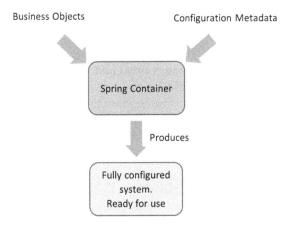

Figure 2.1 – Spring container

Now that you have some understanding of how beans are managed, let's learn more about what a bean is and what it can do.

Defining a bean and its scope

Beans are Java objects that are managed by the IoC containers. The developer supplies the configuration metadata to an IoC container, which then uses the metadata to construct, assemble, and manage the beans. Each bean should have a unique identifier inside a container. A bean can even have more than one identity using an **alias**.

You can define beans using XML, Java, and annotations. Let's declare a simple bean using a Java-based configuration:

```
public class SampleBean {
   public void init() { // initialization logic }
   public void destroy() { // destruction logic }
   // bean code
}

public interface BeanInterface {
   // interface code
}

public class BeanInterfaceImpl implements BeanInterface
               {
```

```
  // bean code
}

@Configuration
public class AppConfig {

  @Bean(initMethod = "init", destroyMethod = "destroy",
    name = {"sampleBean", "sb"})
  @Description("Demonstrate a simple bean")
  public SampleBean sampleBean() {
    return new SampleBean();
  }
  @Bean
  public BeanInterface beanInterface() {
    return new BeanInterfaceImpl();
  }
}
```

In the preceding code, the bean is declared using the AppConfig class. @Configuration is a class-level annotation that indicates that the class contains code for configuration. @Bean is a method-level annotation used to define the bean. You can also pass the bean's initialization and destruction life cycle method using the @Bean annotation attributes as shown in the preceding code.

In general, a bean's name is a class name with its initial letter in lowercase. For example, the bean name of BeanInterface would be beanInterface. However, you can also use the name attribute to define a bean name and its aliases – as we can see in the preceding code, SampleBean has two bean names: sampleBean and sb.

> **Note**
>
> The default methods for destructions are close/shutdown public methods, which are called by the container automatically. However, if you wish to follow a different method, you can use the preceding code. If you don't want the container to call the default destruction method, you can assign an empty string to the destroyMethod attribute (destroyMethod = "").

You can also create a bean using the interfaces shown in the previous code for the BeanInterface bean.

Note that the @Bean annotation should be inside the @Component annotation. The @Component annotation is a generic way to declare a bean. A class annotated with @Configuration lets the method return a bean annotated with @bean. @Configuration is meta-annotated with @Component; therefore, the @Bean annotation works inside it. There are other annotations such as @Controller, @Service, and @Repository, which are also annotated with @Component.

The @Description annotation, as the name suggests, is used to describe a bean. When monitoring tools are used, these descriptions help us to understand the beans at runtime.

The @ComponentScan annotation

The @ComponentScan annotation allows the auto-scanning of beans. It takes a few arguments, such as base packages and their classes. The Spring container then investigates all the classes inside the base package and looks for beans. It scans all classes annotated with @Component, or other annotations that are meta-annotated with @Component, such as @Configuration, @Controller, and so on.

By default, Spring Boot takes the default base package from the class, which has the @ComponentClass annotation. You can use the basePackageClasses attribute to identify which packages should be scanned.

Another way to scan more than two packages is by using the basePackages attribute. It allows you to scan more than one package.

If you want to use more than one @ComponentScan, then you can wrap them inside the @ComponentScans annotation as follows:

```
@Configuration
@ComponentScans({
  @ComponentScan(basePackages = "com.packt.modern.api"),
  @ComponentScan(basePackageClasses = AppConfig.class)
})
class AppConfig { //code }
```

The bean's scope

Spring containers are responsible for creating a bean's instances. How instances will be created by the Spring container is defined by the scope. The default scope is singleton, where only one instance will be created per IoC container, and the same instance will be injected. If you want to create a new instance each time it is requested, you can define the prototype scope for the bean.

The singleton and prototype scopes are available for all Spring-based applications. There are four more scopes available for web applications: request, session, application, and websocket. For these last four scopes, the application context should be web-aware, as Spring Boot-based web applications are web-aware.

The following table contains all the scopes:

Scope	Usage
singleton	Creates a new instance per IoC container. This is the default scope.
prototype	Creates a new instance for each injection (for collaborated beans).
request	Only for web-aware context. A single bean instance will be created for each HTTP request and will be valid throughout the HTTP request life cycle.

Scope	Usage
`session`	Only for web-aware context. A single bean instance will be created for each HTTP session and will be valid throughout the HTTP session life cycle.
`application`	Only for web-aware context. A single instance will be created for the application scope, and be valid throughout the life cycle of servlet-context.
`websocket`	Only for web-aware context. A single instance will be created for each WebSocket session.

Let's see how we can define the singleton and prototype scopes in code:

```
@Configuration
public class AppConfig {
  // no scope is defined so default singleton
  // scope is applied.
  // If you want to define it explicitly, you can
  // do that using
  // @Scope(value =
      ConfigurableBeanFactory.SCOPE_SINGLETON)
  // OR
  // @Scope(ConfigurableBeanFactory.SCOPE_SINGLETON)
  // Here,
  // ConfigurableBeanFactory.SCOPE_SINGLETON
  // is a string constant with value "singleton".
  @Bean
  public SingletonBean singletonBean() {
    return new SingletonBean();
  }

  @Bean
  @Scope(ConfigurableBeanFactory.SCOPE_PROTOTYPE)
  public PrototypeBean prototypeBean() {
    return new PrototypeBean();
  }

  @Bean
  @Scope(value =
      WebApplicationContext.SCOPE_REQUEST,
      proxyMode = ScopedProxyMode.TARGET_CLASS)
  // You need a proxyMode attribute because when
  // web-aware context is instantiated,
```

```
    // you don't have any HTTP request.
    // Therefore, Spring injects the proxy as a
    // dependency and instantiates the bean when the
    // HTTP request is invoked. OR, in short you
    // can simply write @RequestScope, a shortcut
    @RequestScope
    public ReqScopedBean requestScopedBean() {
      return new ReqScopedBean();
    }
}
```

Similarly, you can create a web-aware context-related bean as follows:

```
@Configuration
public class AppConfig {
  @Bean
  @Scope(value =
      WebApplicationContext.SCOPE_REQUEST,
      proxyMode = ScopedProxyMode.TARGET_CLASS)
  // You need a proxyMode attribute because when
  // web-aware context is instantiated, you
  // don't have any HTTP request. Therefore,
  // Spring injects the proxy as a dependency and
  // instantiate the bean when HTTP request is
  // invoked. OR, in short you can annotate with
  // @SessionScope, a shortcut for above
  @SessionScope
  public ReqScopedBean requestScopedBean() {
    return new ReqScopedBean();
  }
  @ApplicationScope
  public ReqScopedBean requestScopedBean() {
    return new ReqScopedBean();
  }
  // here "scopeName" is alias for value
  // interestingly, no shortcut. Also hard coded
  // value for websocket
  @Scope(scopeName = "websocket",
      proxyMode = ScopedProxyMode.TARGET_CLASS)
  public ReqScopedBean requestScopedBean() {
    return new ReqScopedBean();
  }
}
```

We have now succinctly covered beans here. However, you can explore more on this topic in the official Spring documentation (`https://docs.spring.io/spring-framework/docs/current/reference/html/core.html#beans-definition`). For now, let us move on and start learning how to configure beans.

Configuring beans using Java

Before Spring 3, you could only define beans using XML. Spring 3 introduced the `@Configuration`, `@Bean`, `@import`, and `@DependsOn` annotations to configure and define Spring beans using Java.

You have already learned about the `@Configuration` and `@Bean` annotations in the *Defining a bean and its scope* section. Now, you will explore how to use the `@Import` and `@DependsOn` annotations.

The `@Import` annotation is more useful when you develop an application without using autoconfiguration.

The @Import annotation

`@Import` is used for modularizing configurations when you have more than one configuration class. You can import the bean's definitions from other configuration classes, and this is useful when you instantiate the context manually. Spring Boot uses auto-configuration, so you don't need to use `@Import`. However, you would have to use `@Import` to modularize the configurations if you want to instantiate the context manually.

Let's say the `FooConfig` configuration class contains `FooBean` and the `BarConfig` configuration class contains `BarBean`. The `BarConfig` class also imports `FooConfig` using `@Import`:

```
@Configuration
public class FooConfig {
  @Bean
  public FooBean fooBean() {
    return new FooBean();
  }
}
@Configuration
@Import(FooConfig.class)
public class BarConfig {
  @Bean
  public BarBean barBean() {
    return new BarBean();
  }
}
```

Now, while instantiating the container (context), you can just supply `BarConfig` to load both the `FooBean` and `BarBean` definitions in the Spring container, shown as follows:

```
public static void main(String[] args) {
  ApplicationContext appContext = new
    AnnotationConfigApplicationContext(
      BarConfig.class);
  // Get FooBean & BarBean beans from context
  FooBean fooBean =
    appContext.getBean(FooBean.class);
  BarBean barBean =
    appContext.getBean(BarBean.class);
}
```

The @DependsOn annotation

The Spring container manages the bean initialization order. What if you have a bean that depends on another bean? You want to make sure that the required bean is initialized before the bean that depends on it. @DependsOn helps you to achieve this when you configure beans using Java (not through XML).

You will get the `NoSuchBeanDefinitionException` exception if a bean's initialization order is messed up and the Spring container fails to find the dependency as a result.

Let's assume we have a bean called `BazBean` that depends on the `FooBean` and `BarBean` beans. You can make use of the @DependsOn annotation to maintain the initialization order. The Spring container will follow the instructions and initialize both the `FooBean` and `BarBean` beans before creating `BazBean`. Here's how the code would look:

```
@Configuration
public class AppConfig {
  @Bean
  public FooBean fooBean() {
    return new FooBean();
  }
  @Bean
  public BarBean barBean () {
    return new BarBean ();
  }
  @Bean
  @DependsOn({"fooBean","barBean"})
  public BazBean bazBean (){
    return new BazBean ();
  }
}
```

How to code DI

Have a look at the following example. CartService has a dependency on CartRepository. The CartRepository instantiation is done inside the CartService constructor:

```
public class CartService {
  private CartRepository repository;
  public CartService() {
    this.repository = new CartRepositoryImpl();
  }
}
```

We can decouple this dependency in the following way:

```
public class CartService {
  private CartRepository repository;
  public CartService(CartRepository repository) {
    this.repository = repository;
  }
}
```

If you create a bean of the CartRepository implementation, you can easily inject the CartRepository bean using configuration metadata. Before that, let's have a look at the Spring container again.

You have seen how ApplicationContext can be initialized in the *The @Import annotation* subsection of this chapter. When it gets created, it takes all the metadata from the bean's configuration. @Import allows you to have multiple configurations.

Each bean can have its dependencies, that is, a bean may need other objects to work (compositions), as in the CartService example. These dependencies can be defined using constructors, setter methods, or properties. These dependent objects (part of the constructors, setter method arguments, or class properties) are injected by the Spring container (ApplicationContext) using the bean's definition and its scope. We'll next investigate each of these ways to define the DI.

> **Note**
>
> DI virtually makes a class independent of its dependencies. Therefore, a change in the dependency won't impact the class (no code change) as long as the interface contract is not broken. Indeed, you can change the underlying implementation of the dependency or use a different implementation class.

Using a constructor to define a dependency

Now, you'll see how you can inject CartRepository into the CartService constructor. A way to inject a dependency using a constructor is as follows:

```
@Configuration
public class AppConfig {
  @Bean
  public CartRepository cartRepository() {
    return new CartRepositoryImpl();
  }
  @Bean
  public CartService cartService() {
    return new CartService(cartRepository());
  }
}
```

Using a setter method to define a dependency

Now, let's change the CartService class. Instead of having a constructor, use the setter method to instantiate the dependency:

```
public class CartService {
private CartRepository repository;
  public void setCartRepository(CartRepository repository) {
    this.repository = repository;
  }
}
```

Now, you can use the following configuration to inject the dependency:

```
@Configuration
public class AppConfig {
  @Bean
  public CartRepository cartRepository() {
    return new CartRepositoryImpl();
  }
  @Bean
  public CartService cartService() {
    CartService service = new CartService();
    Service.setCartService(cartRepository());
    return service;
  }
}
```

Using a class property to define a dependency

Spring also provides an out-of-the-box solution for injecting a dependency using the @Autowired annotation. It makes code look cleaner. Have a look at the following example:

```
@Service
public class CartService {
  @Autowired
  private CartRepository repository;
}
```

The Spring container will take care of injecting the CartRepository bean. You'll learn more about @Autowired in the next section.

Configuring a bean's metadata using annotations

The Spring Framework provides lots of annotations to configure the metadata for beans. However, we'll focus on the most used annotations: @Autowired, @Qualifier, @Inject, @Resource, @Primary, and @Value.

How to use @Autowired?

The @Autowired annotation allows you to define the configuration in a bean's class itself, instead of writing a separate configuration class annotated with @Configuration.

The @Autowired annotation can be applied to a field (as we saw in the class property-based DI example), constructor, setter, or any method.

The Spring container makes use of reflections to inject the beans annotated with @Autowired. This also makes it more costly than other injection approaches.

Please make a note that applying @Autowired to class members will only work if there is no constructor or setter method to inject the dependent bean.

Here is a code example demonstrating the use of @Autowired to inject dependencies:

```
@Component
public class CartService {
  private CartRepository repository;
  private ARepository aRepository;
  private BRepository bRepository;
  private CRepository cRepository;
  @Autowired // member(field) based auto wiring
  private AnyBean anyBean;
  @Autowired // constructor based autowired
```

```
  public CartService(CartRepository repository) {
    this.repository = repository;
  }
  @Autowired // Setter based auto wiring
  public void setARepository(ARepository aRepo) {
   this.aRepository = aRepo;
  }
  @Autowired // method based auto wiring
  public void xMethod(BRepository bRepo, CRepository cRepo) {
   this.bRepository = bRepo;
   this.cRepository = cRepo;
  }
}
```

@Autowired works based on reflection. However, to remove the ambiguity, matching beans are found and injected using type matching, qualifier matching, or name matching in this order of precedence. These are applicable to both field and setter method injections.

Matching by type

The following example works because type matches take precedence. The code finds the CartService bean and injects it into CartController:

```
@Configuration
public class AppConfig {
  @Bean
  public CartRepository cartRepository() {
    return new CartRepositoryImpl();
  }
  @Bean
  public CartService cartService() {
    CartService service = new CartService();
    service.setCartRepository(cartRepository());
    return service;
  }
}
@Controller
public class CartController {
  @Autowired
  private CartService service;
}
```

Matching by qualifier

Let's assume there is more than one bean of a given type. As a result, the Spring container won't be able to determine the correct bean by type matching:

```
@Configuration
public class AppConfig {
  @Bean
  public CartService cartService1() {
    return new CartServiceImpl1();
  }
  @Bean
  public CartService cartService2() {
    return new CartServiceImpl2();
  }
}
@Controller
public class CartController {
  @Autowired
  private CartService service1;
  @Autowired
  private CartService service2;
}
```

This example, when run, will return NoUniqueBeanDefinitionException. To sort this out, we can make use of the @Qualifier annotation.

If you look closely, you'll find that the configuration class has two beans identified by their method names: cartService1 and cartService2. Or, you can also make use of the value attribute of the @Bean annotation to give it a name/alias. Now, you can use these names to assign the same type to these two different beans using the @Qualifier annotation as shown:

```
@Controller
public class CartController {
  @Autowired
  @Qualifier("cartService1")
  private CartService service1;
  @Autowired
  @Qualifier("cartService2")
  private CartService service2;
}
```

Matching by name

Let's define a service using the @Service annotation, which is a specialized type of @Component. Let's assume we have a component scan in place:

```
@Service(value="cartServc")
public class CartService {
  // code
}
@Controller
public class CartController {
  @Autowired
  private CartService cartServc;
}
```

This code works because the field name of CartController for CartService is the same as was given to the value attribute of the @Service annotation. If you change the field name from cartServc to something else, it will fail with NoUniqueBeanDefinitionException.

There are other annotations: @Inject (JSR-330: https://jcp.org/en/jsr/ detail?id=330) and @Resource (JSR-250: https://jcp.org/en/jsr/ detail?id=250). @Inject also requires the javax.inject library. @Resource and @Inject are similar to @Autowired and can be used for injecting dependencies. Both @Autowired and @Inject have the same execution path precedence (by type, then by qualifier, then by name). However, the @Resource execution path preference is by name (the first preference), then by type, and lastly by qualifier.

What is the purpose of @Primary?

In the previous subsection, we saw that @Qualifier helps you to resolve which type should be used when multiple beans are available for injection. The @Primary annotation allows you to set one of the type's beans as the default. Bean annotation with @Primary will be injected into auto-wired fields:

```
@Configuration
public class AppConfig {
  @Bean
  @Primary
  public CartService cartService1() {
    return new CartServiceImpl1();
  }
  @Bean
  public CartService cartService2() {
    return new CartServiceImpl2();
  }
}
```

```
@Controller
public class CartController {
  @Autowired
  private CartService service;
}
```

In this example, the bean marked with @Primary will be used to inject the dependency into the CartController class for CartService.

When can we use @Value?

Spring supports the use of external property files via <xyz>.properties or <xyz>.yml. Let's say you want to use the value of any property in your code. You can achieve this using the @Value annotation. Let's have a look at the following example code:

```
@Configuration
@PropertySource("classpath:application.properties")
public class AppConfig {}
@Controller
public class CartController {
  @Value("${default.currency}")
  String defaultCurrency;
}
```

The defaultCurrency field would take its value from the default.currency field defined in the application.properties file. If you are using Spring Boot, you don't need to use @PropertySource. You just need to place application.yml or the properties file under the src/main/resources directory.

Now that you have some idea of how a bean's metadata can be configured to achieve different results, let us move on to writing some code for AOP, which we introduced conceptually earlier in this chapter.

Writing code for AOP

We discussed AOP previously in the *Understanding the patterns and paradigms of Spring* section. In simple terms, AOP is a programming paradigm that solves cross-cutting concerns such as logging, transactions, and security. These cross-cutting concerns are known as *aspects* in AOP. They allow you to modularize your code and place cross-cutting concerns in a central location.

The following code captures the time taken by a method to execute:

```
class Test
  public void performSomeTask() {
    long start = System.currentTimeMillis();
```

```
      // Business Logic
      long executionTime =
            System.currentTimeMillis() - start;
      System.out.println("Time taken: " + executionTime + "ms");
   }
}
```

Time calculations are used for monitoring performance. This code captures the execution time (how long the code takes to run). If you have hundreds of methods in your application, then you need to add a time-capturing piece of code in each one of them to monitor the time taken for their execution. But what if you later wanted to modify the time-capturing code? You would have to modify the code in all those places. You don't want to do that. This is where AOP helps you. It makes your cross-cutting code modular.

Let's create an AOP example for capturing the amount of time a method takes to execute. Here, logging monitoring time will be our aspect that will capture the amount of time taken by a method for its execution.

As the first step, you'll define an annotation (TimeMonitor) to target the method. Methods annotated with @TimeMonitor will log the time taken by that method. This will help us to identify the pointcut. The pointcut is defined in the Aspect class's code explanation:

```
@Target(ElementType.METHOD)
@Retention(RetentionPolicy.RUNTIME)
public @interface TimeMonitor {}
```

Next, we need to define the Aspect. Aspects insert additional logic during the execution of the program at a certain point. This point is known as the **join point**. The join point could be a field being modified, a method being called, or an exception being thrown.

The @Aspect annotation is used to mark the class as an Aspect. The time-monitoring aspect has been defined using the following code:

```
@Aspect
@Component
public class TimeMonitorAspect {
   @Around("@annotation(TimeMonitor)")
   public Object logTime(ProceedingJoinPoint joinPoint) throws
Throwable {
      long start = System.nanoTime();
      Object proceed = joinPoint.proceed();
      long exeTime = System.nanoTime() - start;
      System.out.println(joinPoint.getSignature()
         .getName() + " took: " + exeTime + " ns");
      return proceed;
```

```
  }
}
```

@Around is a method annotation that defines Advice. **Advice** is an action taken by the Aspect at a specific time (Joinpoint). Advice could be any of the following:

- @Before: Advice executes before the JoinPoint.

- After: Advice executes after the JoinPoint. It has three subtypes:

  - @After: Advice executes after the JoinPoint irrespective of the method's outcome – successful or failed

  - @AfterReturning: Advice executes after the JoinPoint executes successfully

  - @AfterThrowing: Advice executes after the JoinPoint throws an exception

- @Around: Advice executes before and after the JoinPoint.

TimeMonitorAspect executes at the method level because the MonitorTime Advice target is a method.

@Around also takes an expression argument, @annotation(com.packt.modern.api. TimeMonitor). This predicate expression is known as the **pointcut**, which determines whether the Advice needs to be executed or not. The logTime method will be executed for all the methods annotated with @TimeMonitor. Spring supports the AspectJ expression syntax (https://www. eclipse.org/aspectj/doc/released/quick5.pdf). These expressions are dynamic in nature, allowing flexibility while defining the pointcut.

JoinPoint is added as a method logTime() parameter. With the JoinPoint object, you can capture all the information of the target and proxy. You can capture the method's full signature, class name, method name, arguments, and so on using the JoinPoint object.

That's all we need to implement TimeMonitorAspect. Now you can simply add the @TimeMonitor annotation to log the computed time taken by the methods as shown here:

```
class Test {
  @TimeMonitor
  public void performSomeTask() {
    // Business Logic
  }
}
```

`JoinPoint` also allows you to capture the target object and proxy. You might be wondering what these are. These are created by the Spring AOP module and are important for AOP to work. An `Advice` is applied to the target object. Spring AOP creates a subclass of the target object and overrides the methods and advice is inserted. On the other hand, the proxy is an object that is created after some advice is applied to the target object using the CGLIB or JDK proxy lib. (CGLIB generates code, and Spring makes use of it for generating the proxy classes and related code. The JDK proxy lib is also used for generating proxy classes and instances.)

Now that we have covered some important concepts in Spring, let us get to know Spring Boot, one of the best tools for developing Spring web applications.

Why use Spring Boot?

Nowadays, Spring Boot is the obvious choice for developing state-of-the-art, production-ready web applications specific to Spring. Its website (`https://projects.spring.io/spring-boot/`) also outlines its huge advantages.

Spring Boot is an amazing Spring tool created by **Pivotal** that was released for General Availability in April 2014. It was developed based on the request of SPR-9888 (`https://jira.spring.io/browse/SPR-9888`) with the title *Improved support for containerless web application architectures*.

You might be wondering: why *containerless*? Because today's cloud environment, with its **Platform-as-a-Service (PaaS)** offerings, provides most of the features offered by container-based web architectures, such as reliability, management, and scaling. Therefore, Spring Boot focuses on making itself an ultralight container.

Spring Boot has its own default configurations and also supports auto-configuration to make production-ready web application development simple. **Spring Initializr** (`http://start.spring.io`) is a web-based service where you simply choose your build tools, such as *Maven* or *Gradle*, along with the project metadata, such as the group, artifacts, and any dependencies. Once you fill in the required fields, you simply click on the **Generate Project** button to be provided with a Spring Boot project that you can use for your production application.

On the page at the preceding link, we can see the default **Packaging** option is **Jar**, which we are going to use throughout this book. You would use the **War** packaging option (for **web archive**) if you wanted to deploy the application on a web server such as *WebLogic* or *Tomcat*.

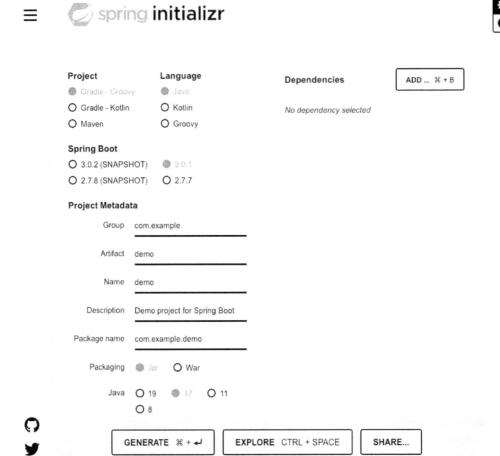

Figure 2.2 – Spring Initializr

In simple words, Spring Initializr does all the configuration for us so we can focus on writing our business logic and APIs.

> **Building and running the code**
>
> You can build the code by running `gradlew clean build` from the root of the project and running the service using `java -jar build/libs/Chapter02-0.0.1-SNAPSHOT.jar`. Make sure to use Java 17 in the path.

In the next section, let's dig into servlet dispatcher, which is responsible for Spring MVC and REST controller features.

Understanding the importance of servlet dispatcher

In the previous chapter, you learned that RESTful web services are developed on top of the HTTP protocol. Java has a Servlets feature to work with HTTP. Servlets allow you to have path mapping that can work at REST endpoints and provides the HTTP method for identification. Servlets also allow you to form different types of response objects, including JSON and XML. However, they offer a somewhat crude way of implementing REST endpoints, as you must still handle the request URI, parse the parameters, and convert JSON/XML and the responses on your own.

Spring MVC comes to your rescue. Spring MVC is based on the **Model-View-Controller** (**MVC**) pattern and has been part of the Spring Framework since its first release. MVC is a well-known design pattern:

- **Model**: Models are Java objects (called POJOs) that contain the application data. They also represent the state of the application.

- **View**: The view is a presentation layer that consists of HTML/JSP/template files. The view renders the data from models and generates the HTML output.

- **Controller**: The controller processes the user requests and builds the model.

DispatcherServlet is part of Spring MVC. It works as a front controller, that is, it handles all the incoming HTTP requests. Spring MVC is a web framework that allows you to develop traditional web applications where UI apps are also part of the backend. However, you'll develop RESTful web services, and the UI will be based on the React JavaScript library; therefore, we'll keep the servlet dispatcher role limited to implementing the REST endpoints using `@RestController`.

Let's have a look at the flow of an example user request in Spring MVC for the REST controller:

1. The user sends the HTTP request, which is received by `DispatcherServlet`.

2. `DispatcherServlet` passes the baton to `HandlerMapping`. `HandlerMapping` does the job of finding the correct controller for the requested URI and passes it back to `DispatcherServlet`.

3. `DispatcherServlet` then makes use of `HandlerAdapter` to handle `Controller`.

4. `HandlerAdapter` calls the appropriate method inside `Controller`.

5. `Controller` then executes the associated business logic and forms the response.

6. Spring makes use of the marshaling/unmarshalling of the request and response objects for JSON/XML conversion from Java and vice versa.

Let's see a visual representation of this process:

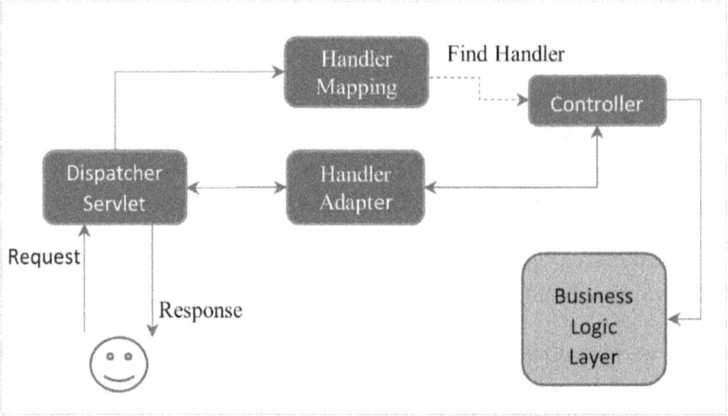

Figure 2.3 – DispatcherServlet

You now have a good understanding of the importance of `DispatcherServlet`, which is key for REST API implementation.

Summary

This chapter helped you learn about Spring's key concepts: beans, DI, and AOP. You also learned how to define the scope of beans and create `ApplicationContext` programmatically, using it to get the beans. You can define beans' configuration metadata using Java and annotations and have learned how to use different beans of the same type.

You also implemented an `Aspect` example, applying a module approach to a cross-cutting concern, and learned the key concepts of the AOP programming paradigm.

Since we are going to implement REST APIs in this book, it is important to understand the servlet dispatcher concept.

In the next chapter, we'll implement our first REST API application using the OpenAPI Specification and use a Spring controller to implement it.

Questions

1. How do you define a bean with the prototype scope?
2. What is the difference between prototype and singleton beans?
3. What is required for a session and request scope to work?

4. What is the relationship between advice and pointcut in terms of AOP?

5. Write an `Aspect` for logging that prints the method and argument names before the method execution, and prints a message with the return type (if any) after the method's successful execution.

Answers

1. By using the `@Scope` annotation as shown:

   ```
   @Scope(ConfigurableBeanFactory.SCOPE_PROTOTYPE)
   ```

2. Beans defined using the `singleton` scope are instantiated only once per Spring container. The same instance is injected every time it is requested, whereas with a bean defined with the `prototype` scope, the container creates a new instance every time the injection is done by the Spring container for the requested bean. In short, a container creates a single bean per container for a singleton-scoped bean, whereas a container creates a new instance every time there is a new injection for prototype-scoped beans.

3. Session and request scopes only work when a web-aware Spring context is used. Other scopes that also need a web-aware context to work are application and `WebSocket` scopes.

4. Advice is an action taken by the `Aspect` at a specific time (`JoinPoint`). Aspects perform the additional logic (advice) at a certain point (`JoinPoint`), such as a method being called, an exception being thrown, and so on.

5. The following code will print the method name and argument names before the method execution, and a message with the return type after the method execution:

   ```
   @Aspect
   @Component
   public class TimeMonitorAspect {
    @Around("@annotation(TimeMonitor)")
    public Object logTime(ProceedingJoinPoint
      joinPoint) throws Throwable {
     System.out.println(String.format("Method
       Name: %s, Arg Name: %s",
       joinPoint.getSignature().getName(),
       Arrays.toString(((CodeSignature)
       joinPoint.getSignature())
       .getParameterNames())));
      Object proceed = joinPoint.proceed();
      System.out.println(String.format("Method
        %s contains the following return type:
        %s",joinPoint.getSignature().getName(),
        ((MethodSignature)
   ```

```
        joinPoint.getSignature()).
        getReturnType().toGenericString()));
    return proceed;
  }
}
```

Further reading

- IoC containers and the DI pattern: `https://martinfowler.com/articles/injection.html`

- The Spring Framework documentation: 6.0.0-M5 was the latest at the time of writing this book: `https://docs.spring.io/spring-framework/docs/6.0.x/reference/html/`

- *Spring Boot 2 Fundamentals*: `https://www.packtpub.com/product/spring-boot-2-fundamentals/9781838821975`

- *Developing Java Applications with Spring and Spring Boot*: `https://www.packtpub.com/product/developing-java-applications-with-spring-and-spring-boot/9781789534757`

3

API Specifications and Implementation

In previous chapters, we learned about the design aspects of REST APIs and the Spring fundamentals required to develop RESTful web services. In this chapter, you'll make use of these two areas to implement REST APIs.

We have chosen a design-first approach for implementation to make our development process understandable for non-technical stakeholders as well. To make this approach possible, we will make use of the **OpenAPI Specification** (**OAS**) to, first, design an API and, later, implement it. We will also learn how to handle errors that occur while serving the request. In this chapter, we will use the example of designing and implementing an API of a sample e-commerce app.

By the end of this chapter, you should be able to design the API specifications and make use of the OpenAPI codegen to generate the code for models and API Java interfaces. You will also know how to write the pseudo-Spring controllers to implement the API Java interfaces and Global Exception Handler for the web service.

We'll cover the following topics as part of this chapter:

- Designing APIs with OAS
- Understanding the basic structure of OAS
- Converting OAS to Spring code
- Implementing the OAS code interfaces
- Adding a Global Exception Handler
- Testing the implementation of the controllers

Technical requirements

You need the following to execute the instructions in this and the following chapters:

- Any Java IDE, such as NetBeans, IntelliJ, or Eclipse

- **Java Development Kit (JDK)** 17

- An internet connection to download the dependencies and Gradle

You can find the code files for this chapter on GitHub at `https://github.com/PacktPublishing/ Modern-API-Development-with-Spring-6-and-Spring-Boot-3/tree/main/ Chapter03`.

Designing APIs with OAS

You can directly start coding the API; however, this approach leads to many issues, such as frequent modifications, difficulty in API management, and difficulty in reviews specifically led by non-technical domain teams. Therefore, you should use the **design-first approach**.

The first question that comes to mind is, how can we design REST APIs? You learned in *Chapter 1*, *RESTful Web Service Fundamentals*, that there is no existing standard to govern REST API implementation. OAS was introduced to solve at least the aspects of the REST API's specification and description. It allows you to write REST APIs in the **YAML Ain't Markup Language** (**YAML**) or **JavaScript Object Notation** (**JSON**) markup languages.

We'll use version 3.0 of OAS (`https://github.com/OAI/OpenAPI-Specification/ blob/main/versions/3.0.3.md`) to implement the e-commerce app REST API. We'll use YAML (pronounced as *yamel*, rhyming with *camel*), which is cleaner and easier to read. YAML is also space-sensitive. It uses space for indentation; for example, it represents the `key: value` pair (pay attention to the space after the colon – `:`). You can read more about YAML at `https://yaml. org/spec/`.

OAS was earlier known as the *Swagger Specification*. Today, OAS-supporting tools are still known as **Swagger tools**. Swagger tools are open source projects that help the overall development life cycle of REST APIs. We'll make use of the following Swagger tools in this chapter:

- **Swagger Editor** (`https://editor.swagger.io/`): This tool is used to design and describe the e-commerce app REST APIs. It allows you to write and preview, at the same time, your REST APIs' design and description. Make sure that you use OAS 3.0. Its beta version is available at `https://editor-next.swagger.io/`.

- **Swagger Codegen** (`https://github.com/swagger-api/swagger-codegen`): This tool is used to generate the Spring-based API models and Java interfaces. You'll use the Gradle plugin (`https://github.com/int128/gradle-swagger-generator-plugin`) to generate code that works on top of Swagger Codegen. There is also an OpenAPI tool Gradle plugin – OpenAPI Generator (`https://github.com/OpenAPITools/openapi-generator/tree/master/modules/openapi-generator-gradle-plugin`). However, we'll opt for the former one because of the open issues count, which is 3.2k (there are multiple for Java/Spring as well) at the time of writing.

- **Swagger UI** (`https://swagger.io/swagger-ui/`): This tool is used to generate the REST API documentation. The same Gradle plugin will be used to generate the API documentation.

Now that you have some idea of how the design-first approach can be used to develop APIs using OAS supporting tools, let's understand the basic structure of OAS.

Understanding the basic structure of OAS

The OpenAPI definition structure can be divided into the following sections (all are keywords and case-sensitive):

- `openapi (version)`
- `info`
- `externalDocs`
- `servers`
- `tags`
- `paths`
- `components`

All the preceding terms are part of `root`. The first three sections (`openapi`, `info`, and `externalDocs`) are used to define the metadata of the API.

You can place an API's definition either in a single file or divided into multiple files. OAS supports both. We'll use a single file to define the sample e-commerce API.

Instead of discussing all the sections theoretically and then writing the e-commerce API definitions, we'll do both together. First, we'll cover each section definition of the e-commerce API, and then we'll discuss why we have used it and what it implies.

The metadata sections of OAS

Let's have a look at the metadata sections of the e-commerce API definitions:

```
openapi: 3.0.3
info:
  title: Sample Ecommerce App
  description: >
```

```
         'This is a ***sample ecommerce app API***.
          You can find
         out more about Swagger at [swagger.io]
            (http://swagger.io).
         Description supports markdown markup. For example,
            you can
         use the `inline code` using back ticks.'
      termsOfService: https://github.com/PacktPublishing/Modern-API-
   Development-with-Spring-6-and-SpringBoot-3/blob/main/LICENSE
      contact:
        name: Packt Support
        url: https://www.packt.com
        email: support@packtpub.com
      license:
        name: MIT
        url: https://github.com/PacktPublishing/Modern-API-
             Development-with-Spring-6-and-Spring-Boot3/blob
             /main/LICENSE
      version: 1.0.0
   externalDocs:
      description: Any document link you want to generate along
                   with API.
      url: http://swagger.io
```

https://github.com/PacktPublishing/Modern-API-Development-with-
Spring-6-and-Spring-Boot-3/tree/main/Chapter03/src/main/resources/
api/openapi.yaml

Now, let us discuss each code section in detail:

- openapi: The openapi section tells us which OAS is used to write the API's definition. OpenAPI uses semantic versioning (https://semver.org/), which means the version will be in the major:minor:patch form. If you look at the openapi metadata value, we use 3.0.3. This reveals that we use major version 3 with patch 3 (the minor version is 0).

- info: The info section contains the metadata about the API. This information is used to generate documentation and can be used by the client. It contains the following fields, out of which only title and version are mandatory fields, with the others being optional:

  - title: The title of the API.

  - description: This is used to describe the API details. As you can see, we can use Markdown (https://spec.commonmark.org/) here. An > (angular bracket) symbol is used to add multi-line values.

- termsOfService: This is a URL that links to the terms of services. Make sure that it follows the proper URL format.

- contact: This is the contact information of the API provider. The email attribute should be the email address of the contact person/organization. Other attributes are name and url. The name attribute represents the name of the contact person or organization. The url attribute provides the link to the contact page. This is an optional field, and all attributes are also optional.

- license: This is the license information. The name attribute is a required field that represents the correct license name, such as MIT. url is optional and provides a link to the license document.

- version: This exposes the API version in the string format.

- externalDocs: This is an optional field that points to extended documentation of the exposed API. It has two attributes – description and url. The description attribute is an optional field that defines a summary of the external documentation. You can use the Markdown syntax for the description. The url attribute is *mandatory* and links to external documentation.

Let's continue building our API definition. We have completed the metadata section, so let's discuss the servers and tags sections.

The servers and tags sections of OAS

After the metadata section, we can now describe the servers and tags sections. Let's have a look at the following code:

```
servers:
  - url: https://ecommerce.swagger.io/v2
tags:
  - name: cart
    description: Everything about cart
    externalDocs:
      description: Find out more (extra document link)
      url: http://swagger.io
  - name: order
    description: Operation about orders
```

https://github.com/PacktPublishing/Modern-API-Development-with-Spring-6-and-Spring-Boot-3/tree/main/Chapter03/src/main/resources/api/openapi.yaml

Now, let us discuss each code section in detail:

- `servers`: The `servers` section is an optional section that contains a list of servers that host the API. If the hosted API document is interactive, then it can be used by Swagger UI to directly call the API and show the response. If it is not provided, then it points to the root (`/`) of the hosted document server. Server URLs are shown using the `url` attribute.

- `tags`: The `tags` section, defined at the root level, contains the collection of tags and their metadata. Tags are used to group the operations performed on the resources. The `tags` metadata contains name, which is a mandatory field, and two additional optional attributes: `description` and `externalDocs`.

The name attribute contains the tag name. We have already discussed the description and `externalDocs` fields in the previous section on metadata.

Let's discuss the last two sections of OAS.

The components section of OAS

If we were going through the structure sequentially, we would have discussed `path` first. However, conceptually, we want to write our models first before we use them in the `path` section. Therefore, we'll discuss the `components` section first, which is used to define the models.

Here is a code snippet from the `components` section of the sample e-commerce app:

```
components:
  schemas:
    Cart:
      description: Shopping Cart of the user
      type: object
      properties:
        customerId:
          description: Id of the customer who possesses
          the cart
          type: string
        items:
          description: Collection of items in cart.
          type: array
          items:
            $ref: '#/components/schemas/Item'
```

https://github.com/PacktPublishing/Modern-API-Development-with-Spring-6-and-Spring-Boot-3/tree/main/Chapter03/src/main/resources/api/openapi.yaml

If you are working with YAML for the first time, you may find it a bit uncomfortable. However, once you go through this section, you'll feel more comfortable with YAML.

Here, we define a model called `Cart`. The `Cart` model is of the `object` type and contains two fields, namely `customerId` (a string) and `items` (an array).

> **The object data type**
>
> You can define any model or field as an object. Once you mark a type as `object`, the next attribute is `properties`, which consists of all the object's fields. For example, the `Cart` model in the previous code will have the following syntax:
>
> ```
> type: object
> properties:
>     <field name>:
>         type: <data type>
> ```

OAS supports six basic data types, which are as follows (all are in lowercase):

- `string`
- `number`
- `integer`
- `boolean`
- `object`
- `array`

Let's discuss the `Cart` model, in which we use the `string`, `object`, and `array` data types. Other data types are `number`, `integer`, and `boolean`. Now, you may be wondering how to define the date, time, and `float` types, and so on. You can do that with the `format` attribute, which you can use along with the `object` type. For example, take a look at the following code:

```
orderDate:
  type: string
  format: date-time
```

In the previous code, `orderDate` is defined with `type string`, but `format` determines what string value it will contain. Since `format` is marked with `date-time`, the `orderDate` field will contain the date and time in the format defined in *RFC 3339, section 5.6* (`https://tools.ietf.org/html/rfc3339#section-5.6`) – for example, `2020-10-22T19:31:58Z`.

There are some other common formats you can use along with types, as follows:

- `type: number` with `format: float`: This will contain the floating-point number

- `type: number` with `format: double`: This will contain the floating-point number with double precision

- `type: integer` with `format: int32`: This will contain the `int` type (signed 32-bit integer)

- `type: integer` with `format: int64`: This will contain the `long` type (signed 64-bit integer)

- `type: string` with `format: date`: This will contain the date as per *RFC 3339* – for example, `2020-10-22`

- `type: string` with `format: byte`: This will contain the Base64-encoded values

- `type: string` with `format: binary`: This will contain the binary data (and can be used for files)

Our `Cart` model's `items` field is an array of the user-defined `Item` type. Here, `Item` is another model and referenced using `$ref`. In fact, all user-defined types are referenced using `$ref`. The `Item` model is also part of the `components/schema` section. Therefore, the value of `$ref` contains an anchor for user-defined types with `#/component/ schemas/{type}`.

`$ref` represents the reference object. It is based on the JSON reference (`https://tools.ietf. org/html/draft-pbryan-zyp-json-ref-03`) and follows the same semantics in YAML. It can refer to an object in the same document or external documents. Therefore, it is used when you have API definitions divided into multiple files. You have already seen one example of its usage in the previous code. Let's see one more example:

```
# Relative Schema Document
$ref: Cart.yaml
# Relative Document with embedded Schema
$ref: definitions.yaml#/Cart
```

There is another caveat to the previous code. If you look closely, you will find two *items* – one is a property of the `Cart` object type and another one is an attribute of the array type. The former one is simple – a field of the `Cart` object. However, the latter belongs to `array` and is a part of the `array` syntax.

```
Array syntax
type: array
items:
    type: <type of object>
```
i. You could have a nested array if you placed the type of object as `array`

ii. You can also refer to the user-defined type using `$ref`, as shown in the code (then, the `type` attribute is not required for `items`)

Let's see what the `Item` model looks like:

```
Item:
  description: Items in shopping cart
  type: object
  properties:
    id:
      description: Item Identifier
      type: string
    quantity:
      description: The item quantity
      type: integer
      format: int32
    unitPrice:
      description: The item's price per unit
      type: number
      format: double
```

https://github.com/PacktPublishing/Modern-API-Development-with-Spring-6-and-Spring-Boot-3/tree/main/Chapter03/src/main/resources/api/openapi.yaml

The `Item` model is also part of the `components/schema` section. We have defined several models used by the e-commerce app API. You can find them in the GitHub code at https://github.com/PacktPublishing/Modern-API-Development-with-Spring-and-Spring-Boot/tree/main/Chapter03/src/main/resources/api/openapi.yaml.

Now, you have learned how you can define models under the `components/schema` section. We'll now discuss how to define an API's endpoints in the `path` section of OAS.

> **Important note**
>
> Like `schemas`, you can also define `requestBodies` (the request payload) and `responses` in the `components` section. This is useful when you have common request bodies and responses.

The path section of OAS

The `path` section is the last section of OAS (sequence-wise, it is second-to-last, but we already discussed `components` in the previous subsection), where we define the endpoints. This is the place where we form the URI and attach the HTTP methods.

Let's write the definition for GET `/api/v1/carts/{customerId}/items`. This API gets the items of the cart associated with a given customer identifier:

```
paths:
  /api/v1/carts/{customerId}:
    get:
      tags:
        - cart
      summary: Returns the shopping cart
      description: Returns the shopping cart of
      given customer
      operationId: getCartByCustomerId
      parameters:
        - name: customerId
          in: path
          description: Customer Identifier
          required: true
          schema:
            type: string
      responses:
        200:
          description: successful operation
          content:
            application/xml:
              schema:
                type: array
                items:
                  $ref: '#/components/schemas/Cart'
            application/json:
              schema:
                type: array
                items:
                  $ref: '#/components/schemas/Cart'
        404:
          description: Given customer ID doesn't exist
          content: {}
```

```
https://github.com/PacktPublishing/Modern-API-Development-with-
Spring-6-and-Spring-Boot-3/tree/main/Chapter03/src/main/resources/
api/openapi.yaml
```

If you just go through the previous code, you can see what the endpoint is, what HTTP method and parameter this API uses, and most importantly, what response you can expect. Let's discuss this in more detail. Here, `v1` represents the version of the API. Each endpoint path (such as `/api/v1/carts/{customerId}/items`) has an HTTP method (such as `POST`) associated with it. The endpoint path always starts with `/`.

Each method can then have seven fields – `tags`, `summary`, `description`, `operationId`, `parameters`, `responses`, and `requestBody`. Let us learn about each of them:

- `tags`: Tags are used to group APIs, as shown in the following screenshot for APIs tagged with `cart`. **Swagger Codegen** makes use of tags to put an API's method in the same class. For example, all the listed `cart` endpoints in the following screenshot will be in `CartsApi.java`:

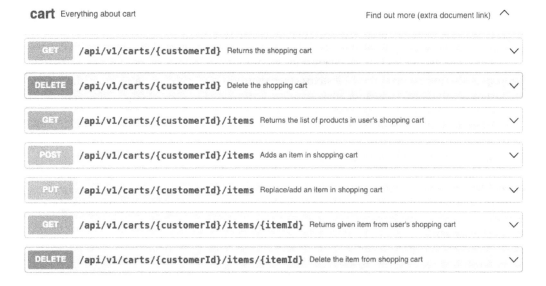

Figure 3.1 – Cart APIs

- `summary` and `description`: The `summary` and `description` sections are the same as we ones we discussed in the *The metadata sections of OAS* section. They contain the given API's operation summary and detailed description, respectively. As usual, you can use Markdown in the description field, as it refers to the same schema.

- `operationId`: This represents the name of the operation. As you can see in the previous code, we have assigned the `getCartByCustomerId` value to it. This same operation name will be used by Swagger Codegen as a method name in the generated API Java interface.

- `Parameters`: If you look closely, you'll find - (a hyphen) in front of the name field. This is used to declare it as an array element. The `parameters` field can contain multiple parameters – in fact, a combination of the `path` and `query` parameters; therefore, it is declared as an array.

   For `path` parameters, you need to make sure that the value of `name`, under `parameters`, is the same as given in `path` inside curly braces.

   The `parameters` field contains the API `query`, `path`, `header`, and `cookie` parameters. In the previous code, we used the `path` parameter (the value of the `in` field). You can change the value to `query` if you want to declare it as a `query` parameter and so on for other parameter types.

   You can mark a field as required or optional using the `required` field inside the `parameters` section, which is a Boolean parameter.

   Finally, you must declare the data type of the parameter, which is where the `schema` field is used.

- `responses`: The `responses` field is a required field for all API operations. This defines the type of responses that can be sent by the API operation when requested. It contains HTTP status codes as the default field. The field must have at least one response, which can be a `default` response or any successful HTTP status code, such as `200`. As the name suggests, the default response will be used when no other response is defined or available in the API operation.

   The response type (such as `200` or `default`) field contains three types of fields – `description`, `content`, and `headers`:

   - The `description` field is used to describe the response.

   - The `headers` field is used to define the header and its value. A `headers` example is shown as follows:

   ```
   responses:
     200:
       description: operation successful
         headers:
           X-RateLimit-Limit:
             schema:
               type: integer
   ```

   - The `content` field, like we have in previous code, defines the type of content that denotes the different media types. We use `application/json`. Similarly, you can define other media types, such as `application/xml`. The `content` type field contains the actual response object that can be defined using the `schema` field, as we have defined an array of the `Item` model inside it.

As mentioned earlier, you can create a reusable response under the `components` section and directly use it with `$ref`.

- `requestBody`: The `requestBody` field is used to define the request payload object. Like the `responses` object, `requestBody` also contains the description and content fields. Content can be defined in a similar fashion to the way it is defined for the `responses` object. You can refer to the previous code of `POST /carts/{customerId}/items` for an example. As a response, you can also create reusable request bodies under the `components` section and directly use them with `$ref`.

Now, you know how to define the API specification using OAS. Great! Here, we just described part of a sample e-commerce app's API. Similarly, you can describe other APIs. You can refer to `openapi.yaml` (`https://github.com/PacktPublishing/Modern-API-Development-with-Spring-6-and-Spring-Boot-3/tree/main/Chapter03/src/main/resources/api/openapi.yaml`) for the complete code of our e-commerce API definitions.

I suggest that you copy the code from `openapi.yaml` and paste it into the editor at `https://editor.swagger.io` to view the API in a nice user interface and play around with it. Make sure to convert the API to OpenAPI version 3 using the **Edit** menu if the default version is not set to 3.0.

We have finished designing our APIs, so now let's generate code using `openapi.yaml` and enjoy the fruits of our hard work.

Converting OAS to Spring code

I am sure you are as excited as I am to start implementing the API. So far, we have learned about the RESTful web service theory and concepts and Spring fundamentals, and also designed our first API specs for a sample e-commerce application.

For this section, you can either clone the Git repository (`https://github.com/PacktPublishing/Modern-API-Development-with-Spring-6-and-Spring-Boot-3`) or start to create a Spring project from scratch using **Spring Initializr** (`https://start.spring.io/`) with the following options:

- **Project**: `Gradle - Groovy`
- **Language**: `Java`
- **Spring Boot**: `3.0.8`

  Or use the 3.X.X available version. Replace the project metadata with your preferred values

- **Packaging**: `Jar`
- **Java**: `17`
- **Dependencies**: `Spring Web`

Once you open the project in your favorite IDE (IntelliJ, Eclipse, or NetBeans), you can add the following extra dependencies required for OpenAPI support under `dependencies` in the `build.gradle` file:

```
swaggerCodegen 'org.openapitools:openapi-generator-cli:6.2.1'
compileOnly 'io.swagger:swagger-annotations:1.6.4'
compileOnly 'org.springframework.boot:spring-boot-starter-
               validation'
compileOnly 'org.openapitools:jackson-databind-nullable:0.2.3'
implementation 'com.fasterxml.jackson.dataformat:jackson-
               dataformat-xml'
implementation 'org.springframework.boot:spring-boot-starter-
               hateoas'
implementation 'io.springfox:springfox-oas:3.0.0'
```

https://github.com/PacktPublishing/Modern-API-Development-with-Spring-6-and-Spring-Boot-3/tree/main/Chapter03/build.gradle

As mentioned earlier, we will use the Swagger plugin for code generation from the API definitions we just wrote. Follow the next seven steps to generate the code.

1. **Adding the Gradle plugin**: To make use of the OpenAPI Generator CLI tool, you can add the Swagger Gradle plugin under `plugins {}` in `build.gradle`, as shown here:

    ```
    plugins {
        …
        …
        id 'org.hidetake.swagger.generator' version '2.19.2'
    }
    ```

2. **Defining the OpenAPI config for code generation**: You need certain configurations, such as what model and API package names OpenAPI Generator's CLI should use, or the library it should use to generate the REST interfaces or date/time-related objects. All these and other configurations can be defined in `config.json` (`/src/main/resources/api/config.json`):

    ```
    {
        "library": "spring-boot",
        "dateLibrary": "java8",
        "hideGenerationTimestamp": true,
        "modelPackage": "com.packt.modern.api.model",
        "apiPackage": "com.packt.modern.api",
        "invokerPackage": "com.packt.modern.api",
        "serializableModel": true,
        "useTags": true,
    ```

```
  "useGzipFeature" : true,
  "hateoas": true,
  "unhandledException": true,
  "useSpringBoot3": true,
  "useSwaggerUI": true,

   ...

   ...

  "importMappings": {
    "ResourceSupport":"org.springframework.hateoas.
        RepresentationModel",
    "Link": "org.springframework.hateoas.Link"
  }
}
```

https://github.com/PacktPublishing/Modern-API-Development-
with-Spring-6-and-Spring-Boot-3/tree/main/Chapter03/src/main/
resources/api/config.json

This configuration sets spring-boot as library – that is, Swagger Codegen will generate
the classes aligned with Spring Boot. You can see that useSpringBoot3 is set to true to
make sure that generated classes are aligned with Spring Boot 3.

All other properties are self-explanatory except importMappings. It contains the mapping
of a type from a YAML file to Java or a type existing in the external library. Therefore, once code
is generated for the importMappings object, it uses the mapped class in the generated code.
If we use Link in any of the models, then the generated models will use the mapped org.
springframework.hateoas.Link class instead of the model defined in the YAML file.

The hateoas configuration property allows us to use the Spring HATEOAS library and add
HATEOAS links.

You can find more information about the configuration at https://github.com/
swagger-api/swagger-codegen#customizing-the-generator.

3. **Defining the OpenAPI Generator ignore file**: You can also add a .gitignore like file
   to ignore certain code you don't want to generate. Add the following line of code to the file
   (/src/main/resources/api/.openapi-generator-ignore):

   ```
   **/*Controller.java
   ```

   We don't want to generate controllers. After the code's addition, only API Java interfaces and
   models will be generated. We'll add controllers manually.

4. Copy the OAS file from https://github.com/PacktPublishing/Modern-API-
   Development-with-Spring-6-and-Spring-Boot-3/blob/main/Chapter03/
   src/main/resources/api/openapi.yaml in /src/main/resources/api.

5. **Defining a swaggerSources task in the Gradle build file**: Now, let's add logic to the swaggerSources task in the `build.gradle` file:

```
swaggerSources {
  def typeMappings = 'URI=URI'
  def importMappings = 'URI=java.net.URI'
  eStore {
    def apiYaml = "${rootDir}/src/main/resources/api/openapi.
      yaml"
    def configJson = "${rootDir}/src/main/resources/api/config.
      json"
    inputFile = file(apiYaml)
    def ignoreFile = file("${rootDir}/src/main/resources/api/.
openapi-generator-ignore")
    code {
      language = 'spring'
      configFile = file(configJson)
      rawOptions = ['--ignore-file-override',
        ignoreFile,
                      '--type-mappings', typeMappings,
                      '--import-mappings',
                        importMappings]
                    as List<String>
      components = [models: true, apis: true, supportingFiles:
        'ApiUtil.java']
      dependsOn validation
    }
  }
}
```

Here, we defined `eStore` (the user-defined name), which contains `inputFile`, pointing to the location of the `openapi.yaml` file. After defining the input, the generator needs to produce the output, which is configured in `code`.

For the code block, `language` is set to Spring (it supports various languages); `configFile` points to `config.json`; `rawOptions` contains an `ignore` file, type mapping, and import mappings; and `components` contains the file flags you want to generate – models and API Java interfaces. Except for `language`, all our other configuration properties are optional in the `code` block.

We only want to generate models and APIs. You can generate other files too, such as clients or test files. `ApiUtil.java` is required in the generated API Java interface otherwise, it will give a compilation error during build time. Therefore, it is added to `components`.

6. **Adding swaggerSources to the compileJava task dependency**: Now we need to add `swaggerSources` to the `compileJava` task as a dependent task.

This task points to the `code` block defined under `eStore`:

```
compileJava.dependsOn swaggerSources.eStore.code
```

Also, you need to add the `generateSwaggerCode` task as a dependency to the `processResources` task:

```
processResources {
    dependsOn(generateSwaggerCode)
}
```

You may get a warning in prior to Gradle 8 versions if you don't define this dependency, and but it will still work. However, this code block is required for the Gradle 8 version.

7. **Adding the generated source code to Gradle sourceSets**: We also need to add the generated source code and resources to `sourceSets`. This makes the generated source code and resources available for development and build:

```
sourceSets.main.java.srcDir "${swaggerSources.eStore.code.
outputDir}/src/main/java"

sourceSets.main.resources.srcDir "${swaggerSources.eStore.
code  .outputDir}/src/main/resources"
```

The source code will be generated in the `/build` directory of the project, such as `Chapter03\build\swagger-code-eStore`. This will append the generated source code and resources to Gradle `sourceSets`.

> **Important note**
>
> You have generated the API Java interfaces and models using the Swagger Codegen utility. Therefore, when you load the project for the first time in your IDE, you may get errors if you don't run your build because IDE won't find the generated Java files (models and API Java interfaces). You can run the build's `gradlew clean build` command to generate these files.

8. **Building the project to generate, compile, and build the code**: The last step is to execute the build. Make sure you have an executable Java code in the `build` path. The Java version should match the version defined in the property of `build.gradle (sourceCompatibility = '17')` or in the IDE settings:

```
$ gradlew clean build
```

Once the build is executed successfully, you can find the generated code in the `build` directory, as shown in the following screenshot:

Figure 3.2 – The OpenAPI-generated code

And that's it. Once you follow all the aforementioned steps, you can successfully generate the API models and API Java interfaces code. In the next section, you'll implement the API Java interfaces generated by OpenAPI Codegen.

Implementing the OAS code interfaces

So far, we have generated code that consists of e-commerce app models and API Java interfaces. These generated interfaces contain all the annotations as per the YAML description provided by us. For example, in `CartApi.java`, `@RequestMapping`, `@PathVariable`, and `@RequestBody` contain the endpoint path (`/api/v1/carts/{customerId}/items`), the value of the `path` variable (such as `{customerId}` in `path`), and the request payload (such as `Item`), respectively. Similarly, generated models contain all the mapping required to support the JSON and XML content types.

Swagger Codegen writes the Spring code for us. We just need to implement the interface and write the business logic inside it. Swagger Codegen generates the API Java interfaces for each of the provided tags. For example, it generates the `CartApi` and `PaymentAPI` Java interfaces for the `cart` and payment tags, respectively. All the paths are clubbed together into a single Java interface based on the given tag. For example, all the APIs with the `cart` tag will be clubbed together into a single Java interface, `CartApi.java`.

Now, we just need to create a class for each of the interfaces and implement it. We'll create `CartController.java` in the `com.packt.modern.api.controllers` package and implement `CartApi`:

```
@RestController
public class CartsController implements CartApi {

  private static final Logger log = LoggerFactory.
getLogger(CartsController.class);

  @Override
  public ResponseEntity<List<Item>> addCartItemsBy
    CustomerId(String customerId, @Valid Item item) {
    log.info("Request for customer ID: {}\nItem: {}",customerId,
item);
    return ok(Collections.EMPTY_LIST);
  }

  @Override
  public ResponseEntity<List<Cart>> getCartByCustomerId(String
customerId) {
    throw new RuntimeException("Manual Exception thrown");
  }
  // Other method implementations (omitted)
}
```

https://github.com/PacktPublishing/Modern-API-Development-with-Spring-6-and-Spring-Boot-3/tree/main/Chapter03/src/main/java/com/packt/modern/api/controllers/CartsController.java

Here, we just implemented the two methods for demonstration purposes. We'll implement the actual business logic in the next chapter.

To add an item (POST `/api/v1/carts/{customerId}/items`) request, we just log the incoming request payload and customer ID inside the `addCartItemsByCustomerId` method. Another method, `getCartByCustomerId`, simply throws an exception. This will allow us to demonstrate the Global Exception Handler in the next section.

Adding a Global Exception Handler

We have multiple controllers that consist of multiple methods. Each method may have checked exceptions or throw runtime exceptions. We should have a centralized place to handle all these errors for better maintainability and modularity and clean code.

Spring provides an AOP feature for this. We just need to write a single class annotated with @ControllerAdvice. Then, we just need to add @ExceptionHandler for each type of exception. This exception handler method will generate user-friendly error messages with other related information.

You can make use of the Project Lombok library if approved by your organization for third-party library usage. This will remove the verbosity of the code for getters, setters, constructors, and so on.

Let's first write the `Error` class in the `exceptions` package that contains all the error information:

```
public class Error {
    private static final long serialVersionUID = 1L;
    private String errorCode;
    private String message;
    private Integer status;
    private String url = "Not available";
    private String reqMethod = "Not available";

    // getters and setters (omitted)
}
```

https://github.com/PacktPublishing/Modern-API-Development-with-Spring-6-and-Spring-Boot-3/tree/main/Chapter03/src/main/java/com/packt/modern/api/exceptions/Error.java

Here, we use the following properties:

- `errorCode`: Application error code, which is different from HTTP error code

- `message`: A short, human-readable summary of the problem

- `status`: An HTTP status code for this occurrence of the problem, set by the origin server

- `url`: A URL of the request that produced the error

- `reqMethod`: A method of the request that produced the error

You can add other fields here if required. The `exceptions` package will contain all the code for user-defined exceptions and global exception handling.

After that, we'll write an enum called `ErrorCode` that will contain all the exception keys, including user-defined errors and their respective error codes:

```
public enum ErrorCode {
  GENERIC_ERROR("PACKT-0001", "The system is unable to
      complete the request. Contact system support."),
  HTTP_MEDIATYPE_NOT_SUPPORTED("PACKT-0002", "Requested
      media type is not supported. Please use
      application/json or application/xml as 'Content-
      Type' header value"),
  HTTP_MESSAGE_NOT_WRITABLE("PACKT-0003", "Missing 'Accept'
      header. Please add 'Accept' header."),
  HTTP_MEDIA_TYPE_NOT_ACCEPTABLE("PACKT-0004", "Requested
      'Accept' header value is not supported. Please use
       application/json or application/xml as 'Accept'
       value"),
  JSON_PARSE_ERROR("PACKT-0005", "Make sure request payload
      should be a valid JSON object."),
  HTTP_MESSAGE_NOT_READABLE("PACKT-0006", "Make sure
      request payload should be a valid JSON or XML
      object according to 'Content-Type'.");

  private String errCode;
  private String errMsgKey;

  ErrorCode(final String errCode, final String errMsgKey) {
    this.errCode = errCode;
    this.errMsgKey = errMsgKey;
  }

  public String getErrCode() {  return errCode;  }
  public String getErrMsgKey() {  return errMsgKey;  }
}
```

https://github.com/PacktPublishing/Modern-API-Development-with-Spring-6-and-Spring-Boot-3/tree/main/Chapter03/src/main/java/com/packt/modern/api/exceptions/ErrorCode.java

Here, we just added a few error code enums with their code and messages. We also just added actual error messages instead of message keys. You can add message keys and add the resource file to `src/main/resources` for internationalization.

Next, you'll add a utility to create an `Error` object, as shown in the following code:

```
public class ErrorUtils {
  private ErrorUtils() {}

  public static Error createError(final String errMsgKey,
      final String errorCode, final Integer httpStatusCode) {
    Error error = new Error();
    error.setMessage(errMsgKey);
    error.setErrorCode(errorCode);
    error.setStatus(httpStatusCode);
    return error;
  }
}
```

Finally, we'll create a class to implement the Global Exception Handler, as shown here:

```
@ControllerAdvice
public class RestApiErrorHandler {
  private final MessageSource messageSource;

  @Autowired
  public RestApiErrorHandler(MessageSource messageSource) {
    this.messageSource = messageSource;
  }


  @ExceptionHandler(Exception.class)
  public ResponseEntity<Error> handleException(HttpServletRequest
request, Exception ex,Locale locale) {
    Error error = ErrorUtils
        .createError(ErrorCode.GENERIC_ERROR.
getErrMsgKey(),  ErrorCode.GENERIC_ERROR.getErrCode(),  HttpStatus.
INTERNAL_SERVER_ERROR.value())    .setUrl(request.getRequestURL().
toString())  .setReqMethod(request.getMethod());
    return new ResponseEntity<>(error, HttpStatus.INTERNAL_SERVER_
ERROR);
  }


  @ExceptionHandler(HttpMediaTypeNotSupportedException.class)
  public ResponseEntity<Error>
      handleHttpMediaTypeNotSupportedException(
        HttpServletRequest request,
        HttpMediaTypeNotSupportedException ex, Locale locale){
    Error error = ErrorUtils
        .createError(ErrorCode.HTTP_MEDIATYPE_NOT_SUPPORTED
```

```
                .getErrMsgKey(),
          ErrorCode.HTTP_MEDIATYPE_NOT_SUPPORTED.getErrCode(),
          HttpStatus.UNSUPPORTED_MEDIA_TYPE.value())
              .setUrl(request.getRequestURL().toString())
              .setReqMethod(request.getMethod());
      return new ResponseEntity<>(
              error, HttpStatus.INTERNAL_SERVER_ERROR);
    }
    // removed code for brevity
  }
```

https://github.com/PacktPublishing/Modern-API-Development-with-
Spring-6-and-Spring-Boot-3/tree/main/Chapter03/src/main/java/com/
packt/modern/api/exceptions/RestApiErrorHandler.java

As you can see, we marked the class with @ControllerAdvice, which enables this class to trace all the request and response processing by the REST controllers and allows us to handle exceptions using @ExceptionHandler.

In the previous code, we handle two exceptions – a generic internal server error exception and HttpMediaTypeNotSupportException. The handling method just populates the Error object using ErrorCode, HttpServletRequest, and HttpStatus. Finally, it returns the error wrapped inside ResponseEntity with the appropriate HTTP status.

In this code, you can add user-defined exceptions too. You can also make use of the Locale instance (a method parameter) and the messageSource class member to support internationalized messages.

Now that we have designed the API and generated the code and implementation, let's now test the implementation in the following subsection.

Testing the implementation of the API

Once the code is ready to run, you can compile and build the artifact using the following command from the root folder of the project:

```
$ ./gradlew clean build
```

The previous command removes the build folder and generates the artifact (the compiled classes and JAR). After the successful build, you can run the application using the following command:

```
$ java -jar build/libs/Chapter03-0.0.1-SNAPSHOT.jar
```

Now, we can perform tests using the curl command:

```
$ curl --request GET 'http://localhost:8080/api/v1/carts/1' --header
'Accept: application/xml'
```

This command calls the GET request for /carts with ID 1. Here, we demand the XML response using the Accept header, and we get the following response:

```
<Error>
  <errorCode>PACKT-0001</errorCode>
  <message>The system is unable to complete the request.
           Contact system support.</message>
  <status>500</status>
  <url>http://localhost:8080/api/v1/carts/1</url>
  <reqMethod>GET</reqMethod>
</Error>
```

If you change the Accept header from application/xml to application/json, you will get the following JSON response:

```
$ curl --request GET 'http://localhost:8080/api/v1/carts/1' --header
'Accept: application/json'
{
  "errorCode":"PACKT-0001",
  "message":"The system is unable to complete the request.
           Contact system support.",
  "status":500,
  "url":"http://localhost:8080/api/v1/carts/1",
  "reqMethod":"GET"
}
```

Similarly, we can also call the API to add an item to the cart, as shown here:

```
$ curl --request POST 'http://localhost:8080/api/v1/carts/1/items' \
  --header 'Content-Type: application/json' \
  --header 'Accept: application/json' \
  --data-raw '{
  "id": "1",
  "quantity": 1,
  "unitPrice": 2.5
  }'
[]
```

Here, we get [] (an empty array) as a response because, in the implementation, we just return the empty collection. You need to provide the Content-Type header in this request because we send the payload (item object) along with the request. You can change Content-Type to application/xml if the payload is written in XML. If the Accept header value is application/xml, it will return the <List/> value. You can remove/change the Content-Type and Accept headers or use the malformed JSON or XML to test the other error response.

This way, we can generate the API description using OpenAPI and then use the generated models and API Java interfaces to implement the APIs.

Summary

In this chapter, we opted for the design-first approach to writing RESTful web services. You learned how to write an API description using OAS and how to generate models and API Java interfaces using the Swagger Codegen tool (using the Gradle plugin). We also implemented a Global Exception Handler to centralize the handling of all the exceptions. Once you have the API Java interfaces, you can write their implementations for business logic. Now, you know how to use OAS and Swagger Codegen to write RESTful APIs. You also now know how to handle exceptions globally.

In the next chapter, we'll implement fully fledged API Java interfaces with business logic with database persistence.

Questions

1. What is OpenAPI and how does it help?

2. How can you define a nested array in a model in a YAML OAS-based file?

3. What annotations do we need to implement a Global Exception Handler?

4. How can you use models or classes written in Java code in your OpenAPI description?

5. Why do we only generate models and API Java interfaces using Swagger Codegen?

Answers

1. OAS was introduced to solve at least a few aspects of a REST API's specification and description. It allows you to write REST APIs in the YAML or JSON markup languages, which allows you to interact with all stakeholders, including those who are non-technical, for review and discussion in the development phase. It also allows you to generate documentation, models, interfaces, clients, and servers in different languages.

2. The array is defined using the following code:

    ```
    type: array
    items:
      type: array
      items:
        type: string
    ```

3. You need a class annotation, `@ControllerAdvice`, and a method annotation, `@ExceptionHandler`, to implement the Global Exception Handler.

4. You can use `--type-mappings` and `--import-mappings` rawOptions in the swaggerSources task of the `build.gradle` file.

5. We only generate the models and API Java interfaces using Swagger Codegen because this allows the complete implementation of controllers by developers only.

Further reading

- OAS 3.0: `https://github.com/OAI/OpenAPI-Specification/blob/master/versions/3.0.3.md`

- The Gradle plugin for OpenAPI Codegen: `https://github.com/int128/gradle-swagger-generator-plugin`

- OAS Code Generator configuration options for Spring: `https://openapi-generator.tech/docs/generators/spring/`

- YAML specifications: `https://yaml.org/spec/`

- Semantic versioning: `https://semver.org/`

4

Writing Business Logic for APIs

You defined API specs using OpenAPI in the previous chapter. API Java interfaces and models were generated by the OpenAPI (Swagger Codegen). In this chapter, you will implement the API's code in terms of both business logic and data persistence. Here, business logic refers to the actual code you are writing for domain functionalities, which in our case comprise operations performed for e-commerce, such as checking out the shopping cart.

You will write services and repositories for implementation and add hypermedia and **entity tags** (**ETags**) to API responses. **Hypermedia As The Engine Of Application State** (**HATEOAS**) will be implemented using Spring and **Hypertext Application Language** (**HAL**). HAL is one of the standards to implement HATEOAS. Others are *Collection+JSON* and *JSON-LD*. You are going to use HAL in this book. You can find a sample of it in the first example in the *Adding ETags to API responses* section denoted by the "`_links`" field. It is worth noting that the code provided only consists of the important lines and not the whole file in the interest of brevity. You can always access the links given after the code to view the complete file.

This chapter covers the following topics:

- Overview of the service design
- Adding a repository component
- Adding service components
- Implementing hypermedia
- Enhancing the controller with a service and HATEOAS
- Adding ETags to API responses
- Testing the APIs

Technical requirements

To execute the instructions in this and the following chapters, you will need any REST API client, such as *Insomnia* or *Postman*.

You can find the code files for this chapter on GitHub at `https://github.com/PacktPublishing/Modern-API-Development-with-Spring-6-and-Spring-Boot-3/tree/main/Chapter04`.

Overview of the service design

We are going to implement a multi-layered architecture that comprises four layers – the presentation layer, application layer, domain layer, and infrastructure layer. Multi-layered architecture is a fundamental building block in the architecture style known as **domain-driven design** (**DDD**). Let's have a brief look at each of these layers:

- **Presentation layer**: This layer represents the **user interface** (**UI**). In *Chapter 7, Designing a User Interface*, you'll develop the UI for a sample e-commerce app.

- **Application layer**: The application layer contains the application logic and maintains and coordinates the overall flow of the application. Just to remind you, it only contains the application logic and *not* the business logic. RESTful web services, async APIs, gRPC APIs, and GraphQL APIs are a part of this layer.

  We already covered REST APIs and controllers in *Chapter 3, API Specifications and Implementation*, which are part of the application layer. We implemented the controllers for demonstration purposes in the previous chapter. In this chapter, we'll implement a controller extensively to serve real data.

- **Domain layer**: This layer contains the business logic and domain information. It contains the state of the business objects, such as `Order` or `Product`. It is responsible for reading/persisting these objects to the infrastructure layer. The domain layer consists of services and repositories too. We'll also be covering these in this chapter.

- **Infrastructure layer**: The infrastructure layer provides support to all other layers. It is responsible for communication, such as interaction with the database, message brokers, and filesystems. Spring Boot works as an infrastructure layer and provides support for communication and interaction with both external and internal systems, such as databases and message brokers.

We'll use the bottom-to-top approach. Let's start implementing the domain layer with the `@Repository` component.

Adding a Repository component

We'll use the bottom-to-top approach to add a @Repository component. Let's start implementing the domain layer with a @Repository component. We'll implement the service and enhance the @Controller component in subsequent sections accordingly. We will code the @Repository component first, then use it in the @Service component using constructor injection. The @Controller component will be enhanced using the @Service component, which will also be injected into the Controller using constructor injection.

The @Repository annotation

Repository components are Java classes marked with the @Repository annotation. This is a special Spring component that is used for interacting with databases.

@Repository is a general-purpose stereotype that represents both DDD's Repository and the **Java Enterprise Edition (JEE) data access object (DAO)** pattern. Developers and teams should handle repository objects based on the underlying approach. In DDD, a repository is a central object that carries references to all the objects and should return the reference of a requested object. We need to have all the required dependencies and configurations in place before we start writing classes marked with @Repository.

You'll use the following libraries as database dependencies:

- **H2 database for persisting data**: We are going to use H2's memory instance; however, you can also use a file-based instance
- **Hibernate object relational mapping (ORM)**: For database object mapping
- **Flyway for database migration**: This helps maintain the database and maintains a database changes history that allows rollbacks, version upgrades, and so on

Let's add these dependencies to the build.gradle file. org.springframework.boot:spring-boot-starter-data-jpa adds all the required JPA dependencies, including Hibernate:

```
implementation 'org.springframework.boot:spring-boot-starter-data-jpa'
implementation 'org.flywaydb:flyway-core'
runtimeOnly   'com.h2database:h2'
```

https://github.com/PacktPublishing/Modern-API-Development-with-Spring-6-and-Spring-Boot-3/tree/main/Chapter04/build.gradle

After adding the dependencies, we can add the configuration related to the database.

Configuring the database and JPA

We also need to modify the `application.properties` file with the following configuration. The configuration file is available at `https://github.com/PacktPublishing/Modern-API-Development-with-Spring-6-and-Spring-Boot-3/tree/main/Chapter04/src/main/resources/application.properties`:

- **Datasource configuration**: The following are the Spring datasource configurations:

```
spring.datasource.name=ecomm
spring.datasource.url=jdbc:h2:mem:ecomm;DB_CLOSE_DELAY=-1;IGNORECASE=TRUE;DATABASE_TO_UPPER=false
spring.datasource.driverClassName=org.h2.Driver
spring.datasource.username=sa
spring.datasource.password=
```

We need to add H2-specific properties to the data source. The URL value suggests that a memory-based H2 database instance will be used.

- **H2 database configuration**: The following are the two H2 database configurations:

```
spring.h2.console.enabled=true
spring.h2.console.settings.web-allow-others=false
```

Here, the H2 console is the H2 web client that allows you to perform different operations on H2, such as viewing tables and executing queries. The H2 console is enabled for local access only; this means you can access the H2 console only on localhost. Also, remote access is disabled by setting `web-allow-others` to `false`.

- **JPA configuration**: The following are the JPA/Hibernate configurations:

```
spring.jpa.properties.hibernate.default_schema=ecomm
spring.jpa.database-platform=org.hibernate.dialect.H2Dialect
spring.jpa.show-sql=true
spring.jpa.format_sql=true
spring.jpa.generate-ddl=false
spring.jpa.hibernate.ddl-auto=none
```

We don't want to generate the DDL or process the SQL file, because we want to use Flyway for database migrations. Therefore, `generate-ddl` is marked with `false` and `ddl-auto` is set to `none`.

- **Flyway configuration**: The following are the Flyway configurations:

```
spring.flyway.url=jdbc:h2:mem:ecomm
spring.flyway.schemas=ecomm
spring.flyway.user=sa
spring.flyway.password=
```

Here, properties that are required for Flyway to connect to the database have been set.

> **Accessing the H2 database**
>
> You can access the H2 database console using `/h2-console`. For example, if your server is running on localhost and port `8080`, then you can access it using `http://localhost:8080/h2-console/`.

You are done with setting up the database configuration. Let's create the database schema and seed data script in the next subsection.

The database and seed data script

Now, we are done configuring the `build.gradle` and `application.properties` files and we can start writing the code. First, we'll add the Flyway database migration script. This script can be written in SQL only. You can place this file in the `db/migration` directory inside the `src/main/resources` directory. We'll follow the Flyway naming convention (`V<version>.<name>.sql`) and create the `V1.0.0__Init.sql` file inside the `db/migration` directory. You can then add the following script to this file:

```
create schema if not exists ecomm;
create TABLE IF NOT EXISTS ecomm.product (
    id uuid NOT NULL DEFAULT random_uuid(),
    name varchar(56) NOT NULL,
    description varchar(200),
    price numeric(16, 4) DEFAULT 0 NOT NULL,
    count numeric(8, 0),
    image_url varchar(40),
    PRIMARY KEY(id)
);
create TABLE IF NOT EXISTS ecomm.tag (
    id uuid NOT NULL DEFAULT random_uuid(),
    name varchar(20),
    PRIMARY KEY(id)
);
-- Other code is removed for brevity
```

https://github.com/PacktPublishing/Modern-API-Development-with-Spring-6-and-Spring-Boot-3/tree/main/Chapter04/src/main/resources/db/migration/V1.0.0__Init.sql

This script creates the `ecomm` schema and adds all the tables required for our sample e-commerce app. It also adds `insert` statements for the seed data.

Adding entities

Now, we can add the entities. An entity is a special object marked with the @Entity annotation that maps directly to the database table using an ORM implementation such as *Hibernate*. Another popular ORM is *EclipseLink*. You can place all entity objects in the com.packt.modern.api. entity package.

Let's create the CartEntity.java file:

```java
@Entity
@Table(name = "cart")
public class CartEntity {
  @Id
  @GeneratedValue
  @Column(name = "ID", updatable = false, nullable = false)
  private UUID id;
  @OneToOne
  @JoinColumn(name = "USER_ID", referencedColumnName = "ID")
  private UserEntity user;
  @ManyToMany( cascade = CascadeType.ALL )
  @JoinTable(
    name = "CART_ITEM",
    joinColumns = @JoinColumn(name = "CART_ID"),
    inverseJoinColumns = @JoinColumn(name = "ITEM_ID")
  )
  private List<ItemEntity> items = Collections.emptyList();
  // Getters/Setter and other codes are removed for brevity
```

https://github.com/PacktPublishing/Modern-API-Development-with-Spring-6-and-Spring-Boot-3/tree/main/Chapter04/src/main/java/com/packt/modern/api/entity/CartEntity.java

Here, the @Entity annotation is part of the jakarta.persistence package, which denotes that it is an entity and should be mapped to the database table. By default, it takes the entity name; however, we are using the @Table annotation to map to the database table. Earlier, the javax. persistence package was part of Oracle. Once Oracle open sourced JEE and handed it over to the Eclipse Foundation, it was legally required to change the name of the package from javax. persistence to jakarta.persistence.

We are also using one-to-one and many-to-many annotations to map the Cart entity to the User entity and Item entity, respectively. The ItemEntity list is also associated with @JoinTable, because we are using the CART_ITEM join table to map the cart and product items based on the CART_ID and ITEM_ID columns in their respective tables.

In `UserEntity`, the `Cart` entity has also been added to maintain the relationship, as shown in the following code block. `FetchType` is marked as `LAZY`, which means the user's cart will be loaded only when explicitly asked. Also, you want to remove the cart if it is not referenced by the user, which can be done by configuring `orphanRemoval` to `true`:

```
@Entity
@Table(name = "user")
public class UserEntity {
// other code
@OneToOne(mappedBy = "user", fetch = FetchType.LAZY, orphanRemoval =
true)
private CartEntity cart;
// other code…
```

https://github.com/PacktPublishing/Modern-API-Development-with-Spring-6-and-Spring-Boot-3/blob/main/Chapter04/src/main/java/com/packt/modern/api/entity/UserEntity.java

All other entities are being added to the entity package located at https://github.com/PacktPublishing/Modern-API-Development-with-Spring-6-and-Spring-Boot-3/blob/main/Chapter04/src/main/java/com/packt/modern/api/entity.

Now, we can add the repositories.

Adding repositories

All the repositories have been added to https://github.com/PacktPublishing/Modern-API-Development-with-Spring-6-and-Spring-Boot-3/blob/main/Chapter04/src/main/java/com/packt/modern/api/repository.

Repositories are simplest to add for CRUD operations, thanks to Spring Data JPA. You just must extend the interfaces with default implementations, such as `CrudRepository`, which provides all the CRUD operation implementations, such as `save`, `saveAll`, `findById`, `findAll`, `findAllById`, `delete`, and `deleteById`. The `save(Entity e)` method is used for both create and update entity operations.

Let's create `CartRepository.java`:

```
public interface CartRepository extends
    CrudRepository<CartEntity, UUID> {
  @Query("select c from CartEntity c join c.user u
    where u.id = :customerId")
  public Optional<CartEntity> findByCustomerId(
      @Param("customerId") UUID customerId);
}
```

https://github.com/PacktPublishing/Modern-API-Development-with-Spring-6-and-Spring-Boot-3/blob/main/Chapter04/src/main/java/com/packt/modern/api/repository/CartRepository.java

The CartRepository interface extends the CrudRepository part of the org.springframework.data.repository package. You can also add methods supported by the JPA query language marked with the @Query annotation (part of the org.springframework.data.jpa.repository package). The query inside the @Query annotation is written in **Java Persistence Query Language** (**JPQL**). JPQL is very similar to SQL; however, here we used the Java class name mapped to a database table instead of using the actual table name. Therefore, we have used CartEntity as the table name instead of Cart.

> **Selecting columns in JPQL**
>
> Similarly, for columns, you should use the variable names given in the class for the fields, instead of using the database table fields. In any case, if you use the database table name or field name and it does not match the class and class members mapped to the actual table, you will get an error.

You must be wondering, *"What if I want to add my own custom method with JPQL or native SQL?"* Well, let me tell you, you can do this too. For orders, we have added a custom interface for this very purpose. First, let's have a look at OrderRepository, which is very similar to CartRepository:

```
@Repository
public interface OrderRepository extends
    CrudRepository<OrderEntity, UUID>, OrderRepositoryExt {
  @Query("select o from OrderEntity o join o.userEntity u
    where u.id = :customerId")
  public Iterable<OrderEntity> findByCustomerId(
      @Param("customerId") UUID customerId);
}
```

https://github.com/PacktPublishing/Modern-API-Development-with-Spring-6-and-Spring-Boot-3/blob/main/Chapter04/src/main/java/com/packt/modern/api/repository/OrderRepository.java

If you look closely, we have extended an extra interface – OrderRepositoryExt. This is our extra interface for the Order repository and consists of the following code:

```
public interface OrderRepositoryExt {
  Optional<OrderEntity> insert(NewOrder m);
}
```

https://github.com/PacktPublishing/Modern-API-Development-with-
Spring-6-and-Spring-Boot-3/blob/main/Chapter04/src/main/java/com/
packt/modern/api/repository/OrderRepositoryExt.java

We already have a save() method for this purpose in CrudRepository; however, we want to use a different implementation. For this purpose, and to demonstrate how you can create your own repository method implementation, we are adding this extra repository interface.

Now, let's create the OrderRepositoryExt interface implementation, as shown here:

```
@Repository
@Transactional
public class OrderRepositoryImpl implements
  OrderRepositoryExt {
  @PersistenceContext
  private EntityManager em;
  private final ItemRepository itemRepo;
  private final CartRepository cRepo;
  private final OrderItemRepository oiRepo;
  public OrderRepositoryImpl(EntityManager em,CartRepository cRepo,
      OrderItemRepository oiRepo) {
    this.em = em;
    this.cRepo = cRepo;
    this.oiRepo= oiRepo;
  }
// rest of the code
```

https://github.com/PacktPublishing/Modern-API-Development-with-
Spring-6-and-Spring-Boot-3/blob/main/Chapter04/src/main/java/com/
packt/modern/api/repository/OrderRepositoryImpl.java

This way, we can also have our own implementation in JPQL/**Hibernate Query Language** (**HQL**), or in native SQL. Here, the @Repository annotation tells the Spring container that this special component is a repository and should be used to interact with the database using the underlying JPA.

It is also marked as @Transactional, which is a special annotation that means that transactions performed by methods in this class will be managed by Spring. It removes all the manual work of adding commits and rollbacks. You can also add this annotation to a specific method inside the class.

We are also using @PersistenceContext for the EntityManager class, which allows us to create and execute the query manually, as shown in the following code:

```
@Override
public Optional<OrderEntity> insert(NewOrder m) {
  Iterable<ItemEntity> dbItems = itemRepo.findByCustomerId(m.
getCustomerId());
```

```java
List<ItemEntity> items = StreamSupport.stream(
            dbItems.spliterator(), false).collect
            (toList());
if (items.size() < 1) {
 throw new ResourceNotFoundException(String.format("There
    is no item found in customer's (ID: %s) cart.",
       m.getCustomerId()));
}
BigDecimal total = BigDecimal.ZERO;
for (ItemEntity i : items) {
  total = (BigDecimal.valueOf(i.getQuantity()).multiply(
     i.getPrice())).add(total);
}
Timestamp orderDate = Timestamp.from(Instant.now());
em.createNativeQuery("""
 INSERT INTO ecomm.orders (address_id, card_id,
   customer_id
 order_date, total, status) VALUES(?, ?, ?, ?, ?, ?)
 """)
.setParameter(1, m.getAddress().getId())
.setParameter(2, m.getCard().getId())
.setParameter(3, m.getCustomerId())
.setParameter(4, orderDate)
.setParameter(5, total)
.setParameter(6, StatusEnum.CREATED.getValue())
.executeUpdate();
Optional<CartEntity> oCart =
 cRepo.findByCustomerId(UUID.fromString
   (m. getCustomerId()));
CartEntity cart = oCart.orElseThrow(() -> new
 ResourceNotFoundException(String.format
("Cart not found for given customer (ID: %s)", m.getCustomerId())));
 itemRepo.deleteCartItemJoinById(cart.getItems().stream()
  .map(i -> i.getId()).collect(toList()), cart. getId());
OrderEntity entity = (OrderEntity)
    em.createNativeQuery("""
SELECT o.* FROM ecomm.orders o WHERE o.customer_id = ? AND
o.order_date >= ?
""", OrderEntity.class)
.setParameter(1, m.getCustomerId())
.setParameter(2, OffsetDateTime.ofInstant(orderDate.
  toInstant(),ZoneId.of("Z")).truncatedTo(ChronoUnit.MICROS))
    .getSingleResult();
```

```
oiRepo.saveAll(cart.getItems().stream()
  .map(i -> new OrderItemEntity().setOrderId
    (entity. getId())
  .setItemId(i.getId())).collect(toList()));
return Optional.of(entity);
}
```

This method basically first fetches the items in the customer's cart. Then, it calculates the order total, creates a new order, and saves it in the database. Next, it removes the items from the cart by removing the mapping because cart items are now part of the order. After that, it saves the mapping of the order and cart items.

Order creation is done using the native SQL query with the prepared statement.

If you look closely, you'll also find that we have used the official *Java 15* feature, **text blocks** (`https://docs.oracle.com/en/java/javase/15/text-blocks/index.html`).

Similarly, you can create a repository for all other entities. All the repositories are available at `https://github.com/PacktPublishing/Modern-API-Development-with-Spring-6-and-Spring-Boot-3/blob/main/Chapter04/src/main/java/com/packt/modern/api/repository`.

Now that we have created the repositories, we can move on to adding services.

Adding a Service component

The `@Service` component is an interface that works between controllers and repositories and is where we'll add the business logic. Though you can directly call repositories from controllers, it is not a good practice as repositories should only be part of the data retrieval and persistence functionalities. Service components also help in sourcing data from various sources, such as databases and other external applications.

Service components are marked with the `@Service` annotation, which is a specialized Spring `@Component` that allows implemented classes to be auto-detected using class-path scanning. Service classes are used to add business logic. Like `Repository`, the `Service` object also represents both DDD's Service and JEE's Business Service Façade patterns. Like `Repository`, it is also a general-purpose stereotype and can be used according to the underlying approach.

First, we'll create the service interface, which is a normal Java interface with all the desired method signatures. This interface will expose all the operations that can be performed by `CartService`:

```
public interface CartService {
   public List<Item> addCartItemsByCustomerId(String customerId, @Valid
Item item);
   public List<Item> addOrReplaceItemsByCustomerId(String customerId, @
Valid Item item);
```

```
    public void deleteCart(String customerId);
    public void deleteItemFromCart(String customerId, String itemId);
    public CartEntity getCartByCustomerId(String customerId);
    public List<Item> getCartItemsByCustomerId(String customerId);
    public Item getCartItemsByItemId(String customerId, String itemId);
}
```

https://github.com/PacktPublishing/Modern-API-Development-with-Spring-6-and-Spring-Boot-3/blob/main/Chapter04/src/main/java/com/packt/modern/api/service/CartService.java

The CartServiceImpl class is annotated with @Service, therefore it would be auto-detected and available for injection. The CartRepository, UserRepository, and ItemService class dependencies are injected using constructor injection.

Let's have a look at one more method implementation of the CartService interface. Check the following code. It adds an item, or updates the price and quantity if the item already exists:

```
@Override
public List<Item> addOrReplaceItemsByCustomerId(
  String customerId, @Valid Item item) {
  // 1
  CartEntity entity = getCartByCustomerId(customerId);
  List<ItemEntity> items = Objects.nonNull
    (entity.getItems())
                ? entity.getItems() : Collections.
                  emptyList();
  AtomicBoolean itemExists = new AtomicBoolean(false);
  // 2
  items.forEach(i -> {
    if(i.getProduct().getId().equals(UUID.fromString(
        item.getId()))) {
     i.setQuantity(item.getQuantity()).
       setPrice(i.getPrice());
     itemExists.set(true);
    }
  });
  if (!itemExists.get()) {
    items.add(itemService.toEntity(item));
  }
  // 3
  return itemService.toModelList(
        repository.save(entity).getItems());
}
```

In the preceding code, we are not managing the application state, but are instead writing the sort of business logic that queries the database, sets the entity object, persists the object, and then returns the model class. Let's have a look at the statement blocks as numbered in the previous code:

1. The method only has `customerId` as a parameter and there is no `Cart` parameter. Therefore, first we get `CartEntity` from the database based on the given `customerId`.

2. The program control iterates through the items retrieved from the `CartEntity` object. If the given item already exists, then the quantity and price are changed. Else, it creates a new `Item` entity from the given `Item` model and then saves it to the `CartEntity` object. The `itemExists` flag is used to find out whether we need to update the existing `Item` or add a new one.

3. Finally, the updated `CartEntity` object is saved in the database. The latest `Item` entity is retrieved from the database, and then gets converted into a model collection and returned to the calling program.

Similarly, you can write `Service` components for others the way you have implemented them for `Cart`. Before we start enhancing the `Controller` classes, we need to add a final frontier to our overall feature.

Implementing hypermedia

We learned about hypermedia and HATEOAS in *Chapter 1*, *RESTful Web Service Fundamentals*. Spring provides state-of-the-art support to HATEOAS using the `org.springframework.boot:spring-boot-starter-hateoas` dependency.

First, we need to make sure that all models returned as part of the API response contain the link field. There are different ways to associate links (that is, the `org.springframework.hateoas.Link` class) with models, either manually or via auto-generation. Spring HATEOAS's links and attributes are implemented according to *RFC-8288* (`https://tools.ietf.org/html/rfc8288`). For example, you can create a self-link manually as follows:

```
import static org.springframework.hateoas.server.mvc.
WebMvcLinkBuilder.linkTo;
import static org.springframework.hateoas.server.mvc.
WebMvcLinkBuilder.methodOn;
// other code blocks…
responseModel.setSelf(linkTo(methodOn(CartController.class)  .
getItemsByUserId(userId,item)).withSelfRel())
```

Here, `responseModel` is a model object that is returned by the API. It has a field called `_self` that is set using the `linkTo` and `methodOn` static methods. The `linkTo` and `methodOn` methods are provided by the Spring HATEOAS library and allow us to generate a self-link for a given controller method.

This can also be done automatically by using Spring HATEOAS's `RepresentationModelAssembler` interface. This interface mainly exposes two methods – `toModel(T model)` and `toCollectionModel(Iterable<? extends T> entities)` – that convert the given entity/entities into Model and `CollectionModel`, respectively.

Spring HATEOAS provides the following classes to enrich the user-defined models with hypermedia. It basically provides a class that contains links and methods to add those to the model:

- `RepresentationModel`: Models/DTOs can extend this to collect the links.

- `EntityModel`: This extends `RepresentationModel` and wraps the domain object (that is, the model) inside it with the content private field. Therefore, it contains the domain model/DTO and the links.

- `CollectionModel`: `CollectionModel` also extends `RepresentationModel`. It wraps the collection of models and provides a way to maintain and store the links.

- `PageModel`: `PageModel` extends `CollectionModel` and provides ways to iterate through the pages, such as `getNextLink()` and `getPreviousLink()`, and through page metadata with `getTotalPages()`, among others.

The default way to work with Spring HATEOAS is to extend `RepresentationModel` with domain models, as shown in the following snippet:

```
public class Cart extends RepresentationModel<Cart>implements
Serializable {
  private static final long serialVersionUID = 1L;
  @JsonProperty("id")
  private String id;
  @JsonProperty("customerId")
  @JacksonXmlProperty(localName = "customerId")
  private String customerId;Implementing hypermedia 101
  @JsonProperty("items")
  @JacksonXmlProperty(localName = "items")
  @Valid
  private List<Item> items = null;
  // Rest of the code is removed for brevity
```

Extending `RepresentationModel` enhances the model with additional methods, including `getLink()`, `hasLink()`, and `add()`.

You know that all these models are being generated by the OpenAPI Codegen; therefore, we need to configure the OpenAPI Codegen to generate new models that support hypermedia, which is done using the following `config.json` file:

```
{
  // ...
```

```
  "apiPackage": "com.packt.modern.api",
  "invokerPackage": "com.packt.modern.api",
  "serializableModel": true,
  "useTags": true,
  "useGzipFeature": true,
  "hateoas": true,
  "unhandledException": true,
  // …
}
```

https://github.com/PacktPublishing/Modern-API-Development-with-Spring-6-and-Spring-Boot-3/blob/main/Chapter04/src/main/resources/api/config.json

Adding the hateoas property and setting it to true would automatically generate models that would extend the RepresentationModel class.

We are halfway there to implement the API business logic. Now, we need to make sure that links will be populated with the appropriate URL automatically. For that purpose, we'll extend the RepresentationModelAssemblerSupport abstract class, which internally implements RepresentationModelAssembler. Let's write the assembler for Cart, as shown in the following code block:

```
@Component
public class CartRepresentationModelAssembler extends
    RepresentationModelAssemblerSupport<CartEntity, Cart> {
  private final ItemService itemService;
  public CartRepresentationModelAssembler(ItemService itemService) {
    super(CartsController.class, Cart.class);
    this.itemService = itemService;
  }
  @Override
  public Cart toModel(CartEntity entity) {
    String uid = Objects.nonNull(entity.getUser()) ?
      entity.getUser().getId().toString() : null;
    String cid = Objects.nonNull(entity.getId()) ?
        entity.getId().toString() : null;
    Cart resource = new Cart();
    BeanUtils.copyProperties(entity, resource);
    resource.id(cid).customerId(uid)
      .items(itemService.toModelList(entity.getItems()));
    resource.add(linkTo(methodOn(CartsController.class)
      .getCartByCustomerId(uid)).withSelfRel());
```

```
        resource.add(linkTo(methodOn(CartsController.class)
          .getCartItemsByCustomerId(uid))
            .withRel("cart-items"));
        return resource;
    }
    public List<Cart> toListModel(
        Iterable<CartEntity> entities) {
        if (Objects.isNull(entities)){return List.of();}
        return StreamSupport.stream(
            entities.spliterator(), false).map(this::toModel)
            .collect(toList());
    }
}
```

https://github.com/PacktPublishing/Modern-API-Development-with-Spring-6-and-Spring-Boot-3/blob/main/Chapter04/src/main/java/com/packt/modern/api/hateoas/CartRepresentationModelAssembler.java

In the previous code, the important part in the Cart assembler is extending RepresentationModelAssemblerSupport and overriding the toModel() method. If you look closely, you'll see that CartController.class, along with the Cart model, is also passed to Rep using the super() call. This allows the assembler to generate the links appropriately as is required for the methodOn method shared earlier. This way, you can generate the link automatically.

You may also need to add additional links to other resource controllers. You can achieve this by writing a bean that implements RepresentationModelProcessor and then override the process() method, as shown here:

```
@Override
public Order process(Order model) {
    model.add(Link.of("/payments/{orderId}").withRel(
        LinkRelation.of("payments")).expand(model.getOrderId()));
    return model;
}
```

You can always refer to https://docs.spring.io/spring-hateoas/docs/current/reference/html/ for more information.

Let's make use of the services and HATEOAS enablers you created in the controller classes in the next section.

Enhancing the controller with a service and HATEOAS

In *Chapter 3, API Specifications and Implementation*, we created the `Controller` class for the Cart API – `CartController` – which just implements the OpenAPI Codegen-generated API specification interface – `CartApi`. It was just a mere block of code without any business logic or data persistence calls.

Now, since we have written the repositories, services, and HATEOAS assemblers, we can enhance the API controller class, as shown here:

```
@RestController
public class CartsController implements CartApi {
  private CartService service;
  private final CartRepresentationModelAssembler assembler;
  public CartsController(CartService service,
    CartRepresentationModelAssembler assembler) {
    this.service = service;
    this.assembler = assembler;
  }
```

https://github.com/PacktPublishing/Modern-API-Development-with-Spring-6-and-Spring-Boot-3/blob/main/Chapter04/src/main/java/com/packt/modern/api/controller/CartsController.java

You can see that `CartService` and `CartRepresentationModelAssembler` are injected using the constructor. The Spring container injects these dependencies at runtime. Then, these can be used as shown in the following code block:

```
@Override
public ResponseEntity<Cart> getCartByCustomerId(  String customerId) {
  return ok(assembler.toModel(service.getCartByCustomerId
    (customerId)));
}
```

https://github.com/PacktPublishing/Modern-API-Development-with-Spring-6-and-Spring-Boot-3/blob/main/Chapter04/src/main/java/com/packt/modern/api/controller/CartsController.java

In the preceding code, you can see that the service retrieves the `Cart` entity based on `customerId` (which internally retrieves it from the repository). This `Cart` entity then gets converted into a model that also contains the hypermedia links made available by Spring HATEOAS's `RepresentationModelAssemblerSupport` class.

The ok () static method of ResponseEntity is used to wrap the returned model that also contains the 200 OK status.

This way, you can also enhance and implement the other controllers. Now, we can also add an ETag to our API responses.

Adding ETags to API responses

An ETag is an HTTP response header that contains a computed hash or equivalent value of the response entity, and a minor change in the entity must change its value. HTTP request objects can then contain the If-None-Match and If-Match headers to receive the conditional responses.

Let's call an API to retrieve the response with an ETag, as shown next:

```
$ curl -v --location --request GET 'http://localhost:8080/ api/v1/
products/6d62d909-f957-430e-8689-b5129c0bb75e' --header 'Content-Type:
application/json' --header 'Accept: application/json'
* … text trimmed
> GET /api/v1/products/6d62d909-f957-430e-8689-b5129c0bb75e HTTP/1.1
> Host: localhost:8080
> User-Agent: curl/7.55.1
> Content-Type: application/json
> Accept: application/json
>
< HTTP/1.1 200
< ETag: "098e97de3b61db55286f5f2812785116f"
< Content-Type: application/json
< Content-Length: 339
<
{
  "_links": {
    "self": {
      "href": "http://localhost:8080/6d62d909-f957-430e
              -8689-b5129c0bb75e"
    }
  },
  "id": "6d62d909-f957-430e-8689-b5129c0bb75e",
  "name": "Antifragile",
  "description": "Antifragile - Things …",
  "imageUrl": "/images/Antifragile.jpg",
  "price": 17.1500,
  "count": 33,
  "tag": ["psychology", "book"]
}
```

Then, you can copy the value from the ETag header to the `If-None-Match` header and send the same request again with the `If-None-Match` header:

```
$ curl -v --location --request GET 'http://localhost:8080/ api/v1/
products/6d62d909-f957-430e-8689-b5129c0bb75e' --header 'Content-
Type: application/json' --header 'Accept: application/json' --header
'If-None-Match: "098e97de3b61db55286f5f2812785116f"'
* … text trimmed
> GET /api/v1/products/6d62d909-f957-430e-8689-b5129c0bb75e HTTP/1.1
> Host: localhost:8080
> User-Agent: curl/7.55.1
> Content-Type: application/json
> Accept: application/jsonAdding ETags to API responses 107
> If-None-Match: "098e97de3b61db55286f5f2812785116f"
>
< HTTP/1.1 304
< ETag: "098e97de3b61db55286f5f2812785116f"
```

You can see that since there is no change to the entity in the database, and it contains the same entity, it sends a `304` (`NOT MODIFIED`) response instead of sending the proper response with `200 OK`.

The easiest way to implement ETags is using Spring's `ShallowEtagHeaderFilter`, as shown here:

```
@Bean
public ShallowEtagHeaderFilter shallowEtagHeaderFilter() {
  return new ShallowEtagHeaderFilter();
}
```

https://github.com/PacktPublishing/Modern-API-Development-with-
Spring-6-and-Spring-Boot-3/blob/main/Chapter04/src/main/java/com/
packt/modern/api/AppConfig.java

For this implementation, Spring calculates the MD5 hash from the cached content written to the response. Next time, when it receives a request with the `If-None-Match` header, it again creates the MD5 hash from the cached content written to the response and then compares these two hashes. If both are the same, it sends the `304 NOT MODIFIED` response. This way, it will save bandwidth, but the same CPU computation will be required.

We can use the HTTP cache control (`org.springframework.http.CacheControl`) class and use the version or a similar attribute that gets updated with each change, if available, to avoid unnecessary CPU computation and for better ETag handling, as shown next:

```
return ResponseEntity.ok()
       .cacheControl(CacheControl.maxAge(5, TimeUnit.DAYS))
       .eTag(prodcut.getModifiedDateInEpoch())
       .body(product);
```

Adding an ETag to the response also allows UI apps to determine whether a page/object refresh is required, or an event needs to be triggered, especially where data changes frequently in applications, such as providing live scores or stock quotes.

Now, you have implemented fully functional APIs. Let's test them next.

Testing the APIs

Now, you must be looking forward to testing. You can find the API client collection at the following location, which is an HTTP Archive file and can be used by Insomnia or Postman API clients. You can import it and then test the APIs:

`https://github.com/PacktPublishing/Modern-API-Development-with-Spring-6-and-Spring-Boot-3/blob/main/Chapter04/Chapter04-API-Collection.har`

> **Building and running the Chapter 4 code**
> You can build the code by running `gradlew clean build` from the root of the project and run the service using `java -jar build/libs/Chapter04-0.0.1-SNAPSHOT.jar`. Make sure to use Java 17 in the path.

Summary

In this chapter, we have learned about database migration using Flyway, maintaining and persisting data using repositories, and writing business logic to services. We have also learned how hypermedia can automatically be added to API responses using Spring HATEOAS assemblers. You have now learned about all the RESTful API development practices, which allows you to use this skill in your day-to-day work involving RESTful API development.

So far, we have written synchronous APIs. In the next chapter, you will learn about async APIs and how to implement them using Spring.

Questions

1. Why is the `@Repository` class used?
2. Is it possible to add extra imports or annotations to Swagger-generated classes or models?
3. How are ETags useful?

Answers

1. Repository classes are marked with @Repository, which is a specialized @Component that makes these classes auto-detectable by package-level auto-scanning and makes them available for injection. Spring provides these classes especially for DDD repositories and the JEE DAO pattern. This is the layer used by the application for interacting with the database – retrieval and persistence as a central repository.

2. It is possible to change the way models and APIs are generated. You must copy the template that you want to modify and then place it in the resources folder. Then, you have to modify the swaggerSources block in the build.gradle file by adding an extra configuration parameter to point to the template source, such as templateDir = file("${rootDir}/ src/main/resources/ templates"). This is the place where you keep modified templates such as api. mustache. This will extend the OpenAPI Codegen templates. You can find all the templates inside the OpenAPI generator JAR file, such as openapi-generator-cli-4.3.1.jar in the \JavaSpring directory. You can copy the one you want to modify in the src/main/resource/templates directory and then play with it. You can make use of the following resources:

    - **JavaSpring templates**: https://github.com/swagger-api/swagger-codegen/ tree/master/modules/swagger-codegen/src/main/resources/JavaSpring

    - **Mustache template variables**: https://github.com/swagger-api/swagger-codegen/wiki/Mustache-Template-Variables

    - **An article explaining implementing a similar approach**: https://arnoldgalovics. com/swagger-codegen-custom-template/

3. ETags help to improve the REST/HTTP client performance and user experience by only re-rendering the page/section when the underlying API response is updated. They also save bandwidth by carrying the response body only when required. CPU utilization can be optimized if the ETag is generated based on values retrieved from the database, for example, the version or date last modified.

Further reading

- Spring HATEOAS: https://docs.spring.io/spring-hateoas/docs/current/ reference/html/

- RFC-8288: https://tools.ietf.org/html/rfc8288

5
Asynchronous API Design

So far, we have developed RESTful web services based on the imperative model, where calls are synchronous. What if you want to make code async and non-blocking? This is what we are going to do in this chapter. You'll learn about asynchronous API design in this chapter, where calls are asynchronous and non-blocking. We'll develop these APIs using Spring **WebFlux**, which is based on Project Reactor (https://projectreactor.io). Reactor is a library for building non-blocking apps on a **Java virtual machine** (**JVM**).

First, we'll walk through the reactive programming fundamentals, and then we'll migrate the existing e-commerce REST API (which we learned about in *Chapter 4, Writing Business Logic for APIs*) to an asynchronous (reactive) API to make things easier by comparing the existing (imperative) way and reactive way of programming. The code will make use of R2DBC for database persistence, which supports reactive programming.

We'll discuss the following topics in this chapter:

- Understanding Reactive Streams
- Exploring Spring WebFlux
- Understanding `DispatcherHandler`
- Controllers
- Functional endpoints
- Implementing reactive APIs for our e-commerce app

By the end of this chapter, you will have learned how to develop and implement reactive APIs and explored async APIs. You will also be able to implement reactive controllers and functional endpoints and make use of R2DBC for database persistence.

Technical requirements

The code for this chapter is available at `https://github.com/PacktPublishing/Modern-API-Development-with-Spring-6-and-Spring-Boot-3/tree/main/Chapter05`.

Understanding Reactive Streams

Normal Java code achieves asynchronicity by using thread pools. Your web server uses a thread pool to serve requests – it assigns a thread to each incoming request. The application uses the thread pool for database connections too. Each database call uses a separate thread and waits for the result. Therefore, each web request and database call uses its own thread. However, there is a wait associated with this and, therefore, these are blocking calls. The thread waits and utilizes the resources until a response is received back from the database or a response object is written. This is kind of a limitation when you scale as you can only use the resources available to the JVM. You overcome this limitation by using a load balancer with other instances of the service, which is a type of horizontal scaling.

In the last decade, there has been a rise in client-server architecture. Lots of IoT-enabled devices, smartphones that have native apps, first-class web apps, and traditional web applications have emerged. Applications not only have third-party services but also have various sources of data, which leads to higher-scale applications. On top of that, microservice-based architecture has increased communication among services themselves. You need lots of resources to serve this higher network communication demand. This makes scaling a necessity. Threads are expensive and not infinite. You don't want to block them for effective utilization. For example, let's say your code is calling the database for data. In this case, the call waits until you get the response in the blocking call. However, a non-blocking call doesn't block anything. It responds only when a response is received from the dependent code (the database in this case). The system can serve other calls during this time. This is where asynchronicity helps. In asynchronous calls, threads become free as soon as a call is done and use a callback utility (common in JavaScript). When data is available at the source, it pushes the data. Project Reactor is based on **Reactive Streams**. Reactive Streams uses the **publisher-subscriber model**, where the source of data, the publisher, pushes the data to the subscriber.

You might be aware that, on the other hand, Node.js uses a single thread to make use of most resources. It is based on an asynchronous non-blocking design, known as an **event loop**.

Reactive APIs are also based on an event loop design and use push-style notifications. If you look closely, Reactive Streams also supports Java stream (a sequence of objects pipelined to produce the desired results by performing various operations) operations, such as `map`, `flatMap`, and `filter`. Internally, Reactive Streams uses a push style, whereas Java streams work according to a pull model; that is, items are pulled from the source, such as a Java collection. In reactive programming, the source (publisher) pushes the data.

In Reactive Streams, streams of data are asynchronous and non-blocking and support backpressure. (Refer to the *Subscriber* subsection of this chapter for an explanation of backpressure.)

There are four basic types of interfaces as per the Reactive Streams specification:

- Publisher
- Subscriber
- Subscription
- Processor

Let's have a look at each of these types.

Publisher

A publisher provides a stream of data to one or more subscribers. A subscriber uses the subscribe() method to subscribe to a publisher. Each subscriber should only subscribe once to a publisher.

Most importantly, the publisher pushes data according to the demand received from subscribers. Reactive Streams are lazy; therefore, the publisher will only push an element if there is a subscriber.

A Publisher interface is defined as follows:

```
package org.reactivestreams;
// T - type of element Publisher sends
public interface Publisher<T> {
  public void subscribe(Subscriber<? super T> s); }
```

Here, the Publisher interface contains the subscribe method. Let's find out about the Subscriber type in the next subsection.

Subscriber

The subscriber consumes the data pushed by the publisher. Publisher-subscriber communication works as follows:

1. When a Subscriber instance is passed to the Publisher.subscribe() method, it triggers the onSubscribe() method. It contains a Subscription parameter, which controls the backpressure, that is, how much data a subscriber demands from the publisher.

2. After the first step, Publisher waits for the Subscription.request(long) call. It only pushes data to Subscriber after the Subscription.request() call is made. This method indicates to the Publisher how many items the subscriber is ready to receive at a time.

   Normally, the publisher pushes the data to the subscriber, irrespective of whether the subscriber can handle it safely or not. However, the subscriber knows best how much data it can handle safely; therefore, in Reactive Streams, Subscriber uses the Subscription instance to communicate the demand for the number of elements to Publisher. This is known as **backpressure** or **flow control**.

You must be wondering, what if `Publisher` asks `Subscriber` to slow down but it can't? In that case, `Publisher` must decide whether to fail, drop, or buffer.

3. Once the demand is made in *step 2*, `Publisher` sends the data notifications and the `onNext()` method is used to consume the data. This method will be triggered until the data notifications are pushed by `Publisher` according to the demand communicated by `Subscription.request()`.

4. At the end, either `onError()` or `onCompletion()` will be triggered as the terminal state. No notification will be sent after one of these invocations has been triggered even if you call `Subscription.request()`. The following are the terminal methods:

   • `onError()` will be invoked the moment any error occurs

   • `onCompletion()` will be invoked when all elements are pushed

The `Subscriber` interface is defined as follows:

```
package org.reactivestreams;
// T - type of element Publisher sends
public interface Subscriber<T> {
  public void onSubscribe(Subscription s);
  public void onNext(T t);
  public void onError(Throwable t);
  public void onComplete();
}
```

Subscription

A subscription is a mediator between the publisher and subscriber. It is the subscriber's responsibility to invoke the `Subscription.subscriber()` method and let the publisher know of the demand. It can be invoked as and when required by the subscriber.

The `cancel()` method asks the publisher to stop sending data notifications and to clean up the resources.

A subscription is defined as follows:

```
package org.reactivestreams;
public interface Subscription {
  public void request(long n);
  public void cancel();
}
```

Processor

The processor is a bridge between the publisher and subscriber and represents the processing stage. It works as both a publisher and subscriber and obeys the contract defined by both. It is defined as follows:

```
package org.reactivestreams;
public interface Processor<T, R>
  extends Subscriber<T>, Publisher<R> {
}
```

Let's have a look at the following example. Here, we are creating `Flux` by using the `Flux.just()` static factory method. `Flux` is a type of publisher in Project Reactor. This publisher contains four integer elements. Then, we use the `reduce` operator (like we do in Java streams) to perform a `sum` operation on it:

```
Flux<Integer> fluxInt = Flux.just(1, 10, 100, 1000).log();
fluxInt.reduce(Integer::sum)
  .subscribe(sum ->
      System.out.printf("Sum is: %d", sum));
```

When you run the previous code, it prints the following output:

```
11:00:38.074 [main] INFO reactor.Flux.Array.1 - |
onSubscribe([Synchronous Fuseable] FluxArray.ArraySubscription)
11:00:38.074 [main] INFO reactor.Flux.Array.1 - |request(unbounded)
11:00:38.084 [main] INFO reactor.Flux.Array.1 - | onNext(1)
11:00:38.084 [main] INFO reactor.Flux.Array.1 - | onNext(10)
11:00:38.084 [main] INFO reactor.Flux.Array.1 - | onNext(100)
11:00:38.084 [main] INFO reactor.Flux.Array.1 - | onNext(1000)
11:00:38.084 [main] INFO reactor.Flux.Array.1 - | onComplete() Sum is:
1111
Process finished with exit code 0
```

Looking at the output, when `Publisher` is subscribed, `Subscriber` sends an unbounded `Subscription.request()`. When the first element is notified, `onNext()` is called, and so on. At the end, when the publisher is done with the push elements, the `onComplete()` event is called. This is how Reactive Streams works.

Now that you have an idea of how Reactive Streams works, let's see how and why Spring makes use of Reactive Streams in the Spring WebFlux module.

Exploring Spring WebFlux

Existing Servlet APIs are blocking APIs. They use input and output streams, which block APIs. Servlet 3.0 containers evolve and use the underlying event loop. Async requests are processed asynchronously but read and write operations still use blocking input/output streams. The *Servlet 3.1* container has evolved further, supporting asynchronicity and having the non-blocking I/O stream APIs. However, there are certain Servlet APIs, such as `request.getParameters()`, which parse the blocking request body and provide synchronous contracts such as `Filter`. The **Spring MVC** framework is based on the Servlet API and Servlet containers.

Therefore, Spring provides **Spring WebFlux**, which is fully non-blocking and provides backpressure functionality. It provides concurrency with a small number of threads and scales with fewer hardware resources. WebFlux provides fluent, functional, and continuation-style APIs to support the declarative composition of asynchronous logic. Writing asynchronous functional code is more complex than writing imperative-style code. However, once you get hands-on with it, you will love it because it allows you to write precise and readable code.

Both Spring WebFlux and Spring MVC can co-exist; however, to ensure the effective use of reactive programming, you should never mix a reactive flow with blocking calls.

Spring WebFlux supports the following features and archetypes:

- The event loop concurrency model
- Both annotated controllers and functional endpoints
- Reactive clients
- Netty and Servlet 3.1 container-based web servers, such as Tomcat, Undertow, and Jetty

Now that you have some idea about WebFlux, you can deep dive into learning how WebFlux works by understanding reactive APIs and Reactor Core. Let's first explore reactive APIs. You'll explore Reactor Core in a subsequent subsection.

Understanding reactive APIs

Spring WebFlux APIs are reactive APIs and accept `Publisher` as the plain input. WebFlux then adapts it to a type supported by a reactive library such as Reactor Core or RxJava. It then processes the input and returns the output in a format supported by the reactive library. This allows WebFlux APIs to be interoperable with other reactive libraries.

By default, Spring WebFlux uses Reactor (`https://projectreactor.io`) as a core dependency. Project Reactor provides the Reactive Streams library. As stated in the previous paragraph, WebFlux accepts the input as `Publisher`, then adapts it to a Reactor type, and then returns it as a `Mono` or `Flux` output.

You know that `Publisher` in Reactive Streams pushes the data to its subscribers based on demand. It can push one or more (possibly infinite) elements. Project Reactor takes it further and provides two `Publisher` implementations, namely `Mono` and `Flux`. `Mono` can return either 0 or 1 to `Subscriber`, whereas `Flux` returns 0 to N elements. Both are abstract classes that implement the `CorePublisher` interface. The `CorePublisher` interface extends the publisher.

Normally, we have the following methods in the repository:

```
public Product findById(UUID id);
public List<Product> getAll();
```

These can be replaced with `Mono` and `Flux`:

```
Public Mono<Product> findById(UUID id);
public Flux<Product> getAll();
```

Streams can be either hot or cold based on whether the source can be restarted or not. The source is restarted if there are multiple subscribers for cold streams, while the same source is used for multiple subscribers in hot streams. Project Reactor streams are, by default, cold. Therefore, once you consume a stream, you can't reuse it until it's restarted. However, Project Reactor allows you to turn a cold stream into a hot one by using `cache()` methods. Both `Mono` and `Flux` abstract classes support cold and hot streams.

Let's understand the cold and hot stream concepts with some examples:

```
Flux<Integer> fluxInt = Flux.just(1, 10, 100).log();
fluxInt.reduce(Integer::sum).subscribe(sum ->        System.out.
printf("Sum is: %d\n", sum));
fluxInt.reduce(Integer::max).subscribe(max ->     System.out.
printf("Maximum is: %d", max));
```

Here, we have created a `Flux` object, `fluxInt`, that contains three numbers. Then, we are performing two operations separately – `sum` and `max`. You can see that there are two subscribers. By default, Project Reactor streams are cold; therefore, when a second subscriber registers, it restarts, as shown in the following output:

```
11:23:35.060 [main] INFO reactor.Flux.Array.1 - |
onSubscribe([Synchronous Fuseable] FluxArray.ArraySubscription)
11:23:35.060 [main] INFO reactor.Flux.Array.1 - | request(unbounded)
11:23:35.060 [main] INFO reactor.Flux.Array.1 - | onNext(1)
11:23:35.060 [main] INFO reactor.Flux.Array.1 - | onNext(10)
11:23:35.060 [main] INFO reactor.Flux.Array.1 - | onNext(100)
11:23:35.060 [main] INFO reactor.Flux.Array.1 - | onComplete()
Sum is: 111
11:23:35.076 [main] INFO reactor.Flux.Array.1 - |
onSubscribe([Synchronous Fuseable] FluxArray.ArraySubscription)
11:23:35.076 [main] INFO reactor.Flux.Array.1 - | request(unbounded)
```

```
11:23:35.076 [main] INFO reactor.Flux.Array.1 - | onNext(1)
11:23:35.076 [main] INFO reactor.Flux.Array.1 - | onNext(10)
11:23:35.076 [main] INFO reactor.Flux.Array.1 - | onNext(100)
11:23:35.076 [main] INFO reactor.Flux.Array.1 - | onComplete()
Maximum is: 100
```

The source is created in the same program, but what if the source is somewhere else, such as in an HTTP request, or you don't want to restart the source? In these cases, you can turn the cold stream into a hot stream by using `cache()`, as shown in the next code block. The only difference between the following code and the previous code is that we have added a `cache()` call to `Flux.just()`:

```
Flux<Integer> fluxInt = Flux.just
    (1, 10, 100).log().cache();
fluxInt.reduce(Integer::sum).subscribe(sum ->    System.out.printf("Sum
is: %d\n", sum));
fluxInt.reduce(Integer::max).subscribe(max ->    System.out.
printf("Maximum is: %d", max));
```

Now, look at the output. The source has not restarted; instead, the same source is used again:

```
11:29:25.665 [main] INFO reactor.Flux.Array.1 - |
onSubscribe([Synchronous Fuseable] FluxArray.ArraySubscription)
11:29:25.665 [main] INFO reactor.Flux.Array.1 - | request(unbounded)
11:29:25.665 [main] INFO reactor.Flux.Array.1 - | onNext(1)
11:29:25.665 [main] INFO reactor.Flux.Array.1 - | onNext(10)
11:29:25.665 [main] INFO reactor.Flux.Array.1 - | onNext(100)
11:29:25.665 [main] INFO reactor.Flux.Array.1 - | onComplete()
Sum is: 111
Maximum is: 100
```

Now that we have got to the crux of reactive APIs, let's see what Spring WebFlux's Reactive Core consists of.

Reactive Core

Reactive Core provides a foundation for developing a reactive web application with Spring. A web application needs three levels of support for serving HTTP web requests:

- Handling web requests by the server using the following:

  - HttpHandler: An interface of the `reactor.core.publisher.Mono` package that is an abstraction of a request/response handler over different HTTP server APIs, such as Netty or Tomcat:

    ```
    public interface HttpHandler {
        Mono<Void> handle(ServerHttpRequest request,
    ```

```
        ServerHttpResponse response);
    }
```

- WebHandler: An interface of the `org.springframework.web.server` package that provides support for user sessions, request and session attributes, a locale and principal for the request, form data, and so on. You can find more information about WebHandler at https://docs.spring.io/spring-framework/docs/current/reference/html/web-reactive.html#webflux-web-handler-api.

- Handling of a web request call by the client using `WebClient`.

- Codecs (`Encoder`, `Decoder`, `HttpMessageWriter`, `HttpMessageReader`, and `DataBuffer`) for the serialization and deserialization of content at both the server and client levels for the request and response.

These components are at the core of Spring WebFlux. WebFlux application configuration also contains the following beans – webHandler (`DispatcherHandler`), `WebFilter`, `WebExceptionHandler`, `HandlerMapping`, `HandlerAdapter`, and `HandlerResultHandler`.

For REST service implementation, there are specific `HandlerAdapter` instances for the following web servers – Tomcat, Jetty, Netty, and Undertow. A web server such as Netty, which supports Reactive Streams, handles the subscriber's demands. However, if the server handler does not support Reactive Streams, then the `org.springframework.http.server.reactive.ServletHttpHandlerAdapter` HTTP HandlerAdapter is used. `HandlerAdapter` handles the adaptation between Reactive Streams and the Servlet 3.1 container async I/O and implements a `Subscriber` class. `HandlerAdapter` uses the OS TCP buffers. OS TCP uses its own backpressure (control flow); that is, when the buffer is full, the OS uses the TCP backpressure to stop incoming elements.

The browser, or any HTTP client, consumes REST APIs using the HTTP protocol. When a request is received by the web server, it forwards it to the Spring WebFlux application. Then, WebFlux builds the reactive pipeline that goes to the controller. `HttpHandler` is an interface between WebFlux and the web server that communicates using the HTTP protocol. If the underlying server supports Reactive Streams, such as Netty, then the subscription is done by the server natively. Else, WebFlux uses `ServletHttpHandlerAdapter` for Servlet 3.1 container-based servers. `ServletHttpHandlerAdapter` then adapts the streams to async I/O Servlet APIs and vice versa. Then, the subscription of Reactive Streams happens with `ServletHttpHandlerAdapter`.

Therefore, in summary, `Mono/Flux` streams are subscribed by WebFlux internal classes, and when the controller sends a `Mono/Flux` stream, these classes convert it into HTTP packets. The HTTP protocol does support event streams. However, for other media types, such as JSON, Spring WebFlux subscribes the `Mono/Flux` streams and waits until `onComplete()` or `onError()` is triggered. Then, it serializes the whole list of elements, or a single element in the case of `Mono`, in one HTTP response.

Spring WebFlux needs a component like `DispatcherServlet` in Spring MVC – a front controller. Let's discuss this in the next section.

Understanding DispatcherHandler

`DispatcherHandler`, a front controller in Spring WebFlux, is the equivalent of `DispatcherServlet` in the Spring MVC framework. `DispatcherHandler` contains an algorithm that makes use of special components – `HandlerMapping` (maps requests to the handler), `HandlerAdapter` (a `DispatcherHandler` helper to invoke a handler mapped to a request), and `HandlerResultHandler` (a palindrome of words, for processing the result and forming results) – for processing requests. The `DispatcherHandler` component is identified by a bean named `webHandler`.

It processes requests in the following way:

1. A web request is received by `DispatcherHandler`.

2. `DispatcherHandler` uses `HandlerMapping` to find a matching handler for the request and uses the first match.

3. It then uses the respective `HandlerAdapter` to process the request, which exposes `HandlerResult` (the value returned by `HandlerAdapter` after processing). The return value could be one of the following – `ResponseEntity`, `ServerResponse`, values returned from `@RestController`, or values (`CharSequence`, view, map, and so on) returned by a view resolver.

4. Then, it makes use of the respective `HandlerResultHandler` to write the response or render a view based on the `HandlerResult` type received from *step 2*. `ResponseEntityResultHandler` is used for `ResponseEntity`, `ServerResponseResultHandler` is used for `ServerResponse`, `ResponseBodyResultHandler` is used for values returned by the `@RestController` or `@ResponseBody` annotated method, and `ViewResolutionResultHandler` is used for values returned by the view resolver.

5. The request is completed.

You can create REST endpoints in Spring WebFlux using either an annotated controller such as Spring MVC or functional endpoints. Let's explore these in the next sections.

Controllers

The Spring team has kept the same annotations for both Spring MVC and Spring WebFlux as these annotations are non-blocking. Therefore, you can use the same annotations we used in previous chapters to create REST controllers. In Spring WebFlux, the annotation runs on Reactive Core and provides a non-blocking flow. However, you, as the developer, have the responsibility of maintaining a fully non-blocking flow and the reactive chain (pipeline). Any blocking calls in a reactive chain will convert the reactive chain into a blocking call.

Let's create a simple REST controller that supports non-blocking and reactive calls:

```
@RestController
public class OrderController {
  @RequestMapping(value = "/api/v1/orders",  method =
    RequestMethod.POST)
  public ResponseEntity<Order> addOrder(
          @RequestBody NewOrder newOrder){
    // …
  }
  @RequestMapping(value = "/api/v1/orders/{id}", method =
    RequestMethod.GET)
  public ResponseEntity<Order>getOrderById(
    @PathVariable("id")  String id){
    // …
  }
}
```

You can see that it uses all the annotations that we have used in Spring MVC:

- @RestController is used to mark a class as a REST controller. Without this, the endpoint won't register, and the request will be returned as NOT FOUND 404.

- @RequestMapping is used to define the path and HTTP method. Here, you can also use @PostMapping with just the path. Similarly, for each of the HTTP methods, there is a respective mapping, such as @GetMapping.

- The @RequestBody annotation marks a parameter as a request body, and an appropriate codec will be used for conversion. Similarly, there are @PathVariable and @RequestParam for the path parameter and query parameter, respectively.

We are going to use an annotation-based model to write the REST endpoints. You'll get a closer look when we implement the e-commerce app controllers using WebFlux. Spring WebFlux also provides a way to write a REST endpoint using a functional programming style, which you'll explore in the next section.

Functional endpoints

The REST controllers we coded using Spring MVC were written in imperative-style programming. Reactive programming, on the other hand, is functional-style programming. Therefore, Spring WebFlux also allows an alternative way to define REST endpoints, using functional endpoints. These also use the same Reactive Core foundation.

Let's see how we can write the same `Order` REST endpoint of the sample e-commerce app, using a functional endpoint:

```
import static org.springframework.http.MediaType.
  APPLICATION_JSON;
import static org.springframework.web.reactive.
  function.server. RequestPredicates.*;
import staticorg.springframework.
web.reactive.function.server. RouterFunctions.route;
// ...
  OrderRepository repository = ...
  OrderHandler handler = new OrderHandler(repository);
  RouterFunction<ServerResponse> route = route()
    .GET("/v1/api/orders/{id}",
          accept(APPLICATION_JSON),
          handler::getOrderById)
    .POST("/v1/api/orders", handler::addOrder)
    .build();
public class OrderHandler {
  public Mono<ServerResponse> addOrder
    (ServerRequest req){
    // ...
  }
  public Mono<ServerResponse> getOrderById(
    ServerRequest req) {
    // ...
  }
}
```

In the previous code, you can see that the `RouterFunctions.route()` builder allows you to write all the REST routes in a single statement using the functional programming style. Then, it uses the method reference of the handler class to process the request, which is the same as the @RequestMapping body of an annotation-based model.

Let's add the following code to the `OrderHandler` methods:

```
public class OrderHandler {
  public Mono<ServerResponse> addOrder(
    ServerRequest req){
    Mono<NewOrder> order = req.bodyToMono(NewOrder.class);
    return ok()
      .build(repository.save(toEntity(order)));
  }
  public Mono<ServerResponse> getOrderById(
```

```
        ServerRequest req) {
      String orderId = req.pathVariable("id");
      return repository
        .getOrderById(UUID.fromString(orderId))
        .flatMap(order -> ok()
            .contentType(APPLICATION_JSON)
            .bodyValue(toModel(order)))
        .switchIfEmpty(ServerResponse.notFound()
        .build());
  }
}
```

Unlike the @RequestMapping() mapping methods in the REST controller, handler methods don't have multiple parameters, such as body, path, or query parameters. They just have a ServerRequest parameter, which can be used to extract the body, path, and query parameters. In the addOrder method, the Order object is extracted using request.bodyToMono(), which parses the request body and then converts it into an Order object. Similarly, the getOrderById() method retrieves the order object identified by the given ID from a server request object by calling request.pathVariable("id").

Now, let's discuss the response. The handler method uses the ServerResponse object in comparison to ResponseEntity in Spring MVC. Therefore, the ok() static method looks like it's from ResponseEntity, but it is from org.springframework.web.reactive.function.server.ServerResponse.ok. The Spring team has tried to keep the API as similar as possible to Spring MVC; however, the underlying implementation differs and provides a non-blocking reactive interface.

The last point about these handler methods is the way a response is written. It uses a functional style instead of an imperative style and makes sure that the reactive chain does not break. The repository returns the Mono object (a publisher) in both cases as a response wrapped inside ServerResponse.

You can find interesting code in the getOrderById() handler method. It performs a flatMap operation on the received Mono object from the repository. It converts it from an entity into a model, then wraps it in a ServerResponse object, and returns the response. You must be wondering what happens if the repository returns null. The repository returns Mono as per the contract, which is similar in nature to the Java Optional class. Therefore, the Mono object can be empty but not null, as per the contract. If the repository returns an empty Mono, then the switchIfEmpty() operator will be used and a NOT FOUND 404 response will be sent.

In the case of an error, there are different error operators that can be used, such as doOnError() or onErrorReturn().

We have discussed the logic flow using the Mono type; the same explanation will apply if you use the Flux type in place of the Mono type.

We have discussed a lot of theory relating to reactive, asynchronous, and non-blocking programming in a Spring context. Let's jump into coding and migrate the e-commerce API developed in *Chapter 4, Writing Business Logic for APIs*, to a reactive API.

Implementing reactive APIs for our e-commerce app

Now that you have an idea of how Reactive Streams works, we can go ahead and implement REST APIs that are asynchronous and non-blocking.

You'll recall that we are following the design-first approach, so we need the API design specification first. However, we can reuse the e-commerce API specification we created previously in *Chapter 3, API Specifications and Implementation*.

OpenAPI Codegen is used to generate the API interface/contract that creates the Spring MVC-compliant API Java interfaces. Let's see what changes we need to do to generate the reactive API interfaces.

Changing OpenAPI Codegen for reactive APIs

You need to tweak a few OpenAPI Codegen configurations to generate Spring WebFlux-compliant Java interfaces, as shown next:

```
{
    "library": "spring-boot",
    "dateLibrary": "java8",
    "hideGenerationTimestamp": true,
    "modelPackage": "com.packt.modern.api.model",
    "apiPackage": "com.packt.modern.api",
    "invokerPackage": "com.packt.modern.api",
    "serializableModel": true,
    "useTags": true,
    "useGzipFeature" : true,
    "reactive": true,
    "interfaceOnly": true,
    …
    …
}
```

https://github.com/PacktPublishing/Modern-API-Development-with-Spring-6-and-Spring-Boot-3/blob/main/Chapter05/src/main/resources/api/config.json

Reactive API support is only there if you opt for `spring-boot` as the library. Also, you need to set the `reactive` flag to `true`. By default, the `reactive` flag is `false`.

Now, you can run the following command:

```
$ gradlew clean generateSwaggerCode
```

This will generate Reactive Streams-compliant Java interfaces, which are annotation-based REST controller interfaces. When you open any API interface, you'll find Mono/Flux reactor types in it, as shown in the following code block for the OrderAPI interface:

```
@Operation(
  operationId = "addOrder",
  summary = "Creates a new order for the …",
  tags = { "order" },
  responses = {
    @ApiResponse(responseCode = "201",
      description = "Order added successfully",
      content = {
        @Content(mediaType = "application/xml",
          schema = @Schema(
            implementation = Order.class)),
        @Content(mediaType = "application/json",
          schema = @Schema(
            implementation = Order.class))
      }),
      @ApiResponse(responseCode = "406",
        description = "If payment not authorized")
  }
)
@RequestMapping(
  method = RequestMethod.POST,
  value = "/api/v1/orders",
  produces = { "application/xml",
               "application/json" },
  consumes = { "application/xml",
               "application/json" }
)
default Mono<ResponseEntity<Order>> addOrder(
 @Parameter(name = "NewOrder", description =
     "New Order Request object")
    @Valid @RequestBody(required = false)
        Mono<NewOrder> newOrder,
 @Parameter(hidden = true)
  final ServerWebExchange exg) throws Exception {
```

You will have observed another change: an additional parameter, `ServerWebExchange`, is also required for reactive controllers.

Now, when you compile your code, you may find compilation errors because we haven't yet added the dependencies required for reactive support. Let's learn how to add them in the next section.

Adding Reactive dependencies to build.xml

First, we'll remove `spring-boot-starter-web` as we don't need Spring MVC now. Second, we'll add `spring-boot-starter-webflux` and `reactor-test` for Spring WebFlux and Reactor support tests, respectively. Once these dependencies are added successfully, you should not see any compilation errors in the OpenAPI-generated code.

You can add the required reactive dependencies to `build.gradle`, as shown next:

```
implementation 'org.springframework.boot:
  spring-boot-starter-webflux'

testImplementation('org.springframework.boot:
  spring-boot-starter-test')
testImplementation 'io.projectreactor:reactor-test'
```

`https://github.com/PacktPublishing/Modern-API-Development-with-Spring-6-and-Spring-Boot-3/blob/main/Chapter05/build.gradle`

We need to have a complete reactive pipeline from the REST controller to the database. However, existing JDBC and Hibernate dependencies only support blocking calls. JDBC is a fully blocking API. Hibernate is also blocking. Therefore, we need to have reactive dependencies for the database.

Hibernate Reactive (`https://github.com/hibernate/hibernate-reactive`) was released after the first edition of this book. Hibernate Reactive supports PostgreSQL, MySQL/MariaDB, Db2 11.5+, CockroachDB 22.1+, MS SQL Server 2019+, and Oracle Database 21+. Hibernate Reactive does not support H2 at the time of writing. Therefore, we will simply use Spring Data, a Spring framework that provides the `spring-data-r2dbc` library for working with Reactive Streams.

Many NoSQL databases, such as MongoDB, already provide a reactive database driver. An R2DBC-based driver should be used for relational databases in place of JDBC for fully non-blocking/reactive API calls. **R2DBC** stands for **Reactive Relational Database Connectivity**. R2DBC is a reactive API open specification that establishes a **Service Provider Interface** (**SPI**) for database drivers. Almost all the popular relational databases support R2DBC drivers – H2, Oracle Database, MySQL, MariaDB, SQL Server, PostgreSQL, and R2DBC Proxy.

Let's add the R2DBC dependencies for Spring Data and H2 to the `build.gradle` file:

```
implementation 'org.springframework.boot:spring-boot-starter-data-r2dbc'

implementation 'com.h2database:h2'
runtimeOnly 'io.r2dbc:r2dbc-h2'
```

Now, we can write end-to-end (from the controller to the repository) code without any compilation errors. Let's add global exception handling before we jump into writing an implementation for API Java interfaces.

Handling exceptions

We'll add the global exception handler the way it was added in Spring MVC in *Chapter 3, API Specifications and Implementation*. Before that, you must be wondering how to handle exceptions in a reactive pipeline. Reactive pipelines are a flow of streams and you can't add exception handling the way you do in imperative code. You need to raise the error in a pipeline flow only.

Check out the following code:

```
.flatMap(card -> {
  if (Objects.isNull(card.getId())) {
    return service.registerCard(mono)
      .map(ce -> status(HttpStatus.CREATED)
        .body(assembler.entityToModel(
          ce, exchange)));
  } else {
    return Mono.error(() -> new
      CardAlreadyExistsException(
        " for user with ID - " + d.getId()));
  }
})
```

Here, a `flatMap` operation is performed. An error should be thrown if `card` is not valid, that is, if `card` does not have the requested ID. Here, `Mono.error()` is used because the pipeline expects `Mono` as a returned object. Similarly, you can use `Flux.error()` if `Flux` is expected as the returned type.

Let's assume you are expecting an object from a service or repository call, but instead, you receive an empty object. Then, you can use the `switchIfEmpty()` operator, as shown in the following code:

```
Mono<List<String>> monoIds =
  itemRepo.findByCustomerId( customerId)
    .switchIfEmpty(Mono.error(new
```

```
    ResourceNotFoundException(". No items
        found in Cart of customer with Id - " +
            customerId)))
    .map(i -> i.getId().toString())
    .collectList().cache();
```

Here, the code expects a `Mono` object of the `List` type from the `item` repository. However, if the returned object is empty, then it simply throws the `ResourceNotFoundException`. `switchIfEmpty()` exception and accepts the alternate `Mono` instance.

By now, you might have a question about the type of exception. It throws a runtime exception. See the `ResourceNotFoundException` class declaration here:

```
public class ResourceNotFoundException
    extends RuntimeException
```

Similarly, you can also use `onErrorReturn()`, `onErrorResume()`, or similar error operators from Reactive Streams. Look at the use of `onErrorReturn()` in the next code block:

```
return service.getCartByCustomerId(customerId)
    .map(cart -> assembler
      .itemfromEntities(cart.getItems().stream()
        .filter(i -> i.getProductId().toString()
          .equals(itemId.trim())).collect(toList()))
        .get(0)).map(ResponseEntity::ok)
    .onErrorReturn(notFound().build())
```

All exceptions should be handled, and an error response should be sent to the user. We'll have a look at the global exception handler first, in the following section.

Handling global exceptions for controllers

We created a global exception handler using `@ControllerAdvice` in Spring MVC. We'll take a slightly different route for handling errors in Spring WebFlux. First, we'll create the `ApiErrorAttributes` class, which can also be used in Spring MVC. This class extends `DefaultErrorAttributes`, which is a default implementation of the `ErrorAttributes` interface. The `ErrorAttributes` interface provides a way to handle maps, a map of error fields, and their values. These error attributes can then be used to display an error to the user or for logging.

The following attributes are provided by the `DefaultErrorAttributes` class:

- `timestamp`: The time that the error was captured

- `status`: The status code

- error: An error description

- exception: The class name of the root exception (if configured)

- message: The exception message (if configured)

- errors: Any ObjectError from a BindingResult exception (if configured)

- trace: The exception stack trace (if configured)

- path: The URL path when the exception was raised

- requestId: The unique ID associated with the current request

We have added two default values to the status and message – an internal server error and a generic error message (The system is unable to complete the request. Contact system support.), respectively – to ApiErrorAttributes, as shown next:

```
@Component
public class ApiErrorAttributes
  extends DefaultErrorAttributes {
  private HttpStatus status =
      HttpStatus.INTERNAL_SERVER_ERROR;
  private String message =
    ErrorCode.GENERIC_ERROR.getErrMsgKey();

  @Override
  public Map<String, Object>
    getErrorAttributes( ServerRequest request,
      ErrorAttributeOptions options) {
    var attributes =
      super.getErrorAttributes(request, options);
    attributes.put("status", status);
    attributes.put("message", message);
    attributes.put("code", ErrorCode.
      GENERIC_ERROR.getErrCode());
    return attributes;
  }
  // Getters and setters
}
```

https://github.com/PacktPublishing/Modern-API-Development-with-Spring-6-and-Spring-Boot-3/blob/main/Chapter05/src/main/java/com/packt/modern/api/exception/ApiErrorAttributes.java

Now, we can use this `ApiErrorAttributes` class in a custom global exception handler class. We'll create the `ApiErrorWebExceptionHandler` class, which extends the `AbstractErrorWebExceptionHandler` abstract class.

The `AbstractErrorWebExceptionHandler` class implements the `ErrorWebExcepti onHandler` and `InitializingBean` interfaces. `ErrorWebExceptionHandler` is a functional interface that extends the `WebExceptionHandler` interface, which indicates that `WebExceptionHandler` is used for rendering exceptions. `WebExceptionHandler` is a contract for handling exceptions when server exchange processing takes place.

The `InitializingBean` interface is a part of the Spring core framework. It is used by components that react when all properties are populated. It can also be used to check whether all the mandatory properties are set.

Now that we have studied the basics, let's jump into writing the `ApiErrorAttributes` class:

```java
@Component
@Order(-2)
public class ApiErrorWebExceptionHandler extends
  AbstractErrorWebExceptionHandler {

 public ApiErrorWebExceptionHandler(
    ApiErrorAttributes errorAttributes,
    ApplicationContext appCon,
    ServerCodecConfigurer serverCodecConfigurer){
   super(errorAttributes,
     new WebProperties().getResources(),appCon);
   super.setMessageWriters(
     serverCodecConfigurer.getWriters());
   super.setMessageReaders(
     serverCodecConfigurer.getReaders());
  }
 @Override
 protected RouterFunction<ServerResponse>
   getRoutingFunction(ErrorAttributes errA) {
   return RouterFunctions.route(
     RequestPredicates.all(),
       this::renderErrorResponse);
 }
```

https://github.com/PacktPublishing/Modern-API-Development-with-Spring-6-and-Spring-Boot-3/blob/main/Chapter05/src/main/java/com/packt/modern/api/exception/ApiErrorWebExceptionHandler.java

The first important observation about this code is that we have added the `@Order` annotation, which tells us the preference of execution. `ResponseStatusExceptionHandler` is placed at the 0 index by the Spring Framework and `DefaultErrorWebExceptionHandler` is ordered at the -1 index. Both are exception handlers like the one we have created. If you don't give an order of precedence to `ApiErrorWebExceptionHandler` superseding both, then it won't ever execute. Therefore, the order of precedence is set at -2.

Next, this class overrides the `getRoutingFunction()` method, which calls the privately defined `renderErrorResponse()` method, where we have our own custom implementation for error handling, as shown next:

```java
private Mono<ServerResponse> renderErrorResponse(
    ServerRequest request) {
  Map<String, Object> errorPropertiesMap =
    getErrorAttributes(request,
      ErrorAttributeOptions.defaults());
  Throwable throwable = (Throwable) request
    .attribute("org.springframework.boot.web
              .reactive.error
              .DefaultErrorAttributes.ERROR")
    .orElseThrow(() -> new IllegalStateException
    ("Missing exception attribute in ServerWebExchange"));
  ErrorCode errorCode = ErrorCode.GENERIC_ERROR;
  if (throwable instanceof
      IllegalArgumentException || throwable
      instanceof DataIntegrityViolationException
      || throwable instanceof
      ServerWebInputException) {
    errorCode = ILLEGAL_ARGUMENT_EXCEPTION;
  } else if (throwable instanceof
    CustomerNotFoundException) {
    errorCode = CUSTOMER_NOT_FOUND;
  } else if (throwable instanceof
    ResourceNotFoundException) {
    errorCode = RESOURCE_NOT_FOUND;
  } // other else-if
  // …
}
```

https://github.com/PacktPublishing/Modern-API-Development-with-Spring-6-and-Spring-Boot-3/blob/main/Chapter05/src/main/java/com/packt/modern/api/exception/ApiErrorWebExceptionHandler.java

Here, first, we extract the error attributes in `errorPropertiesMap`. This will be used when we form the error response. Next, we capture the occurred exception using `throwable`. Then, we check the type of the exception and assign an appropriate code to it. We keep the default as `GenericError`, which is nothing more than `InternalServerError`.

Next, we use a `switch` statement to form an error response based on the raised exception, as shown here:

```
switch (errorCode) {
  case ILLEGAL_ARGUMENT_EXCEPTION ->{errorPropertiesMap.put
    ("status", HttpStatus.BAD_REQUEST);
    errorPropertiesMap.put("code",
      ILLEGAL_ARGUMENT_EXCEPTION.getErrCode());
    errorPropertiesMap.put("error",
      ILLEGAL_ARGUMENT_EXCEPTION);
    errorPropertiesMap.put("message", String
      .format("%s %s",
      ILLEGAL_ARGUMENT_EXCEPTION.getErrMsgKey(),
      throwable.getMessage()));
    return ServerResponse.status(
        HttpStatus.BAD_REQUEST)
      .contentType(MediaType.APPLICATION_JSON)
      .body(BodyInserters.fromValue(
        errorPropertiesMap));
  }
  case CUSTOMER_NOT_FOUND -> {
    errorPropertiesMap.put("status",
      HttpStatus.NOT_FOUND);
    errorPropertiesMap.put("code",
      CUSTOMER_NOT_FOUND.getErrCode());
    errorPropertiesMap.put("error",
      CUSTOMER_NOT_FOUND);
    errorPropertiesMap.put("message", String
      .format("%s %s",
      CUSTOMER_NOT_FOUND.getErrMsgKey(),
      throwable.getMessage()));
    return ServerResponse.status(
        HttpStatus.NOT_FOUND)
      .contentType(MediaType.APPLICATION_JSON)
      .body(BodyInserters.fromValue(
        errorPropertiesMap));
  }
  case RESOURCE_NOT_FOUND -> {
    // rest of the code …
}
```

Probably in the next version of Java, we will be able to combine the `if-else` and `switch` blocks to make the code more concise. You can also create a separate method that takes `errorPropertiesMap` as an argument and returns the formed server response based on it. Then, you can use `switch`.

Custom application exception classes, such as `CustomerNotFoundException`, and other exception-handling-supported classes, such as `ErrorCode` and `Error`, are being used from the existing code (from *Chapter 4, Writing Business Logic for APIs*).

Now that we have studied exception handling, we can concentrate on HATEOAS.

Adding hypermedia links to an API response

HATEOAS support for reactive APIs exists and is a bit like what we did in the previous chapter using Spring MVC. We create these assemblers again for HATEOAS support. We also use the HATEOAS assembler classes to convert a model into an entity and vice versa.

Spring WebFlux provides the `ReactiveRepresentationModelAssembler` interface for forming hypermedia links. We would override its `toModel()` method to add the links to response models.

Here, we will do some groundwork for populating the links. We will create an `HateoasSupport` interface with a single default method, as shown next:

```
public interface HateoasSupport {
  default UriComponentsBuilder
    getUriComponentBuilder(@Nullable
            ServerWebExchange exchange) {
    if (exchange == null) {
      return UriComponentsBuilder.fromPath("/");
    }
    ServerHttpRequest request = exchange.getRequest();
    PathContainer contextPath = request.getPath().
      contextPath();

    return UriComponentsBuilder
        .fromHttpRequest(request)
        .replacePath(contextPath.toString())
        .replaceQuery("");
  }
}
```

```
https://github.com/PacktPublishing/Modern-API-Development-with-
Spring-6-and-Spring-Boot-3/blob/main/Chapter05/src/main/java/com/
packt/modern/api/hateoas/HateoasSupport.java
```

Here, this class contains a single default method, getUriComponentBuilder(), which accepts ServerWebExchange as an argument and returns the UriComponentsBuilder instance. This instance can then be used to extract the server URI that would be used to add the links with a protocol, host, and port. If you remember, the ServerWebExchange argument was added to controller methods. This interface is used to get the HTTP request, response, and other attributes.

Now, we can use these two interfaces – HateoasSupport and ReactiveRepresentation ModelAssembler – to define the representation model assemblers.

Let's define the address's representational model assembler, as shown next:

```
@Component
public class AddressRepresentationModelAssembler
    implements ReactiveRepresentationModelAssembler
        <AddressEntity, Address>, HateoasSupport {

    private static String serverUri = null;
    private String getServerUri(
            @Nullable ServerWebExchange exch) {
      if (Strings.isBlank(serverUri)) {
        serverUri = getUriComponentBuilder
          (exch).toUriString();
      }
      return serverUri;
    }
```

https://github.com/PacktPublishing/Modern-API-Development-with-Spring-6-and-Spring-Boot-3/blob/main/Chapter05/src/main/java/com/packt/modern/api/hateoas/AddressRepresentationModelAssembler.java

Here, we have defined another private method, getServerUri(), which extracts the server URI from UriComponentBuilder, which itself is returned from the default getUriComponentBuilder() method of the HateoasSupport interface.

Now, we can override the toModel() method, as shown in the following code block:

AddressRepresentationModelAssembler.java

```
@Override
public Mono<Address> toModel(AddressEntity entity,
    ServerWebExchange exch) {
    return Mono.just(entityToModel(entity, exch));
}
public Address entityToModel(AddressEntity entity,
```

```
  ServerWebExchange exch) {
  Address resource = new Address();
  if(Objects.isNull(entity)) {
    return resource;
  }
  BeanUtils.copyProperties(entity, resource);
  resource.setId(entity.getId().toString());
  String serverUri = getServerUri(exchange);
  resource.add(Link.of(String.format(
      "%s/api/v1/addresses", serverUri))
      .withRel("addresses"));
  resource.add(Link.of(String.format(
      "%s/api/v1/addresses/%s",serverUri,
      entity.getId())).withSelfRel());
  return resource;
}
```

The `toModel()` method returns the `Mono<Address>` object with hypermedia links formed from the `AddressEntity` instance using the `entityToModel()` method.

`entityToModel()` copies the properties from the entity instance to the model instance. Most importantly, it adds hypermedia links to the model using the `resource.add()` method. The `add()` method takes the `org.springframework.hateoas.Link` instance as an argument. Then, we use the `Link` class's `of()` static factory method to form the link. You can see that a server URI is used here to add it to the link. You can form as many links as you want and add these to the resource using the `add()` method.

The `ReactiveRepresentationModelAssembler` interface provides the `toCollectionModel()` method with a default implementation that returns the `Mono<CollectionModel<D>>` collection model. However, we can also add the `toListModel()` method, as shown here, which returns the list of addresses using `Flux`:

AddressRepresentationModelAssembler.java

```
public Flux<Address> toListModel(
        Flux<AddressEntity> ent,
        ServerWebExchange exchange) {
  if (Objects.isNull(ent)) {
    return Flux.empty();
  }
  return Flux.from(ent.map(e ->
          entityToModel(e, exchange)));
}
```

This method internally uses the `entityToModel()` method. Similarly, you can create a representation model assembler for other API models. You can find all these models at `https://github.com/PacktPublishing/Modern-API-Development-with-Spring-6-and-Spring-Boot-3/tree/main/Chapter05/src/main/java/com/packt/modern/api/hateoas`.

Now that we are done with the basic code infrastructure, we can develop the API implementation based on the interfaces generated by OpenAPI Codegen. Here, we'll first develop the repositories that will be consumed by the services. At the end, we'll write the controller implementation. Let's start with the repositories.

Defining an entity

Entities are defined in more or less the same way as we defined and used them in *Chapter 4, Writing Business Logic for APIs*. However, instead of using Hibernate mappings and JPA, we'll use Spring Data annotations, as shown here:

```java
@Table("ecomm.orders")
public class OrderEntity {
  @Id
  @Column("id")
  private UUID id;

  @Column("customer_id")
  private UUID customerId;

  @Column("address_id")
  private UUID addressId;

  @Column("card_id")
  private UUID cardId;

  @Column("order_date")
  private Timestamp orderDate;

  // other fields mapped to table columns

  private UUID cartId;

  private UserEntity userEntity;

  private AddressEntity addressEntity;

  private PaymentEntity paymentEntity;
```

```
private List<ShipmentEntity> shipments = new ArrayList<>();

// other entities fields and getters/setters
```

https://github.com/PacktPublishing/Modern-API-Development-with-Spring-6-and-Spring-Boot-3/blob/main/Chapter05/src/main/java/com/packt/modern/api/entity/OrderEntity.java

Here, because we are using Spring Data in place of Hibernate, we use Spring Data annotations, namely @Table, to associate an entity class to a table name, and @Column for mapping a field to a column of the table. As is obvious, @Id is used as the identifier column. Similarly, you can define the other entities.

After defining the entities, let's add repositories in the next subsection.

Adding repositories

A repository is an interface between our application code and database. It is the same as the respository you used in Spring MVC. However, we are writing the code using the reactive paradigm. Therefore, it is necessary to have repositories that use an R2DBC-/reactive-based driver and return instances of reactive types on top of Reactive Streams. This is the reason why we can't use JDBC.

Spring Data R2DBC provides different repositories for Reactor and RxJava, for example, ReactiveCrudRepository and ReactiveSortingRepository and RxJava2CrudRepository and RxJava3CrudRepository. Also, you can write your own custom implementation.

We are going to use ReactiveCrudRepository and also write a custom implementation.

We'll write repositories for the Order entity. For other entities, you can find the repositories at https://github.com/PacktPublishing/Modern-API-Development-with-Spring-6-and-Spring-Boot-3/tree/main/Chapter05/src/main/java/com/packt/modern/api/repository.

First, let's write the CRUD repository for the Order entity, as shown next:

```
@Repository
public interface OrderRepository extends
    ReactiveCrudRepository<OrderEntity, UUID>,
        OrderRepositoryExt {

  @Query("select o.* from ecomm.orders o join
          ecomm.\"user\" u on o.customer_id =
          u.id where u.id = :cusId")
  Flux<OrderEntity> findByCustomerId(UUID cusId);
}
```

```
https://github.com/PacktPublishing/Modern-API-Development-with-
Spring-6-and-Spring-Boot-3/blob/main/Chapter05/src/main/java/com/
packt/modern/api/repository/OrderRepository.java
```

This is as simple as shown. The `OrderRepository` interface extends `ReactiveCrudRepository` and our own custom repository interface, `OrderRepositoryExt`.

We'll discuss `OrderRepositoryExt` a bit later; let's discuss `OrderRepository` first. We have added one extra method, `findByCustomerId()`, to the `OrderRepository` interface, which finds the order by the given customer ID. The `ReactiveCrudRepository` interface and the `Query()` annotation are part of the Spring Data R2DBC library. `Query()` consumes native SQL queries, unlike the repository we created in the previous chapter.

We can also write our own custom repository. Let's write a simple contract for it, as shown next:

```java
public interface OrderRepositoryExt {
  Mono<OrderEntity> insert(Mono<NewOrder> m);
  Mono<OrderEntity> updateMapping(OrderEntity e);
}
```

```
https://github.com/PacktPublishing/Modern-API-Development-with-
Spring-6-and-Spring-Boot-3/blob/main/Chapter05/src/main/java/com/
packt/modern/api/repository/OrderRepositoryExt.java
```

Here, we have written two method signatures – the first one inserts a new order record in the database and the second one updates the order item and cart item mapping. The idea is that once an order is placed, items should be removed from the cart and added to the order. If you want, you can also combine both operations.

Let's first define the `OrderRepositoryExtImpl` class, which extends the `OrderRepositoryExt` interface, as shown in the following code block:

```java
@Repository
public class OrderRepositoryExtImpl implements OrderRepositoryExt {

    private ConnectionFactory connectionFactory;
    private DatabaseClient dbClient;
    private ItemRepository itemRepo;
    private CartRepository cartRepo;
    private OrderItemRepository oiRepo;

    public OrderRepositoryExtImpl(ConnectionFactory
        connectionFactory, ItemRepository itemRepo,
        OrderItemRepository oiRepo, CartRepository
        cartRepo, DatabaseClient dbClient) {
```

```
      this.itemRepo = itemRepo;
      this.connectionFactory = connectionFactory;
      this.oiRepo = oiRepo;
      this.cartRepo = cartRepo;
      this.dbClient = dbClient;
   }
```

https://github.com/PacktPublishing/Modern-API-Development-with-Spring-6-and-Spring-Boot-3/blob/main/Chapter05/src/main/java/com/packt/modern/api/repository/OrderRepositoryExtImpl.java

We have just defined a few class properties and added these properties in the constructor as an argument for constructor-based dependency injection.

As per the contract, it receives Mono<NewOrder>. Therefore, we need to add a method that converts a model into an entity to the OrderRepositoryExtImpl class. We also need an extra argument as CartEntity contains the cart items. Here it is:

OrderRepositoryExtImpl.java

```java
private OrderEntity toEntity(NewOrder order, CartEntity c) {
   OrderEntity orderEntity = new OrderEntity();
   BeanUtils.copyProperties(order, orderEntity);
   orderEntity.setUserEntity(c.getUser());
   orderEntity.setCartId(c.getId());
   orderEntity.setItems(c.getItems())
      .setCustomerId(UUID.fromString(order.getCustomerId()))
      .setAddressId(UUID.fromString
        (order.getAddress().getId()))
      .setOrderDate(Timestamp.from(Instant.now()))
      .setTotal(c.getItems().stream()
      .collect(Collectors.toMap(k ->
        k.getProductId(), v ->
          BigDecimal.valueOf(v.getQuantity())
          .multiply(v.getPrice()))))
      .values().stream().reduce(
        BigDecimal::add).orElse(BigDecimal.ZERO));
   return orderEntity;
}
```

This method is straightforward except for the code where the total is set. The total is calculated using the stream. Let's break it down to understand it:

1.   First, it takes the items from CartEntity.

2.   Then, it creates streams from items.

3. It creates a map with the key as the product ID and the value as the product of the quantity and price.

4. It takes the value from the map and converts it into a stream.

5. It performs the reduce operation by adding a method to `BigDecimal`. It then gives the total amount.

6. If values are not present, then it simply returns 0.

After the `toEntity()` method, we also need another mapper that reads rows from the database and converts them into `OrderEntity`. For this purpose, we'll write `BiFunction`, which is a part of the `java.util.function` package:

OrderRepositoryExtImpl.java

```java
class OrderMapper implements BiFunction<Row,Object,
    OrderEntity> {
  @Override
  public OrderEntity apply(Row row, Object o) {
    OrderEntity oe = new OrderEntity();
    return oe.setId(row.get("id", UUID.class))
        .setCustomerId(
            row.get("customer_id", UUID.class))
        .setAddressId(
            row.get("address_id", UUID.class))
        .setCardId(
            row.get("card_id", UUID.class))
        .setOrderDate(Timestamp.from(
            ZonedDateTime.of(
            (LocalDateTime) row.get("order_date"),
            ZoneId.of("Z")).toInstant()))
        .setTotal(
            row.get("total", BigDecimal.class))
        .setPaymentId(
            row.get("payment_id", UUID.class))
        .setShipmentId(
            row.get("shipment_id", UUID.class))
        .setStatus(StatusEnum.fromValue(
            row.get("status", String.class)));
  }
}
```

Here, we have overridden the `apply()` method, which returns `OrderEntity`, by mapping properties from the row to `OrderEntity`. The second parameter of the `apply()` method is not used because it contains metadata that we don't need.

Let's first implement the `updateMapping()` method from the `OrderRepositoryExt` interface:

OrderRepositoryExtImpl.java

```java
public Mono<OrderEntity> updateMapping(OrderEntity
  orderEntity) {
  return oiRepo.saveAll(orderEntity.getItems()
    .stream().map(i -> new OrderItemEntity()
      .setOrderId(orderEntity.getId())
      .setItemId(i.getId())).collect(toList()))
      .then(
        itemRepo.deleteCartItemJoinById(
          orderEntity.getItems().stream()
            .map(i -> i.getId())
            .collect(toList()),
          orderEntity.getCartId())
            .then(Mono.just(orderEntity))
      );
}
```

Here, we have created a pipeline of Reactive Streams and performed two back-to-back database operations. First, it creates the order item mapping using `OrderItemRepository`, and then it removes the cart item mapping using `ItemRepository`.

Java streams are used to create an input list of `OrderItemEntity` instances in the first operation, and a list of item IDs in the second operation.

So far, we have made use of `ReactiveCrudRepository` methods. Let's implement a custom method using an entity template, as shown next:

OrderRepositoryExtImpl.java

```java
@Override
public Mono<OrderEntity> insert(Mono<NewOrder> mdl) {
  AtomicReference<UUID> orderId =new AtomicReference<>();
  Mono<List<ItemEntity>> itemEntities =
      mdl.flatMap(m ->
        itemRepo.findByCustomerId(
          UUID.fromString(m.getCustomerId()))
        .collectList().cache());
```

```
Mono<CartEntity> cartEntity =
    mdl.flatMap(m ->
        cartRepo.findByCustomerId(
            UUID.fromString(m.getCustomerId())))
        .cache();
cartEntity = Mono.zip(cartEntity, itemEntities,
        (c, i) -> {
            if (i.size() < 1) {
                throw new ResourceNotFoundException(
                String.format("There is no item found
                in customer's (ID:%s) cart.",
                    c.getUser().getId())));
        }
    return c.setItems(i);
}).cache();
```

Here, we override the `insert()` method from the `OrderRepositoryExt` interface. The `insert()` method is filled with fluent, functional, and reactive APIs. The `insert()` method receives a `NewOrder` model `Mono` instance as an argument that contains the payload for creating a new order. Spring Data R2DBC does not allow fetching nested entities. However, you can write a custom repository for `Cart` in the same way you have written for `Order` that can fetch `Cart` and its items together.

We are using `ReactiveCrudRepository` for `Cart` and `Item` entities. Therefore, we are fetching them one by one. First, we use the item repository to fetch the cart items based on the given customer ID. `Customer` has a one-to-one mapping with `Cart`. Then, we fetch the `Cart` entity using `CartRepository` by using the customer ID.

We get the two separate Mono objects – Mono<List<ItemEntity>> and Mono<CartEntity>. Now, we need to combine them. Mono has a `zip()` operator, which allows you to take two Mono objects and then use the Java `BiFunction` to merge them. `zip()` returns a new Mono object only when both the given Mono objects produce the item. `zip()` is polymorphic and therefore other forms are also available.

We have the cart and its items, plus the `NewOrder` payload. Let's insert these items into a database, as shown in the next code block:

OrderRepositoryExtImpl.java

```
R2dbcEntityTemplate template = new R2dbcEntityTemplate
  (connectionFactory);
Mono<OrderEntity> orderEntity = Mono.zip(mdl,
    cartEntity, (m, c) -> toEntity(m, c)).cache();
return orderEntity.flatMap(oe -> dbClient.sql("""
```

```
INSERT INTO ecomm.orders (address_id,
card_id, customer_id, order_date, total,
status) VALUES($1, $2, $3, $4, $5, $6)
""")
.bind("$1", Parameter.fromOrEmpty(
    oe.getAddressId(), UUID.class))
.bind("$2", Parameter.fromOrEmpty(
    oe.getCardId(), UUID.class))
.bind("$3", Parameter.fromOrEmpty(
    oe.getCustomerId(), UUID.class))
.bind("$4",OffsetDateTime.ofInstant(
    oe.getOrderDate().toInstant(), ZoneId.of(
      "Z")).truncatedTo(ChronoUnit.MICROS))
.bind("$5", oe.getTotal())
.bind("$6", StatusEnum.CREATED.getValue())
  .map(new OrderMapper()::apply).one())
.then(orderEntity.flatMap(x ->
    template.selectOne(
      query(where("customer_id").is(
          x.getCustomerId()).and("order_date")
          .greaterThanOrEquals(OffsetDateTime.
          ofInstant(x.getOrderDate().
            toInstant(),ZoneId.of("Z"))
            .truncatedTo(ChronoUnit.MICROS))),
      OrderEntity.class).map(t -> x.setId(
          t.getId()).setStatus(t.getStatus())))
));
```

Here, we again use `Mono.zip()` to create an `OrderEntity` instance. Now, we can use values from this instance to insert into the `orders` table.

There are two ways to interact with the database to run SQL queries – by using either `DatabaseClient` or `R2dbcEntityTemplate`. Now, `DatabaseClient` is a lightweight implementation that uses the `sql()` method to deal with SQL directly, whereas `R2dbcEntityTemplate` provides a fluent API for CRUD operations. We have used both classes to demonstrate their usage.

First, we use `DatabaseClient.sql()` to insert the new order into the `orders` table. We use `OrderMapper` to map the row returned from the database to the entity. Then, we use the `then()` reactive operator to select the newly inserted record and then map it back to `orderEntity` using the `R2dbcEntityTemplate.selectOne()` method.

Similarly, you can create repositories for other entities. Now, we can use these repositories in services. Let's define services in the next subsection.

Adding services

Let's add a service for `Order`. There is no change in the `OrderService` interface, as shown next. You just need to make sure that interface method signatures have reactive types as returned types to keep the non-blocking flow in place:

```
public interface OrderService {
  Mono<OrderEntity> addOrder(@Valid Mono<NewOrder>
    newOrder);
  Mono<OrderEntity> updateMapping(@Valid OrderEntity
    orderEntity);
  Flux<OrderEntity> getOrdersByCustomerId(@NotNull @Valid
    String customerId);
  Mono<OrderEntity> getByOrderId(String id);
}
```

https://github.com/PacktPublishing/Modern-API-Development-with-Spring-6-and-Spring-Boot-3/blob/main/Chapter05/src/main/java/com/packt/modern/api/service/OrderService.java

Next, you are going to implement each of these four methods described in `OrderService`. Let's first implement the first two methods of `OrderService` in the following way:

```
@Override
public Mono<OrderEntity> addOrder(@Valid Mono<NewOrder>
  newOrder) {
  return repository.insert(newOrder);
}

@Override
public Mono<OrderEntity> updateMapping(
  @Valid OrderEntity orderEntity) {
  return repository.updateMapping(orderEntity);
}
```

https://github.com/PacktPublishing/Modern-API-Development-with-Spring-6-and-Spring-Boot-3/blob/main/Chapter05/src/main/java/com/packt/modern/api/service/OrderServiceImpl.java

The first two are straightforward; we just use the `OrderRepository` instance to call the respective methods. In an idle scenario, the overridden `updateMapping` method will trigger the rest of the process after updating the mappings:

1. Initiate the payment.

2. Once the payment is authorized, change the status to `paid`.

3. Initiate the shipment and changed the status to `Shipment Initiated` or `Shipped`.

As our application is not a real-world app and we are writing for learning purposes, we are not writing the code for executing all three steps. For simplicity, we are just updating the mapping.

Let's implement the third one (`getOrdersByCustomerId`). This is a bit tricky, as shown next:

OrderServiceImpl.java

```java
private BiFunction<OrderEntity, List<ItemEntity>, OrderEntity>
biOrderItems = (o, fi) -> o
    .setItems(fi);
@Override
public Flux<OrderEntity> getOrdersByCustomerId(
    String customerId) {
 return repository.findByCustomerId(UUID
  .fromString(customerId)).flatMap(order ->
   Mono.just(order)
    .zipWith(userRepo.findById(order.getCustomerId()))
    .map(t -> t.getT1().setUserEntity(t.getT2()))
    .zipWith(addRepo.findById(order.getAddressId()))
    .map(t ->
        t.getT1().setAddressEntity(t.getT2()))
    .zipWith(cardRepo.findById(
       order.getCardId() != null
       ? order.getCardId() : UUID.fromString(
        "0a59ba9f-629e-4445-8129-b9bce1985d6a"))
           .defaultIfEmpty(new CardEntity()))
    .map(t -> t.getT1().setCardEntity(t.getT2()))
    .zipWith(itemRepo.findByCustomerId(
       order.getCustomerId()).collectList(),biOrderItems)
  );
}
```

The previous method looks complicated, but it's not. What you are doing here is basically fetching data from multiple repositories and then populating the nested entities inside `OrderEntity`. This is done with the `zipWith()` operator by using either the `map()` operator alongside it or `BiFunction` as a separate argument.

The preceding method first fetches the orders by using the customer ID, then flat maps the orders to populate its nested entities such as `Customer`, `Order`, and `Items`. Therefore, we are using `zipWith()` inside the `flatMap()` operator. If you observe the first `zipWith()`, it fetches the user entity and then sets the nested user entity's property using the `map()` operator. Similarly, other nested entities are populated.

In the last `zipWith()` operator, we are using `BiFunction biOrderItems` to set the `item` entities in the `OrderEntity` instance.

The same algorithm is used to implement the last method (`getOrderById`) of the `OrderService` interface, as shown in the following code:

OrderServiceImpl.java

```java
@Override
public Mono<OrderEntity> getByOrderId(String id) {
  return repository.findById(UUID.fromString(id))
    .flatMap(order ->
      Mono.just(order)
        .zipWith(userRepo.findById(order.getCustomerId()))
        .map(t -> t.getT1().setUserEntity(t.getT2()))
        .zipWith(addRepo.findById(order.getAddressId()))
        .map(t -> t.getT1().setAddressEntity(t.getT2()))
        .zipWith(cardRepo.findById(order.getCardId()))
        .map(t -> t.getT1().setCardEntity(t.getT2()))
        .zipWith(itemRepo.findByCustomerId
            (order.getCustomerId()).collectList()
               ,biOrderItems)
    );
}
```

So far, you have used the `zipWith()` operator to merge different objects. You may find another way to merge two Mono instances using the `Mono.zip()` operator, as shown next:

```java
private BiFunction<CartEntity, List<ItemEntity>, CartEntity>
cartItemBiFun = (c, i) -> c
     .setItems(i);
@Override
public Mono<CartEntity> getCartByCustomerId(String
  customerId) {
  Mono<CartEntity> cart = repository.findByCustomerId(
    UUID.fromString(customerId))
      .subscribeOn(Schedulers.boundedElastic());
  Mono<UserEntity> user = userRepo.findById(
    UUID.fromString(customerId))
      .subscribeOn(Schedulers.boundedElastic());
  cart = Mono.zip(cart, user, cartUserBiFun);
  Flux<ItemEntity> items =
      itemRepo.findByCustomerId(
        UUID.fromString(customerId))
```

```
        .subscribeOn(Schedulers.boundedElastic());
    return Mono.zip(cart, items.collectList(),
        cartItemBiFun);
}
```

https://github.com/PacktPublishing/Modern-API-Development-with-Spring-6-and-Spring-Boot-3/blob/main/Chapter05/src/main/java/com/packt/modern/api/service/CardServiceImpl.java

This example is taken from the CartServiceImpl class. Here, we make two separate calls – one using the cart repository and another one from the item repository. As a result, these two calls produce two Mono instances and merge them using the Mono.zip() operator. This we call directly using Mono; the previous example was used on Mono/Flux instances with the zipWith() operator.

Using similar techniques, the remaining services have been created. You can find them at https://github.com/PacktPublishing/Modern-API-Development-with-Spring-6-and-Spring-Boot-3/tree/main/Chapter05/src/main/java/com/packt/modern/api/service.

You have implemented async services that allow you to perform async operations, including database calls. Now, you can consume these service classes in controllers. Let's move our focus on to the last development subsection (controllers) of our reactive API implementation.

Adding controller implementations

REST controller interfaces are already generated by the OpenAPI Codegen tool. We can now create an implementation of those interfaces. The only different thing while implementing the reactive controller is having the reactive pipelines to call the services and assemblers. You should also only return ResponseEntity objects wrapped in either Mono or Flux based on the generated contract.

Let's implement OrderApi, which is the controller interface for the Orders REST API:

```
@RestController
public class OrderController implements OrderApi {
  private final OrderRepresentationModelAssembler
    assembler;
  private OrderService service;
  public OrderController(OrderService service,
   OrderRepresentationModelAssembler assembler) {
    this.service = service;
    this.assembler = assembler;
  }
```

https://github.com/PacktPublishing/Modern-API-Development-with-Spring-6-and-Spring-Boot-3/blob/main/Chapter05/src/main/java/com/packt/modern/api/controller/OrderController.java

Here, `@RestController` is a trick that combines `@Controller` and `@ResponseBody`. These are the same annotations we used in *Chapter 4, Writing Business Logic for APIs*, to create the REST controller. However, the methods have different signatures now to apply the reactive pipelines. Make sure you don't break the reactive chain of calls or add any blocking calls. If you do, either the REST call will not be fully non-blocking or you may see undesired results.

We use constructor-based dependency injection to inject the order service and assembler. Let's add the method implementations:

OrderController.java

```java
@Override
public Mono<ResponseEntity<Order>>
    addOrder(@Valid Mono<NewOrder> newOrder,
        ServerWebExchange exchange) {
  return service.addOrder(newOrder.cache())
    .zipWhen(x -> service.updateMapping(x))
    .map(t -> status(HttpStatus.CREATED)
      .body(assembler.entityToModel(
          t.getT2(), exchange)))
    .defaultIfEmpty(notFound().build());
}
```

Both the method argument and return type are reactive types (`Mono`), used as a wrapper. Reactive controllers also have an extra parameter, `ServerWebExchange`, which we discussed earlier.

In this method, we simply pass the `newOrder` instance to the service. We have used `cache()` because we need to subscribe to it more than once. We get the newly created `EntityOrder` through the `addOrder()` call. Then, we use the `zipWhen()` operator, which performs the `updateMapping` operation using the newly created order entity. At the end, we send the `Order` object by wrapping it inside `ResponseEntity`. Also, it returns NOT FOUND 404 when an empty instance is returned.

Let's have a look at other method implementations of the `order` API interface:

OrderController.java

```java
@Override
public Mono<ResponseEntity<Flux<Order>>>
    getOrdersByCustomerId(@NotNull
```

```
        @Valid String customerId, ServerWebExchange
          exchange) {
    return Mono
        .just(ok(assembler.toListModel(service
          .getOrdersByCustomerId(customerId),
            exchange)));
  }


  @Override
  public Mono<ResponseEntity<Order>>
    getByOrderId(String id, ServerWebExchange
      exchange) {
    return service.getByOrderId(id).map(o ->
      assembler.entityToModel(o, exchange))
        .map(ResponseEntity::ok)
        .defaultIfEmpty(notFound().build());
  }
```

In the previous code, both methods are kind of similar in nature; the service returns OrderEntity based on the given customer ID and order ID, respectively. It then gets converted into a model and is wrapped inside ResponseEntity and Mono.

Similarly, other REST controllers are implemented using the same approach. You can find the rest of them at https://github.com/PacktPublishing/Modern-API-Development-with-Spring-6-and-Spring-Boot-3/tree/main/Chapter05/src/main/java/com/packt/modern/api/controller.

We are almost done with the reactive API implementation. Let's look into some of the other minor changes.

Adding H2 Console to an application

The H2 Console app is not available by default in Spring WebFlux the way it is available in Spring MVC. However, you can add it by defining the bean on your own, as follows:

```
@Component
public class H2ConsoleComponent {
    private Server webServer;

    @Value("${modern.api.h2.console.port:8081}")
    Integer h2ConsolePort;

    @EventListener(ContextRefreshedEvent.class)
    public void start()
        throws java.sql.SQLException {
```

```
        this.webServer = org.h2.tools.Server
          .createWebServer("-webPort",
            h2ConsolePort.toString(), "-
              tcpAllowOthers").start();
    }

    @EventListener(ContextClosedEvent.class)
    public void stop() {
      this.webServer.stop();
    }
}
```

https://github.com/PacktPublishing/Modern-API-Development-with-
Spring-6-and-Spring-Boot-3/blob/main/Chapter05/src/main/java/com/
packt/modern/api/H2ConsoleComponent.java

The previous code (H2ConsoleComponent) is straightforward; we have added the
start() and stop() methods, which are executed on ContextRefreshEvent and
ContextStopEvent, respectively. ContextRefreshEvent is an application event that gets
fired when ApplicationContext is refreshed or initialized. ContextStopEvent is also an
application event that gets fired when ApplicationContext is closed.

The start() method creates the web server using the H2 library and starts it on a given port. The
stop() method stops the H2 web server, that is, the H2 Console app.

You need a different port to execute H2 Console, which can be configured by adding the modern.
api.h2.console.port=8081 property to the application.properties file. The
h2ConsolePort property is annotated with @Value("${modern.api.h2.console.
port:8081}"); therefore, the value configured in application.properties will be picked and
assigned to h2ConsolePort when the H2ConsoleComponent bean is initialized by the Spring
Framework. The value 8081 will be assigned if the property is not defined in the application.
properties file.

Since we are discussing application.properties, let's have a look at some of the other changes.

Adding application configuration

We are going to use Flyway for database migration. Let's add the configuration required for it:

```
spring.flyway.url=jdbc:h2:file:./data/ecomm;AUTO_SERVER=TRUE;DB_CLOSE_
DELAY=-1;IGNORECASE=TRUE;DATABASE_TO_UPPER=FALSE
spring.flyway.schemas=ecomm
spring.flyway.user=
spring.flyway.password=
```

```
https://github.com/PacktPublishing/Modern-API-Development-with-
Spring-6-and-Spring-Boot-3/blob/main/Chapter05/src/main/resources/
application.properties
```

You must be wondering why we are using JDBC here, instead of R2DBC. This is because Flyway hasn't yet started supporting R2DBC (at the time of writing). You can change it to R2DBC once support is added.

We have specified the ecomm schema and set a blank username and password.

Similarly, you can add the Spring Data configuration in the application.properties file:

```
spring.r2dbc.url=r2dbc:h2:file://././data/ecomm?options=AUTO_
SERVER=TRUE;DB_CLOSE_DELAY=-1;IGNORECASE=TRUE;DATABASE_TO_
UPPER=FALSE;;TRUNCATE_LARGE_LENGTH=TRUE;DB_CLOSE_ON_EXIT=FALSE
spring.r2dbc.driver=io.r2dbc:r2dbc-h2
spring.r2dbc.name=
spring.r2dbc.password=
```

Spring Data supports R2DBC; therefore, we are using an R2DBC-based URL. We have set io.r2dbc:r2dbc-h2 for the driver to H2 and set a blank username and password.

Similarly, we have added the following logging properties to logback-spring.xml to add debug statements to the console for Spring R2DBC and H2:

```
<logger name="org.springframework.r2dbc"
      level="debug" additivity="false">
   <appender-ref ref="STDOUT"/>
</logger>
<logger name="reactor.core" level="debug"
   additivity="false">
   <appender-ref ref="STDOUT"/>
</logger>
<logger name="io.r2dbc.h2" level="debug"
    additivity="false">
   <appender-ref ref="STDOUT"/>
</logger>
```

This concludes our implementation of reactive RESTful APIs. Now, you can test them.

Testing reactive APIs

Now, you must be looking forward to testing. You can find the API client collection at the following location. You can import it and then test the APIs using any API client that supports the HAR-type file import:

```
https://github.com/PacktPublishing/Modern-API-Development-with-Spring-
6-and-Spring-Boot-3/blob/main/Chapter05/Chapter05-API-Collection.har
```

> **Building and running the Chapter 05 code**
>
> You can build the code by running `gradlew clean build` from the root of the project and run the service using `java -jar build/libs/Chapter05-0.0.1-SNAPSHOT.jar`. Make sure to use Java 17 in the path.

Summary

I hope you enjoyed learning about reactive API development with an asynchronous, non-blocking, and functional paradigm. At first glance, you may find it complicated if you are not very familiar with the fluent and functional paradigm, but with practice, you'll start writing only functional-style code. Definitely, familiarity with Java streams and functions will help you to grasp the concepts easily.

Now that you have reached the end of this chapter, you have the skills to write functional and reactive code. You can write reactive, asynchronous, and non-blocking code and REST APIs. You also learned about R2DBC, which will become more solid and enhanced in the future as long as reactive programming continues to be used.

In the next chapter, we'll explore the security aspect of RESTful service development.

Questions

1. Do you really need the reactive paradigm for application development?
2. Are there any disadvantages to using the reactive paradigm?
3. Who plays the role of the subscriber in the case of an HTTP request in Spring WebFlux?

Answers

1. Yes, it is required only if you need vertical scaling. In the cloud, you pay to use the resources, and reactive applications help you to use resources optimally. This is a new way of achieving scale. You need a small number of threads compared to non-reactive applications. The cost of connection to a database, I/O, or any external source is a callback; therefore, reactive-based applications do not require much memory. However, while reactive programming is superior in terms of vertical scaling, you should continue using your existing or non-reactive applications. Even Spring recommends that. There is no new or old style; both can co-exist. However, when you need scaling for any special component or application, you can go the reactive way. A few years back, Netflix replaced the Zuul API gateway with the reactive Zuul2 API gateway. This helped them to achieve scale. However, they still have/use non-reactive applications.

2. There are pros and cons to everything. Reactive is no exception. Reactive code is not easy to write compared to the imperative style. It is very difficult to debug because it does not use a single thread. However, if you have developers who are proficient in the reactive paradigm, this isn't an issue.

3. WebFlux internal classes subscribe to Mono/Flux streams sent by the controller and convert them into HTTP packets. The HTTP protocol does support event streams. However, for other media types, such as JSON, Spring WebFlux subscribes Mono/Flux streams and waits till onComplete() or onError() is triggered. Then, it serializes the whole list of elements, or a single element in the case of Mono, in one HTTP response. You can learn more about it in the *Reactive Core* section.

Further reading

- Project Reactor: `https://projectreactor.io`

- Spring Reactive documentation: `https://docs.spring.io/spring-framework/docs/current/reference/html/web-reactive.html`

- Spring Data R2DBC – reference documentation: `https://docs.spring.io/spring-data/r2dbc/docs/current/reference/html/`

- *Hands-On Reactive Programming in Spring 5*: `https://www.packtpub.com/product/hands-on-reactive-programming-in-spring-5/9781787284951`

- *Learn Java 17 Programming – Second Edition*: `https://www.packtpub.com/product/learn-java-17-programming-second-edition/9781803241432`

Part 2 –
Security, UI, Testing,
and Deployment

In this part, you will learn how to secure REST APIs with JWTs and Spring Security. After completing this part, you will also be able to authorize REST endpoints based on user roles. You will learn how APIs are consumed by the UI app, and you will learn how to automate the unit testing and integration testing of APIs. By the end of this part, you will be able to containerize the built app and then deploy it in a Kubernetes cluster.

This part contains the following chapters:

- *Chapter 6, Securing REST Endpoints Using Authorization and Authentication*
- *Chapter 7, Designing a User Interface*
- *Chapter 8, Testing APIs*
- *Chapter 9, Deployment of Web Services*

6

Securing REST Endpoints Using Authorization and Authentication

In previous chapters, we developed a RESTful web service using imperative and reactive coding styles. Now, you'll learn how you can secure these REST endpoints using Spring Security. You'll implement token-based authentication and authorization for REST endpoints. A successful authentication provides two types of tokens – a **JavaScript Object Notation (JSON) Web Token (JWT)** as an access token, and a refresh token in response. This JWT-based access token is then used to access the secured **Uniform Resource Locators (URLs)**. A refresh token is used to request a new JWT if the existing JWT has expired, and a valid request token provides a new JWT to use.

You'll associate users with roles such as *admin* and *user*. These roles will be used as authorization to make sure that REST endpoints can only be accessed if a user holds certain roles. We'll also briefly discuss **cross-site request forgery (CSRF)** and **cross-origin resource sharing (CORS)**.

The topics of this chapter are divided into the following sections:

- Implementing authentication using Spring Security and JWTs
- Securing REST APIs with JWTs
- Configuring CORS and CSRF
- Understanding authorization
- Testing security

By the end of the chapter, you will know how to implement authentication and authorization using Spring Security and protect your web service from CORS and CSRF attacks.

Technical requirements

The code for this chapter is available at `https://github.com/PacktPublishing/Modern-API-Development-with-Spring-6-and-Spring-Boot-3/tree/main/Chapter06`.

Implementing authentication using Spring Security and JWT

Spring Security is a framework consisting of a collection of libraries that allow you to implement enterprise application security without worrying about writing boilerplate code. In this chapter, we will use the Spring Security framework to implement token-based (JWT) authentication and authorization. Throughout the course of this chapter, you will also learn about CORS and CSRF configuration.

It's useful to know that Spring Security also provides support for opaque tokens, just like it does for JWTs. The main difference between them is how information is read from the token. You can't read the information from an opaque token the way you can with a JWT – only the issuer is aware of how to do this.

> **Note**
>
> A token is a string of characters such as
>
> `5rm1tc1obfshrm2354lu9dlt5reqm1ddjchqh81 7rbk37q95b768bib0j`
> `f44df6suk1638sf78cef7 hfo1g4ap3bkighbnk7inr68ke780744fpej0gtd`
> `9qflm999o8q`.
>
> It allows you to call secured HTTP endpoints or resources that are stateless by using various authorization flows.

You learned about `DispatcherServlet` in *Chapter 2, Spring Concepts and REST APIs*. This is an interface between a client request and the REST controller. Therefore, if you want to place logic for token-based authentication and authorization, you will have to do this before a request reaches `DispatcherServlet`. Spring Security libraries provide the servlet with pre-filters (as a part of the filter chain), which are processed before the request reaches `DispatcherServlet`. A **pre-filter** is a servlet filter that is processed before it reaches the actual servlet, which in Spring Security's case is `DispatcherServlet`. Similarly, **post-filters** get processed after a request has been processed by the servlet/controller.

There are two ways you can implement token-based (JWT) authentication – by using either `spring-boot-starter-security` or `spring-boot-starter-oauth2-resource-server`. We will use the latter because it does the boilerplate configuration for us.

The former contains the following libraries:

- `spring-security-core`
- `spring-security-config`
- `spring-security-web`

`spring-boot-starter-oauth2-resource-server` provides the following, along with all three preceding **Java ARchive (JAR)** files:

- `spring-security-oauth2-core`
- `spring-security-oauth2-jose`
- `spring-security-oauth2-resource-server`

When you start this chapter's code, you will find the following log. You can see that, by default, `DefaultSecurityFilterChain` is `auto-configured`. The `log` statement lists the configured filters in `DefaultSecurityFilterChain`, as shown in the following log block:

```
INFO [Chapter06,,,] [null] [null] [null] [null] 31975 ---
[            main] o.s.s.web.DefaultSecurityFilterChain     :
Will secure any request with [org.springframework.
security.web.session.DisableEncodeUrlFilter@781dac73, org.
springframework.security.web.context.request.async.
WebAsyncManagerIntegrationFilter@3a4e524, org.springframework.
security.web.context.SecurityContextHolderFilter@22048bd6, org.
springframework.security.web.header.HeaderWriterFilter@5844a2d1, org.
springframework.web.filter.CorsFilter@5e67a490, org.springframework.
security.web.csrf.CsrfFilter@7b95bdb0, org.springframework.
security.web.authentication.logout.LogoutFilter@6bcdd6e4, org.
springframework.security.oauth2.server.resource.web.authentication.
BearerTokenAuthenticationFilter@6826b70f, org.springframework.
security.web.savedrequest.RequestCacheAwareFilter@2e2f20b8,
org.springframework.security.web.servletapi.
SecurityContextHolderAwareRequestFilter@504497fa, org.springframework.
security.web.authentication.AnonymousAuthenticationFilter@2dac2e1b,
org.springframework.security.web.session.
SessionManagementFilter@4af7dd6a, org.springframework.security.web.
access.ExceptionTranslationFilter@401317a0, org.springframework.
security.web.access.intercept.FilterSecurityInterceptor@1ad1c363
```

Therefore, when a client fires an HTTP request, it will go through all the following security filters one after the other before reaching the REST controller (although the order may vary based on the authentication outcome) for `DefaultSecurityFilterChain`:

1. `WebAsyncManagerIntegrationFilter`
2. `SecurityContextPersistenceFilter`
3. `HeaderWriterFilter`

4. `CorsFilter`

5. `CsrfFilter`

6. `LogoutFilter`

7. `BearerTokenAuthenticationFilter`

8. `RequestCacheAwareFilter`

9. `SecurityContextHolderAwareRequestFilter`

10. `AnonymousAuthenticationFilter`

11. `SessionManagementFilter`

12. `ExceptionTranslationFilter`

13. `FilterSecurityInterceptor`

This filter chain may change in future releases. Also, the security filter chain will be different if you just used `spring-boot-starter-security` or changed the configuration. You can find all the filters available in `springSecurityFilterChain` at `https://docs.spring.io/spring-security/reference/servlet/architecture.html#servlet-security-filters`.

Now, you know about the different filters and their order in the default security chain. Next, let's add the required dependencies, making use of the Spring OAuth 2.0 resource server for authentication in the following subsections.

Adding the required Gradle dependencies

Let's add the following dependencies to the `build.gradle` file, as shown next:

```
implementation 'org.springframework.boot:spring-boot-starter-oauth2-resource-server'
implementation 'com.auth0:java-jwt:4.3.0'
```

`https://github.com/PacktPublishing/Modern-API-Development-with-Spring-6-and-Spring-Boot-3/blob/main/Chapter06/build.gradle`

The Spring Boot Starter OAuth 2 resource server dependency will add the following JARs:

- `spring-security-core`
- `spring-security-config`
- `spring-security-web`
- `spring-security-cropto`
- `spring-security-oauth2-core`
- `spring-security-oauth2-jose`
- `spring-security-oauth2-resource-server`

For JWT implementation, we will use the `java-jwt` library from `auth0.com`.

We will now explore how to code these two filters – through login and token-based authentication.

Authentication using the OAuth 2.0 resource server

The Spring Security OAuth 2.0 resource server allows you to implement authentication and authorization using `BearerTokenAuthenticationFilter`. This contains the bearer token authentication logic. However, you still need to write the REST endpoint to generate the token. Let's explore how the authentication flow works in the OAuth2.0 resource server. Take a look at the following diagram:

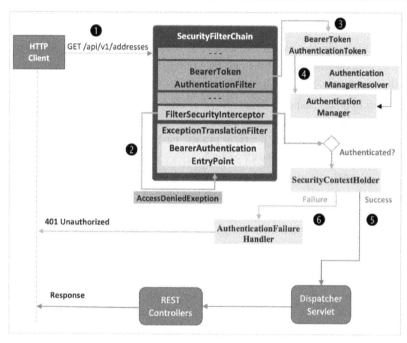

Figure 6.1 – A token authentication flow using the OAuth 2.0 resource server

Let's understand the flow depicted in *Figure 6.1*:

1. The client sends a GET HTTP request to `/api/v1/addresses`.

2. `BearerTokenAuthenticationFilter` comes into action. If the request doesn't contain the `Authorization` header, then `BearerTokenAuthenticationFilter` does not authenticate the request since it did not find the bearer token. It passes the call to `FilterSecurityInterceptor`, which does the authorization. It throws an `AccessDeniedException` exception (marked as **2** in *Figure 6.1*). `ExceptionTranslationFilter` springs into action. Control is moved to `BearerTokenAuthenticationEntryPoint`, which responds with a `401`

Unauthorized status and a WWW-Authenticate header, with a Bearer value. If the client receives a WWW-Authenticate header with a Bearer value in response, it means it must retry with the Authorization header that holds the valid bearer token. At this stage, the request cache is NullRequestCache (that is, empty) due to security reasons – the client can replay the request.

3. Let's assume the HTTP request contains an Authorization header. It extracts the Authorization header from the HTTP request and, apparently, the token from the Authorization header. It creates an instance of BearerTokenAuthentication Token using the token value. BearerTokenAuthenticationToken is a type of AbstractAuthenticationToken class that implements an Authentication interface, representing the token/principal for the authenticated request.

4. The HTTP request is passed to AuthenticationManagerResolver, which provides the AuthenticationManager based on the configuration. AuthenticationManager verifies the BearerTokenAuthenticationToken token.

5. If authentication is successful, then Authentication is set on the SecurityContext instance. This instance is then passed to SecurityContextHolder.setContext(). The request is passed to the remaining filters for processing, then routes to DispatcherServlet, and then, finally, to AddressController.

6. If authentication fails, then SecurityContextHolder.clearContext() is called to clear the context value. ExceptionTranslationFilter springs into action. Control is moved to BearerTokenAuthenticationEntryPoint, which responds with a 401 Unauthorized status and a WWW-Authenticate header, with a value that contains the appropriate error message, such as the following:

```
Bearer error="invalid_token", error_description="An error
occurred while attempting to decode the Jwt: Jwt expired at
2022-12-14T17:23:30Z", error_uri="https://tools.ietf. org/html/
rfc6750#section-3.1".
```

Now that you have learned about the complete authentication flow using the OAuth 2.0 resource server, let's learn the fundamentals of JWT.

Exploring the structure of JWT

You need authority in the form of permissions or rights to carry out any activity or access any information in general. This authority is known as a claim with respect to JWT. A claim is represented as a key-value pair. The key contains the claim name and the value contains the claim, which can be a valid JSON value. A claim can also be metadata about the JWT.

> **How to pronounce JWT**
>
> As per `https://tools.ietf.org/html/rfc7519`, the suggested pronunciation of *JWT* is the same as the English word *jot*.

A JWT is an encoded string that contains a set of claims. These claims are either digitally signed by a **JSON Web Signature** (**JWS**) or encrypted by **JSON Web Encryption** (**JWE**). *JWT is a self-contained way to transmit claims securely between parties.* The links to these **Request for Comments** (**RFC**)-proposed standards are provided in the *Further reading* section of this chapter.

A JWT is an encoded string such as `aaa.bbb.ccc`, consisting of the following three parts, separated by dots (`.`):

- Header
- Payload
- Signature

A few websites, such as `https://jwt.io`, allow you to view the content of a JWT and generate one.

Let's have a look at the following sample JWT string. You can paste it into one of the aforementioned websites to decode the content:

```
eyJhbGciOiJIUzI1NiIsInR5cCI6IkpXVCJ9.
eyJzdWIiOiIxMjM0NTY3ODkwIiwibmFtZSI6IkpvaG4gRG9lIiwiaWF0IjoxNTE2MjM5
MDIyfQ.SflKxwRJSMeKKF2QT4fwpMeJf36POk6yJV_adQssw5c
```

This sample token demonstrates how a JWT is formed and divided into three parts using dots.

Header

A **header** consists of a Base64URL-encoded JSON string, normally containing two key-value pairs – a type of token (with a `typ` key) and a signing algorithm (with an `alg` key).

This sample JWT string contains the following header:

```
{
  "alg": "HS256",
  "typ": "JWT"
}
```

The preceding header contains the `typ` and `alg` fields, representing the type and algorithm, respectively.

Payload

A **payload** is the second part of the JWT, which contains the claims and also comprises a Base64URL-encoded JSON string. There are three types of claims – registered, public, and private. These are outlined as follows:

- **Registered claims**: A few claims are registered on the **Internet Assigned Numbers Authority (IANA) JWT Claims** registry; therefore, these claims are known as registered claims. These are not mandatory but are recommended. Some registered claims are listed here:

  - **Issuer claim** (`iss` key): This claim identifies the principal who issued a token

  - **Subject claim** (`sub` key): This should be a unique value that represents the subject of the JWT

  - **Expiration time claim** (`exp` key): This is a numeric value representing the expiration time on or after which a JWT should be rejected

  - **Issued at claim** (`iat` key): This claim identifies the time at which a JWT is issued

  - **JWT ID claim** (`jti` key): This claim represents the unique identifier for a JWT

  - **Audience claim** (`aud` key): This claim identifies the recipients, which JWT is intended for

  - **Not before claim** (`nbf` key): This represents the time before which a JWT must be rejected

- **Public claims**: These are defined by JWT issuers and must not collide with registered claims. Therefore, these should either be registered with the IANA JWT Claims registry or defined as a URI with a collision-resistant namespace.

- **Private claims**: These are custom claims defined and used by the issuer and audience. They are neither registered nor public.

Here is a sample JWT string containing a payload:

```
{
  "sub": "scott2",
  "roles": [
    "ADMIN"
  ],
  "iss": "Modern API Development with Spring and
    Spring Boot",
  "exp": 1676526792,
  "iat": 1676525892
}
```

The preceding payload contains `sub` (subject), `iss` (issuer), `roles` (custom claim roles), `exp` (expires), `iat` (issued at), and `jti` (JWT ID) fields.

Signature

A **signature** is also a Base64-encoded string and makes up the third part of a JWT-encoded string. A signature is there to protect the content of the JWT. The content is visible but cannot be modified if the token is signed. A Base64-encoded header and payload are passed to the signature's algorithm, along with either a secret or a public key to make the token a signed token. If you wish to include any sensitive or secret information in the payload, then it's better to encrypt it before assigning it to the payload.

A signature makes sure that the content is not modified once it is received. The use of a public/private key enhances the security step by verifying the sender.

You can use a combination of both a JWT and JWE. However, the recommended way is to first encrypt the payload using JWE and then sign it.

We'll use public/private keys to sign the token in the next section. Let's jump into coding in the next section.

Securing REST APIs with JWT

In this section, you'll secure the REST endpoints exposed in *Chapter 4, Writing Business Logic for APIs*. Therefore, we'll use the code from *Chapter 4* and enhance it to secure the APIs.

The REST APIs should be protected using the following techniques:

- No secure API should be accessed without a JWT.
- A JWT can be generated using sign-in/sign-up or a refresh token.
- A JWT and a refresh token should only be provided for a valid user's username/password combination or a valid user sign-up.
- The password should be stored in an encoded format using a `bcrypt` strong hashing function.
- The JWT should be signed with **Rivest-Shamir-Adleman** (**RSA**) keys with a strong algorithm.

RSA

RSA is an algorithm approved by the **Federal Information Processing Standards** (**FIPS**) (FIPS 186) for digital signatures and in **Special Publication** (**SP**) (SP800-56B) for key establishment.

- Claims in the payload should not store sensitive or secured information. If they do, then they should be encrypted.
- You should be able to authorize API access for certain roles.

We need to include new APIs for the authorization flow. Let's add them first.

Adding new APIs

You will enhance the existing APIs by adding four new APIs – sign-up, sign-in, sign-out, and a refresh token. The sign-up, sign-in, and sign-out operations are self-explanatory.

The refresh token provides a new access token (JWT) once the existing token expires. This is the reason why the sign-up/sign-in API provides two types of tokens – an access token and a refresh token as a part of its response. The JWT access token self-expires; therefore, a sign-out operation would only remove the refresh token.

Let's add these APIs to the `openapi.yaml` document next.

Apart from adding the new APIs, you also need to add a new user tag for these APIs that will expose all these APIs through the `UserApi` interface. Let's first add a sign-up endpoint.

Sign-up endpoint

Add the following specification for the sign-up endpoint in `openapi.yaml`:

```
/api/v1/users:
  post:
    tags:
      - user
    summary: Signup the a new customer (user)
    description: Creates a new customer (user)
    operationId: signUp
    requestBody:
      content:
        application/xml:
          schema:
            $ref: '#/components/schemas/User'
        application/json:
          schema:
            $ref: '#/components/schemas/User'
    responses:
      201:
        description: For successful user creation.
        content:
          application/xml:
            schema:
              $ref: '#/components/schemas/SignedInUser'
          application/json:
            schema:
              $ref: '#/components/schemas/SignedInUser'
```

```
https://github.com/PacktPublishing/Modern-API-Development-with-
Spring-6-and-Spring-Boot-3/blob/main/Chapter06/src/main/resources/
api/openapi.yaml
```

The sign-up API call returns the new `SignedInUser` model. This contains `accessToken`, `refreshToken`, `username`, and user ID fields. The code to add the model is shown in the following snippet:

```
SignedInUser:
  description: Signed-in user information
  type: object
  properties:
    refreshToken:
      description: Refresh Token
      type: string
    accessToken:
      description: JWT Token aka access token
      type: string
    username:
      description: User Name
      type: string
    userId:
      description: User Identifier
      type: string
```

Now, let's add the sign-in endpoint.

Sign-in endpoint definition

Add the following specification for the sign-in endpoint to `openapi.yaml`:

```
/api/v1/auth/token:
  post:
    tags:
      - user
    summary: Signin the customer (user)
    description: Generates the JWT and refresh token
    operationId: signIn
    requestBody:
      content:
        application/xml:
          schema:
            $ref: '#/components/schemas/SignInReq'
        application/json:
          schema:
```

```
                    $ref: '#/components/schemas/SignInReq'
       responses:
         200:
           description: Returns the access and refresh token.
           content:
             application/xml:
               schema:
                 $ref: '#/components/schemas/SignedInUser'
             application/json:
               schema:
                 $ref: '#/components/schemas/SignedInUser'
```

The sign-in API uses the new request payload object – `SignInReq`. The object just contains the username and password fields. Let's add it, as follows:

```
SignInReq:
  description: Request body for Sign-in
  type: object
  properties:
    username:
      description: username of the User
      type: string
    password:
      description: password of the User
      type: string
```

Now, let's add the sign-out endpoint.

Sign-out endpoint

Add the following specification for the sign-out endpoint to `openapi.yaml`:

```
# Under the /api/v1/auth/token
delete:
  tags:
    - user
  summary: Signouts the customer (user)
  description: Signouts the customer (user).
  operationId: signOut
  requestBody:
    content:
      application/xml:
        schema:
          $ref: '#/components/schemas/RefreshToken'
      application/json:
```

```
      schema:
        $ref: '#/components/schemas/RefreshToken'
  responses:
    202:
      description: Accepts the request for logout.
```

In an ideal scenario, you should remove the refresh token of a user received from the request. You can fetch the user ID from the token itself and then use that ID to remove the refresh token from the USER_TOKEN table. This endpoint requires you to send a valid access token.

We have opted for an easy way to remove the token, which is for it to be sent by the user as a payload. Therefore, this endpoint needs the following new model, `RefreshToken`. Here is the code to add the model:

```
RefreshToken:
  description: Contains the refresh token
  type: object
  properties:
    refreshToken:
      description: Refresh Token
      type: string
```

Finally, let's add an endpoint to refresh the access token.

Refresh token endpoint

Add the following specification for the refresh token endpoint to `openapi.yaml`:

```
/api/v1/auth/token/refresh:
  post:
    tags:
      - user
    summary: Provides new JWT based on valid refresh token.
    description: Provides JWT based on valid refresh token.
    operationId: getAccessToken
    requestBody:
      content:
        application/json:
          schema:
            $ref: '#/components/schemas/RefreshToken'
    responses:
      200:
        description: For successful operation.
        content:
```

```
        application/json:
          schema:
            $ref: '#/components/schemas/SignedInUser'
```

Here, we have used an exception by defining the refresh endpoint, in terms of forming a URI that represents the refresh token resources. Ideally, a POST call generates the new resource defined in URI. However, this endpoint generates the access token in place of the refresh token inside the response object, SignedInUser.

In the existing code, we don't have a table to store the refresh token. Therefore, let's add one.

Storing the refresh token using a database table

You can modify the Flyway database script to add a new table, as shown in the following code snippet:

```
create TABLE IF NOT EXISTS ecomm.user_token (
    id uuid NOT NULL DEFAULT random_uuid(),
    refresh_token varchar(128),
    user_id uuid NOT NULL,
    PRIMARY KEY(id),
    FOREIGN KEY (user_id)
        REFERENCES ecomm."user"(id)
);
```

https://github.com/PacktPublishing/Modern-API-Development-with-
Spring-6-and-Spring-Boot-3/blob/main/Chapter06/src/main/resources/
db/migration/V1.0.0__Init.sql

Here, the table contains three fields – id, refresh_token, and user_id – for storing the row identifier (primary key), the refresh token, and the user's ID, respectively. Also, we have put the table name user in double quotation marks because the H2 database also makes use of the term "user".

Now, you have completed the API specification modification for authentication and authorization. Next, let's start writing the implementation code for JWT-based authentication.

Implementing the JWT manager

Let's add a constant class that contains all the constants related to the security functionality before we implement the JWT manager class, as shown in the following code snippet:

```
public class Constants {
    public static final String ENCODER_ID = "bcrypt";
    public static final String API_URL_PREFIX = "/api/v1/**";
    public static final String H2_URL_PREFIX = "/h2-console/**";
    public static final String SIGNUP_URL = "/api/v1/users";
```

```
   public static final String TOKEN_URL = "/api/v1/auth/token";
   public static final String REFRESH_URL =
                                   "/api/v1/auth/token/refresh";
   public static final String PRODUCTS_URL =
      "/api/v1/products/**";
   public static final String AUTHORIZATION =
      "Authorization";
   public static final String TOKEN_PREFIX = "Bearer ";
   public static final String SECRET_KEY = "SECRET_KEY";
   public static final long EXPIRATION_TIME = 900_000;
   public static final String ROLE_CLAIM = "roles";
   public static final String AUTHORITY_PREFIX = "ROLE_";
 }
```

https://github.com/PacktPublishing/Modern-API-Development-with-Spring-6-and-Spring-Boot-3/blob/main/Chapter06/src/main/java/com/packt/modern/api/security/Constants.java

These constants are self-explanatory, except the EXPIRATION_TIME long value (900_000), which represents 15 minutes as a time unit.

Now, we can define the JWT manager class – JwtManager. **JwtManager** is a custom class that is responsible for generating a new JWT. It uses the java-jwt library from auth0.com. We will use public/private keys to sign the token. Let's define this class, as follows:

```
@Component
public class JwtManager {
  private final RSAPrivateKey privateKey;
  private final RSAPublicKey publicKey;
  public JwtManager(@Lazy RSAPrivateKey privateKey,
    @Lazy RSAPublicKey publicKey) {
    this.privateKey = privateKey;
    this.publicKey = publicKey;
  }
  public String create(UserDetails principal) {
    final long now = System.currentTimeMillis();
    return JWT.create()
        .withIssuer("Modern API Development with Spring…")
        .withSubject(principal.getUsername())
        .withClaim(
            ROLE_CLAIM,
            principal.getAuthorities().stream()
                .map(GrantedAuthority::getAuthority)
                .collect(toList()))
```

```
                    .withIssuedAt(new Date(now))
                    .withExpiresAt(new Date(now + EXPIRATION_TIME))
                    .sign(Algorithm.RSA256(publicKey, privateKey));
        }
    }
```

https://github.com/PacktPublishing/Modern-API-Development-with-Spring-6-and-Spring-Boot-3/blob/main/Chapter06/src/main/java/com/packt/modern/api/security/JwtManager.java

Here, JWT is a class from the java-jwt library that provides a fluent API to generate the token. It adds issuer (iss), subject (sub), issued at (iat), and expired at (exp) claims.

It also adds a custom claim, ROLE_CLAIM (roles), which is populated using authorities from UserDetails. This is an interface provided by Spring Security. You can use the org. springframework.security. core.userdetails.User.builder() method to create a UserBuilder class. UserBuilder is a final builder class that allows you to build an instance of UserDetails.

Finally, this method (JwtManager.create()) signs the JWT, using SHA256 with the RSA algorithm by calling the sign operation, which uses the provided public and private RSA keys. The JWT header specifies an HS256 value for the algorithm (alg) claim.

Signing is done using the public and private RSA keys. Let's add the code for RSA key management in our sample e-commerce application.

Generating the public/private keys

You can use JDK's keytool to create a key store and generate public/private keys, as shown in the following code snippet:

```
$ keytool -genkey -alias "jwt-sign-key" -keyalg RSA -keystore jwt-keystore.jks -keysize 4096
Enter keystore password:
Re-enter new password:
What is your first and last name?
[Unknown]: Modern API Development
What is the name of your organizational unit?
[Unknown]: Org Unit
What is the name of your organization?
[Unknown]: Packt
What is the name of your City or Locality?
[Unknown]: City
What is the name of your State or Province?
[Unknown]: State
```

```
What is the two-letter country code for this unit?
[Unknown]: IN
Securing REST APIs with JWT 191
Is CN=Modern API Development, OU=Org Unit, O=Packt, L=City, ST=State,
C=IN correct?
[no]: yes
Generating 4,096 bit RSA key pair and self-signed certificate
(SHA384withRSA) with a validity of 90 days
for: CN=Modern API Development, OU=Org Unit, O=Packt, L=City,
ST=State, C=IN
```

The generated key store should be placed under the `src/main/resources` directory.

> **Important note**
>
> Public/private keys are valid only for 90 days from the time they are generated. Therefore, make sure that you create a new set of public/private keys before you run this chapter's code.

Required values used in the `keytool` command should also be configured in the `application.properties` file, as shown here:

```
app.security.jwt.keystore-location=jwt-keystore.jks
app.security.jwt.keystore-password=password
app.security.jwt.key-alias=jwt-sign-key
app.security.jwt.private-key-passphrase=password
```

`https://github.com/PacktPublishing/Modern-API-Development-with-Spring-6-and-Spring-Boot-3/blob/main/Chapter06/src/main/resources/application.properties`

Now, we can configure the key store and public/private keys in the security configuration class.

Configuring the key store and keys

Let's add a `SecurityConfig` configuration class to configure the security relation configurations. This class extends the `WebSecurityConfigurerAdapter` class. Here's the code to do this:

```
@Configuration
@EnableWebSecurity
@EnableGlobalMethodSecurity(prePostEnabled = true)
public class SecurityConfig {
    @Value("${app.security.jwt.keystore-location}")
    private String keyStorePath;
    @Value("${app.security.jwt.keystore-password}")
    private String keyStorePassword;
    @Value("${app.security.jwt.key-alias}")
```

```
  private String keyAlias;
  @Value("${app.security.jwt.private-key-passphrase}")
  private String privateKeyPassphrase;
  …
  …
}
```

https://github.com/PacktPublishing/Modern-API-Development-with-Spring-6-and-Spring-Boot-3/blob/main/Chapter06/src/main/java/com/packt/modern/api/security/SecurityConfig.java

Here, we have added all the properties defined in application.properties.

Now, we can make use of the properties defined in application.properties to configure the KeyStore, RSAPrivateKey, and RSAPublicKey beans in the security configuration class, as shown in the following few subsections.

KeyStore bean

You can create a new bean for KeyStore by adding the following method and annotating it with @ Bean in SecurityConfig.java:

```
@Bean
public KeyStore keyStore() {
  try {
    KeyStore keyStore =
            KeyStore.getInstance(KeyStore.getDefaultType());
    InputStream resStream = Thread.currentThread().
      getContextClassLoader().getResourceAsStream
        (keyStorePath);
    keyStore.load(resStream, keyStorePassword.
      toCharArray());
    return keyStore;
  } catch (IOException | CertificateException |
    NoSuchAlgorithmException | KeyStoreException e) {
    LOG.error("Unable to load keystore: {}",
      keyStorePath, e);
  }
  throw new IllegalArgumentException
    ("Can't load keystore");
}
```

This creates a KeyStore instance, using the KeyStore class from the java.security package. It loads the key store from the src/main/resources package and uses the password configuration in the application.properties file.

Let's define the RSAPrivateKey bean next.

RSAPrivateKey bean

You can create a new bean for RSAPrivateKey by adding the following method and annotating it with @Bean in SecurityConfig.java:

```
@Bean
public RSAPrivateKey jwtSigningKey(KeyStore keyStore) {
  try {
    Key key = keyStore.getKey(keyAlias,
                        privateKeyPassphrase.toCharArray());
    if (key instanceof RSAPrivateKey) {
      return (RSAPrivateKey) key;
    }
  } catch (UnrecoverableKeyException |
    NoSuchAlgorithmException | KeyStoreException e) {
    LOG.error("key from keystore: {}", keyStorePath, e);
  }
  throw new IllegalArgumentException("Cant load
    private key");
}
```

This method uses a key alias and a private key password to retrieve the private key, which is used to return the RSAPrivateKey bean.

Let's define the RSAPublicKey bean next.

RSAPublicKey bean

You can create a new bean for RSAPublicKey by adding the following method and annotating it with @Bean in SecurityConfig.java:

```
@Bean
public RSAPublicKey jwtValidationKey(KeyStore keyStore) {
 try {
  Certificate certificate = keyStore.getCertificate
    (keyAlias);
  PublicKey publicKey = certificate.getPublicKey();
  if (publicKey instanceof RSAPublicKey) {
    return (RSAPublicKey) publicKey;
  }
 } catch (KeyStoreException e) {
   LOG.error("key from keystore: {}", keyStorePath, e);
 }
```

```
    throw new IllegalArgumentException("Can't load public key");
  }
```

Again, a key alias is used to retrieve the certificate from the key store. Then, the public key is retrieved from the certificate and returned.

As you know, JwtManager uses these public and private RSA keys to sign the JWT; therefore, a JWT decoder should use the same public key to decode the token. The Spring OAuth 2.0 resource server uses the org.springframework.security.oauth2.jwt. JwtDecoder interface to decode the token. Therefore, we need to create an instance of the JwtDecoder implementation and set the same public key in it to decode the token.

The Spring OAuth 2.0 resource server provides a NimbusJwtDecoder implementation class of JwtDecoder. Let's now create a bean of it with the public key.

JwtDecoder bean

You can create a new bean for JwtDecoder by adding the following method and annotating it with @Bean in SecurityConfig.java:

```
@Bean
public JwtDecoder jwtDecoder(RSAPublicKey rsaPublicKey) {
    return NimbusJwtDecoder.withPublicKey(rsaPublicKey).build();
}
```

You have defined all the beans required to sign the JWT token. Now, you can implement the newly added REST APIs.

Implementing new APIs

Let's implement the APIs exposed using UserApi. This is a code part that was autogenerated using OpenAPI Codegen. First, you need to define a new entity mapped to the user_token table.

Coding user token functionality

You can create UserTokenEntity based on the user_token table, as shown in the following code snippet:

```
@Entity
@Table(name = "user_token")
public class UserTokenEntity {
    @Id
    @GeneratedValue
    @Column(name = "ID", updatable = false, nullable = false)
    private UUID id;
```

```
@NotNull(message = "Refresh token is required.")
@Basic(optional = false)
@Column(name = "refresh_token")
private String refreshToken;
@ManyToOne(fetch = FetchType.LAZY)
private UserEntity user;
...

...
}
```

https://github.com/PacktPublishing/Modern-API-Development-with-
Spring-6-and-Spring-Boot-3/blob/main/Chapter06/src/main/java/com/
packt/modern/api/entity/UserTokenEntity.java

Similarly, we can expose the following CRUD repository for UserTokenEntity with the following
two methods – deleteByUserId(), which will remove the UserToken table record based on
a given user ID, and findByRefreshToken(), which will find the UserToken table record
based on a given refresh token. The code is illustrated in the following snippet:

```
public interface UserTokenRepository extends
    CrudRepository<UserTokenEntity, UUID> {
  Optional<UserTokenEntity> findByRefreshToken
    (StringrefreshToken);
  Optional<UserTokenEntity> deleteByUserId(UUID userId);
}
```

https://github.com/PacktPublishing/Modern-API-Development-with-
Spring-6-and-Spring-Boot-3/blob/main/Chapter06/src/main/java/com/
packt/modern/api/repository/UserTokenRepository.java

You have defined both the entity and its repository. Now, you will add new operations in UserService
that will consume these new classes.

Enhancing the UserService class

We also need to add new methods to UserService for the UserApi interface. Let's add new
methods to the service, as follows:

```
UserEntity findUserByUsername(String username);
Optional<SignedInUser> createUser(User user);
SignedInUser getSignedInUser(UserEntity userEntity);
Optional<SignedInUser> getAccessToken(RefreshToken refToken);
void removeRefreshToken(RefreshToken refreshToken);
```

https://github.com/PacktPublishing/Modern-API-Development-with-Spring-6-and-Spring-Boot-3/blob/main/Chapter06/src/main/java/com/packt/modern/api/service/UserService.java

Here, each method performs a specific operation, as outlined here:

- findUserByUsername(): This finds and returns a user based on a given username.

- createUser(): This adds a new signed-up user to the database.

- getSignedInUser(): This creates a new model instance of SignedInUser that holds the refresh token, access token (JWT), user ID, and username.

- getAccessToken(): This generates and returns a new access token (JWT) for a given valid refresh token.

- removeRefreshToken(): This removes the refresh token from the database. It is called when the user wants to sign out.

Let's implement each of these methods in the UserServiceImpl class.

Implementing findUserByUsername()

First, you can add the implementation for findUserByUsername() in UserServiceImpl class, as shown in the following code snippet:

```
public UserEntity findUserByUsername(String username) {
  if (Strings.isBlank(username)) {
    throw new UsernameNotFoundException("Invalid user.");
  }
  final String uname = username.trim();
  Optional<UserEntity> oUserEntity =
     repository.findByUsername(uname);
  UserEntity userEntity = oUserEntity.orElseThrow
    (() -> new UsernameNotFoundException(String.format(
      "Given user(%s) not found.", uname)));
  return userEntity;
}
```

https://github.com/PacktPublishing/Modern-API-Development-with-Spring-6-and-Spring-Boot-3/blob/main/Chapter06/src/main/java/com/packt/modern/api/service/UserServiceImpl.java

This is a straightforward operation. You query the database based on a given username. If the username is found, then it returns the user; otherwise, it throws a UsernameNotFoundException exception.

createUser() implementation

Next, you can add the implementation for the `createUser()` method to the `UserServiceImpl` class, as shown in the following code snippet:

```
@Transactional
public Optional<SignedInUser> createUser(User user) {
  Integer count = repository.findByUsernameOrEmail(
    user.getUsername(), user.getEmail());
  if (count > 0) {
    throw new GenericAlreadyExistsException
      ("Use different username and email.");
  }
  UserEntity userEntity = repository.save(toEntity(user));
  return Optional.of(createSignedUserWithRefreshToken(
    userEntity));
}
```

Here, we first check whether an existing user was assigned the same username or email in the sign-up request. If there was, an exception is simply raised; otherwise, a new user is created in the database and a `SignedInUser` instance is returned with refresh and access tokens, using the `createSignedUserWithRefreshToken()` method.

First, we can add a private `createSignedUserWithRefreshToken()` method in the `UserServiceImpl` class, as shown in the following code snippet:

```
private SignedInUser createSignedUserWithRefreshToken(
    UserEntity userEntity) {
  return createSignedInUser(userEntity)
    .refreshToken(createRefreshToken(userEntity));
}
```

This also uses another private method, `createSignedInUser()`, which returns `SignedInUser`; then, it adds the refresh token by calling the `createRefreshToken()` method.

Let's define the two `createSignedInUser()` and `createRefreshToken()` private methods in the `UserServiceImpl` class, as shown in the following code snippet:

```
private SignedInUser createSignedInUser(UserEntity uEntity) {
  String token = tokenManager.create(
          org.springframework.security.core.userdetails
            .User.builder()
          .username(userEntity.getUsername())
          .password(userEntity.getPassword())
          .authorities(Objects.nonNull
```

```
                    (userEntity.getRole()) ?
                     userEntity.getRole().name() : "")
                  .build());
       return new SignedInUser()
           .username(userEntity.getUsername())
           .accessToken(token)
           .userId(userEntity.getId().toString());
   }
   private String createRefreshToken(UserEntity user) {
     String token = RandomHolder.randomKey(128);
     userTokenRepository.save(new UserTokenEntity().
       setRefreshToken(token).setUser(user));
     return token;
   }
```

Here, `tokenManager` is used in the `createSignedIn()` method to create the JWT. `tokenManager` is an instance of `JwtManager`. The `User.builder()` method is used to create a `UserBuilder` class. `UserBuilder`, which is a final builder class, is used to create an instance of `UserDetails`. The `JwtManager.create()` method uses this `UserDetails` instance to create a token.

The `createRefreshToken()` method uses the `RandomHolder` private static class to generate a refresh token. This token is not a JWT; we can use a longer-lasting valid token, such as one valid for a day, as a refresh token. Saving a JWT as a refresh token in the database removes the sole purpose of using the JWT because it expires by the configured time, and it should not be stored in the database, as it automatically becomes invalid. Therefore, we should think carefully before using a JWT as a refresh token and then saving it in the database.

Let's add the `RandomHolder` private static class to the `UserServiceImpl` class, as shown in the following code snippet:

```
// https://stackoverflow.com/a/31214709/109354
private static class RandomHolder {
  static final Random random = new SecureRandom();
  public static String randomKey(int length) {
    return String.format("%"+length+"s",new BigInteger
      (length * 5 /*base32, 2^5*/,random).toString(32)).
      replace('\u0020', '0');
  }
}
```

This class uses a `SecureRandom` instance to generate a random `BigInteger` instance. Then, this random `BigInteger` value is converted into a string with a radix size of 32. Finally, the space is replaced with 0 if found in a converted string.

You can also use the `org.apache.commons.lang3.RandomStringUtils.randomAlphanumeric()` method, or use any other secured random key generator, to generate a refresh token.

We also need to modify the `UserRepository` class to add a new method that returns the count of users with a given username or email.

getSignedInUser() implementation

The implementation of the `getSignedInUser()` method is straightforward. Add it to the `UserServiceImpl` class, as shown in the following code snippet:

```
@Transactional
public SignedInUser getSignedInUser(UserEntity userEntity) {
    userTokenRepository.deleteByUserId(userEntity.getId());
    return createSignedUserWithRefreshToken(userEntity);
}
```

Here, this method first removes the existing token from the database associated with the given user, and then it returns the new instance of `SignedInUser` that was created using `createSignedUserWithRefreshToken()`, defined previously in the *createUser() implementation* subsection.

getAccessToken() implementation

The implementation of the `getAccessToken()` method is, again, straightforward. Add it to the `UserServiceImpl` class, as shown in the following code snippet:

```
public Optional<SignedInUser> getAccessToken(
    RefreshToken refreshToken) {
    return userTokenRepository
        .findByRefreshToken(refreshToken.getRefreshToken())
        .map(ut ->
            Optional.of(createSignedInUser(ut.getUser())
                .refreshToken(refreshToken.getRefreshToken())))
        .orElseThrow(() ->
            new InvalidRefreshTokenException
                ("Invalid token."));
}
```

First, the `getAccessToke()` method finds the user's token entity using the `UserTokenRepository` instance. Then, it populates the `SignedInUser` POJO using the retrieved `UserToken` entity. The `createSignedInUser()` method does not populate the refresh token; therefore, we assign the same refresh token back. If it does find the user token entry in the database based on the refresh token, it throws an exception.

Also, you can add a validation for time that will remove/invalidate the refresh token, which has not been added here for simplicity.

You can also add a time validation logic for the refresh token – for example, storing the refresh token creation time in the database and using the configured valid time for refresh token validation, which is a kind of expiration logic for JWTs.

removeRefreshToken() implementation

You can add the removeRefreshToken() method to the UserServiceImpl class, as shown in the following code snippet:

```
public void removeRefreshToken(RefreshToken refreshToken) {
  userTokenRepository
    .findByRefreshToken(refreshToken.getRefreshToken())
    .ifPresentOrElse(
      userTokenRepository::delete,
      () -> {
        throw new InvalidRefreshTokenException
          ("Invalid token.");
      });
}
```

First, the method finds the given refresh token in the database. If this is not found, then it throws an exception. If the given refresh token is found in the database, then it deletes it.

You have implemented all the extra methods added to the UserService class. Now, you will add additional methods in UserRespository too in the following section.

Enhancing the UserRepository class

Let's add the findByUsername() and findByUsernameOrEmail() methods to UserRepository, as follows:

```
public interface UserRepository extends
    CrudRepository<UserEntity, UUID> {
  Optional<UserEntity> findByUsername(String username);
  @Query( value = "select count(u.*) from ecomm.\"user\" u
          where u.username = :username or u.email = :email",
          nativeQuery = true)
  Integer findByUsernameOrEmail(String username,
    String email);
}
```

https://github.com/PacktPublishing/Modern-API-Development-with-Spring-6-and-Spring-Boot-3/blob/main/Chapter06/src/main/java/com/packt/modern/api/repository/UserRepository.java

The findByUsernameOrEmail method returns a count of the records matching the given username or email.

You are now ready to implement the new APIs added to the UserApi interface to write the REST controllers. Let's do that next.

Implementing the REST controllers

In the previous section, you developed and enhanced the services and repositories required to implement the APIs defined in the UserApi interface, generated by OpenAPI Codegen. The only pending dependency is PasswordEncoder. PasswordEncoder is required to encode the password before storing and matching the given password in the sign-in request.

Adding a bean for PasswordEncoder

You should expose the PasswordEncoder bean because Spring Security needs to know which encoding you want to use for password encoding, as well as for decoding the passwords. Let's add a PasswordEncoder bean to AppConfig, as follows:

```
@Bean
public PasswordEncoder passwordEncoder() {
  Map<String, PasswordEncoder> encoders =
      Map.of(
          ENCODER_ID, new BCryptPasswordEncoder(),
          "pbkdf2", Pbkdf2PasswordEncoder.
            defaultsForSpringSecurity_v5_8(),
          "scrypt", ScryptPasswordEncoder
            .defaultsForSpringSecurity_v5_8());
  return new DelegatingPasswordEncoder
    (ENCODER_ID, encoders);
}
```

https://github.com/PacktPublishing/Modern-API-Development-with-Spring-6-and-Spring-Boot-3/blob/main/Chapter06/src/main/java/com/packt/modern/api/AppConfig.java

Here, you can directly create a new instance of BcryptPasswordEncoder and return it for bcrypt encoding. However, the use of DelegatingPasswordEncoder not only allows you to support existing passwords but also facilitates migration to a new, better encoder if one is available in the future. This code uses Bcrypt as a default password encoder, which is the best among the currently available encoders.

For `DelegatingPasswordEncoder` to work, you need to add a hashing algorithm prefix such as `{bcrypt}` to encoded passwords – for example, add `{bcrypt}$2a$10$neR0EcYY5./tLVp4litNyuBy/ kfrTsqEv8hiyqEKX0TXIQQwC/5Rm` to the persistent store if you already have a hashed password in the database, or if you're adding any seed/test users to the database script. The new password will store the password with a prefix anyway, as configured in the `DelegatingPasswordEncoder` constructor. You have passed `bcrypt` into the constructor; therefore, all new passwords will be stored with a `{bcrypt}` prefix.

`PasswordEncoder` reads the password from the persistence store and removes the prefix before matching. It uses the same prefix to find out which encoder it needs to use for matching. Now, you can start implementing the new APIs based on `UserApi`.

Implementing the Controller class

First, create a new `AuthController` class, as shown in the following code snippet:

```
@RestController
public class AuthController implements UserApi {
  private final UserService service;
  private final PasswordEncoder passwordEncoder;

  public AuthController(UserService s, PasswordEncoder e) {
    this.service = s;
    this.passwordEncoder = e;
  }
  ...
  ...
}
```

https://github.com/PacktPublishing/Modern-API-Development-with-Spring-6-and-Spring-Boot-3/blob/main/Chapter06/src/main/java/com/packt/modern/api/controller/AuthController.java

The `AuthController` class is annotated with `@RestController` to mark it as a REST controller. Then, it uses two beans, `UserService` and `PasswordEncoder`, which will be injected at the time of the `AuthController` construction.

First, let's add the sign-in operation, as follows:

```
public ResponseEntity<SignedInUser> signIn(@Valid SignInReq signInReq)
{
  UserEntity userEntity = service
      .findUserByUsername(signInReq.getUsername());
  if (passwordEncoder.matches(signInReq.getPassword(),
      userEntity.getPassword())) {
```

```
      return ok(service.getSignedInUser(userEntity));
   }
   throw new InsufficientAuthenticationException
     ("Unauthrzed");
}
```

The operation first finds the user and matches the password using the `PasswordEncoder` instance. If everything goes through successfully, it returns the `SignedInUser` instance with refresh and access tokens; otherwise, it throws an exception.

Let's add other operations to `AuthController`, as follows:

```
public ResponseEntity<Void> signOut(
  @Valid RefreshToken refreshToken) {
  service.removeRefreshToken(refreshToken);
  return accepted().build();
}
public ResponseEntity<SignedInUser> signUp
   (@Valid User user) {
 return status(HttpStatus.CREATED)
    .body(service.createUser(user).get());
}
public ResponseEntity<SignedInUser> getAccessToken(
   @Valid RefreshToken refreshToken) {
  return ok(service.getAccessToken(refreshToken)
    .orElseThrow(InvalidRefreshTokenException::new));
}
```

All operations such as `signOut()`, `signUp()`, and `getAccessToken()` are straightforward, as outlined here:

- `signOut()` uses the user service to remove the given refresh token. Ideally, you would like to get the user ID from the logged-in user's request and remove the refresh token, based on the retrieved user ID from the request.

- `signUp()` creates a valid new user and returns the `SignedInUser` instance as a response. Here, we haven't added the validation of the payload for simplicity. In a real-world application, you must validate the request payload.

- `getAccessToken()` returns `SignedInUser` with a new access token if the given refresh token is valid.

We have finished coding the controllers. Let's configure security in the next subsection.

Configuring web-based security

This is the last puzzle to sort out the authentication and authorization piece. The SecurityConfig class is also annotated with @EnableWebSecurity. With the new version of Spring Security, you now don't need to extend WebSecurityConfigurerAdapter and override the configure() method, as we did in the last edition of this book. Instead, you now create a bean that returns the configured instance of the SecurityFilterChain class.

The method (filterChain) that returns SecurityFilterChain takes HttpSecurity as a parameter. HttpSecurity contains DSL (fluent methods). You can make use of these methods to configure web-based security, such as which web paths to allow and which method to allow. Let's make the following configurations using these fluent methods to return the SecurityFilterChain instance, as shown in the following code snippet from SecurityConfig.java:

```
@Bean
protected SecurityFilterChain filterChain(HttpSecurity http) throws
Exception {
  http.httpBasic().disable()
      .formLogin().disable()
      .csrf().ignoringRequestMatchers(API_URL_PREFIX)
          .ignoringRequestMatchers(toH2Console())
      .and()
      .headers().frameOptions().sameOrigin()
      .and()
      .cors()
      .and()
      .authorizeHttpRequests(req ->
          req.requestMatchers(toH2Console()).permitAll()
            .requestMatchers(new AntPathRequestMatcher(
                TOKEN_URL, HttpMethod.POST.name())))
                .permitAll()
            .requestMatchers(new AntPathRequestMatcher(
                TOKEN_URL, HttpMethod.DELETE.name())).
                permitAll()
            .requestMatchers(new AntPathRequestMatcher(
                SIGNUP_URL, HttpMethod.POST.name())))
                .permitAll()
            .requestMatchers(new AntPathRequestMatcher(
                REFRESH_URL, HttpMethod.POST.name())).
                permitAll()
            .requestMatchers(new AntPathRequestMatcher(
                PRODUCTS_URL, HttpMethod.GET.name())).
                permitAll()
            .requestMatchers("/api/v1/addresses/**")
```

```
        .hasAuthority(RoleEnum.ADMIN.getAuthority())
      .anyRequest().authenticated())
    .oauth2ResourceServer(oauth2ResourceServer ->
      oauth2ResourceServer.jwt(jwt ->
        jwt.jwtAuthenticationConverter(
          getJwtAuthenticationConverter())))
    .sessionManagement().sessionCreationPolicy(
      SessionCreationPolicy.STATELESS);
  return http.build();
}
```

https://github.com/PacktPublishing/Modern-API-Development-with-Spring-6-and-Spring-Boot-3/blob/main/Chapter06/src/main/java/com/packt/modern/api/security/SecurityConfig.java

Here, you configure the following security settings:

1. First of all, you disable the basic authentication and form login using the disable() method.

2. Then, you ignore the CSRF configuration for the API base path and H2 console URLs.

3. Then, you set the headers setting for frame options that have the same origin to allow the H2 console application to work fine in the browser. The H2 console UI is based on HTML frames. The H2 console UI won't display in browsers because, by default, the security header (X-Frame-Options) is not sent with permission to allow frames with the same origin. Therefore, you need to configure headers().frameOptions().sameOrigin().

4. Then, you enable the CORS setting. You'll learn more about this in the next section.

5. Then, you configure the authorization of the request, which takes the request object as a parameter. You use this request object to restrict access, based on URL patterns, by using the requestMatchers() method and an instance of the AntPathRequestMatcher class:

    *   Configure URL patterns and respective HTTP methods using **Ant matchers**, which allows you to use Ant (build tool) pattern-matching styles. You can also use mvcMatchers(), which uses the same pattern-matching style as a Spring **Model-View-Controller** (**MVC**) and works well with MVC-based Spring applications.

    *   Static method toH2Console() is a utility that provides a matcher that includes the H2 console location.

    *   The /api/v1/addresses/** pattern has been configured to be accessed only by the user who has the ADMIN role, by calling hasAuthority() and passing the admin authority in it. You'll learn more about it in the *Understanding authorization* section of this chapter.

6. All URLs, except those configured explicitly by authorizeHttpRequests(), should be allowed by any authenticated user (by using anyRequest(). authenticated()).

- Enable JWT bearer token support for the OAuth 2.0 resource server (`oauth2ResourceServer.jwt()`).

- Enable the `STATELESS` session creation policy (that is, `sessionManagement().sessionCreationPolicy` won't create any `HTTPSession`)

7. Finally, the `filterChain` method returns `SecurityFilterChain` by building the instance from the configured `HttpSecurity` instance (the `http.build()` call).

In this section, you learned how to configure Spring security for authentication and authorization. Next, we will learn about the CORS and CSRF.

Configuring CORS and CSRF

Browsers restrict cross-origin requests from scripts for security reasons. For example, a call from `http://mydomain.com` to `http://mydomain-2.com` can't be made using a script. Also, an origin not only indicates a domain but also includes a scheme and a port.

Before hitting any endpoint, the browser sends a pre-flight request using the HTTP method option to check whether the server will permit the actual request. This request contains the following headers:

- The actual request's headers (`Access-Control-Request-Headers`).

- A header containing the actual request's HTTP method (`Access-Control- Request-Method`).

- An `Origin` header that contains the requesting origin (scheme, domain, and port).

- If the response from the server is successful, then only the browser allows the actual request to fire. The server responds with other headers, such as `Access- Control-Allow-Origin`, which contains the allowed origins (an asterisk * value means any origin), `Access-Control-Allow-Methods` (allowed methods), `Access-Control-Allow-Headers` (allowed headers), and `Access-Control-Max-Age` (allowed time in seconds).

You can configure CORS to take care of cross-origin requests. For that, you need to make the following two changes:

- Add a `CorsConfigurationSource` bean that takes care of the CORS configuration using a `CorsConfiguration` instance.

- Add the `cors()` method to `HTTPSecurity` in the `configure()` method. The `cors()` method uses `CorsFilter` if a `corsFilter` bean is added; otherwise, it uses `CorsConfigurationSource`. If neither is configured, then it uses the Spring MVC pattern inspector handler.

Let's now add the `CorsConfigurationSource` bean to the `SecurityConfig` class.

The default permitted values (new CorsConfiguraton(). applyPermitDefaultValues()) configure CORS for any origin (*), all headers, and simple methods (GET, HEAD, and POST), which have an allowed max age of 30 minutes.

You need to allow mostly all of the HTTP methods, including the DELETE method, and you need more custom configuration; therefore, we will use the following bean definition in SecurityConfig. java:

```
@Bean
CorsConfigurationSource corsConfigurationSource() {
  CorsConfiguration configuration = new CorsConfiguration();
  configuration.setAllowedOrigins(List.of("*"));
  configuration.setAllowedMethods(Arrays.asList("HEAD",
    "GET", "PUT", "POST", "DELETE", "PATCH"));
  // For CORS response headers
  configuration.addAllowedOrigin("*");
  configuration.addAllowedHeader("*");
  configuration.addAllowedMethod("*");
  UrlBasedCorsConfigurationSource source = new
    UrlBasedCorsConfigurationSource();
  source.registerCorsConfiguration("/**", configuration);
  return source;
}
```

Here, you create a CorsConfiguration instance using the default constructor and then set the allowed origins, allowed methods, and response headers. Finally, you pass it as an argument while registering it to the UrlBasedCorsConfigurationSource instance and returning it.

In the previous section, inside the SecurityChainFilter method annotated with @Bean, you have configured CSRF using the csrf() DSL. We have applied CSRF protection to all URLs, except URLs starting with /api/v1 and the /h2-console H2 database console URLs. You can change the configuration based on your requirement.

Let's first understand what CSRF/XSRF is. **CSRF** or **XSRF** stands for **cross-site request forgery**, which is a web security vulnerability. To understand how this vulnerability comes into effect, let's assume you are a bank customer and are currently signed in to your account online. While you are logged in, you may receive an email and click on a link in it, or on any other malicious website's link, that contains a malicious script. This script can then send a request to your bank for a fund transfer. The bank then transfers the funds to a perpetrator's account because the bank thinks that the request has been sent by you, as you are signed in. Hackers can use this vulnerability similarly for different hacking activities.

To prevent such attacks, the application sends new unique CSRF tokens associated with the signed-in user for each new request. These tokens are stored in hidden form fields. When a user submits a form, the same token should be sent back with the request. The application then verifies the CSRF token and only processes the request if the verification is successful. This works because malicious scripts can't read the token due to the same origin policy.

However, if a perpetrator also tricks you into revealing the CSRF token, then it is very difficult to prevent such attacks. You can disable CSRF protection for this web service by using `csrf().disable()` because we only expose REST endpoints.

Now, let's move on to the final section, where you will configure the authorization based on the user's role.

Understanding authorization

Your valid username/password or access token for authentication gives you access to secure resources, such as URLs, web resources, or secure web pages. Authorization is one step ahead; it allows you to configure access security further with scopes such as read, write, or roles such as Admin, User, and Manager. Spring Security allows you to configure any custom authority.

We will configure three types of roles for our sample e-commerce app – namely, Customer (user), Admin, and Customer Support Representative (CSR). Obviously, each user will have their own specific authority. For example, a user can place an order and buy stuff online but should not be able to access the CSR or admin resources. Similarly, a CSR should not be able to have access to admin-only resources. A security configuration that allows authority or role-based access to resources is known as authorization. A failed authentication should return an HTTP 401 status (unauthorized), and a failed authorization should return an HTTP 403 status (forbidden), which means the user is authenticated but does not have the required authority/role to access the resource.

Let's introduce these three roles in a sample e-commerce app, as shown in the following code snippet:

```
public enum RoleEnum implements GrantedAuthority {
  USER(Const.USER), ADMIN(Const.ADMIN), CSR(Const.CSR);
  private String authority;
  RoleEnum(String authority) {
    this.authority = authority;
  }
  @JsonCreator
  public static RoleEnum fromAuthority(String authority) {
    for (RoleEnum b : RoleEnum.values()) {
      if (b.authority.equals(authority)) {
        return b;
      }
    }
  }
  @Override
  public String toString() {
    return String.valueOf(authority);
  }
```

```
  @Override
  @JsonValue
  public String getAuthority() {
    return authority;
  }
  public class Const {
    public static final String ADMIN = "ROLE_ADMIN";
    public static final String USER = "ROLE_USER";
    public static final String CSR = "ROLE_CSR";
  }
}
```

https://github.com/PacktPublishing/Modern-API-Development-with-Spring-6-and-Spring-Boot-3/blob/main/Chapter06/src/main/java/com/packt/modern/api/entity/RoleEnum.java

Here, we declared an enum that implements Spring Security's `GrantedAuthority` interface to override the `getAuthority()` method. `GrantedAuthority` is an authority granted to an `Authentication` (interface) object. As you know, `BearerTokenAuthenticationToken` is a type of `AbstractAuthenticationToken` class that implements the authentication interface, which represents the token/principal for an authenticated request. We have used the string constants for the user's roles in this enum, as we need these when we configure the role-based restriction at a method level.

Let's discuss the role and authority in detail.

Role and authority

An authority can be assigned for fine-grained control, whereas roles should be applied to large sets of permissions. A role is an authority that has the ROLE_ prefix. This prefix is configurable in Spring Security.

Spring Security provides the `hasRole()` and `hasAuthority()` methods to apply role- and authority-based restrictions. `hasRole()` and `hasAuthority()` are almost identical, but the `hasRole()` method maps with Authority without the ROLE_ prefix. If you use `hasRole('ADMIN')`, your Admin enum must be ROLE_ADMIN instead of ADMIN because a role is an authority and should have a ROLE_ prefix, whereas if you use `hasAuthority('ADMIN')`, your Admin enum must be only ADMIN.

The OAuth 2.0 resource server, by default, populates authorities based on the scope (scp) claim. If you provide access to a user's resources, such as order history for integration with another application, then you can limit an application's access to a user's account before granting access to other applications for third-party integration. Third-party applications can request one or more scopes; this information is then presented to the user on the consent screen, and the access token issued to the application will be limited to the scopes granted. However, in this chapter, we haven't provided OAuth 2.0 authorization flows and will limit security access to REST endpoints.

If the JWT contains a claim with the name *scope* (`scp`), then Spring Security will use the value in that claim to construct the authorities by prefixing each value with `SCOPE_`. For example, if a payload contains a `scp=["READ","WRITE"]` claim, this means that an `Authority` list will consist of `SCOPE_READ` and `SCOPE_WRITE`.

We need to change the default authority mapping behavior because a scope (`scp`) claim is the default authority for the OAuth2.0 resource server in Spring. We can do that by adding a custom authentication converter to `JwtConfigurer` in `OAuth2ResourceServer` in your security configuration. Let's add a method that returns the converter, as follows:

```
private Converter<Jwt, AbstractAuthenticationToken>
    getJwtAuthenticationConverter() {
  JwtGrantedAuthoritiesConverter authorityConverter =
    new JwtGrantedAuthoritiesConverter();
  authorityConverter.setAuthorityPrefix(AUTHORITY_PREFIX);
  authorityConverter.setAuthoritiesClaimName(ROLE_CLAIM);
  JwtAuthenticationConverter converter =
    new JwtAuthenticationConverter();
  converter.setJwtGrantedAuthoritiesConverter(authorityConverter);
  return converter;
}
```

https://github.com/PacktPublishing/Modern-API-Development-with-Spring-6-and-Spring-Boot-3/blob/main/Chapter06/src/main/java/com/packt/modern/api/security/SecurityConfig.java

Here, we first create a new instance of `JwtGrantedAuthorityConverter` and then assign an authority prefix (`ROLE_`) and authority claim name (the key of the claim in JWT) as `roles`.

Now, we can use this private method to configure the OAuth 2.0 resource server. You can now modify the existing configuration with the following code. We can also add configuration to add role-based restrictions to the `POST /api/v1/addresses` API call, in the following code snippet in `SecurityConfig.java`:

```
.requestMatchers("/api/v1/addresses/**")
    .hasAuthority(RoleEnum.ADMIN.getAuthority())
.anyRequest().authenticated())
.oauth2ResourceServer(oauth2ResourceServer ->
  oauth2ResourceServer.jwt(jwt ->
    jwt.jwtAuthenticationConverter(
    getJwtAuthenticationConverter()))))
```

After setting this configuration to add an address (POST /api/v1/addresses), it now requires both authentication and authorization. This means the logged-in user must have the ADMIN role to call this endpoint successfully. Also, we changed the default claim from scope to role.

Now, we can proceed further with method-level, role-based restrictions. Spring Security provides a feature that allows you to place authority- and role-based restrictions on public methods of Spring beans, using a set of annotations such as @PreAuthorize, @Secured, and @RolesAllowed. By default, these are disabled; therefore, you need to enable them explicitly.

Let's enable these by adding the @EnableGlobalMethodSecurity(prePostEnabled = true) annotation to the Spring Security configuration class, as follows:

```
@Configuration
@EnableWebSecurity
@EnableGlobalMethodSecurity(prePostEnabled = true)
public class SecurityConfig {
```

Now, you can use the @PreAuthorize (the given access-control expression would be evaluated before the method invocation) and @PostAuthorize (the given access-control expression would be evaluated after the method invocation) annotations to place restrictions on public methods of Spring beans because you have set the prePostEnabled property to true when enabling the global method-level security.

@EnableGlobalMethodSecurity also supports the following properties:

- securedEnabled: This allows you to use @Secured annotation on public methods.
- jsr250Enabled: This allows you to use JSR-250 annotations such as @RolesAllowed, which can be applied to both public classes and methods. As the name suggests, you can use a list of roles for access restrictions.

@PreAuthorize and @PostAuthorize are more powerful than the other security annotations because not only can they be configured for authorities/roles but also for any valid **Spring Expression Language (SpEL)** expression:

```
For demonstration purposes, let's add the @PreAuthorize annotation to
the deleteAddressesById() method, which is associated with DELETE /v1/
auth/addresses/{id} in AddressController, as shown in the following
code snippet:
@PreAuthorize("hasRole('" + Const.ADMIN + "')")
@Override
public ResponseEntity<Void> deleteAddressesById(String id) {
  service.deleteAddressesById(id);
  return accepted().build();
}
```

Let's break down the preceding code snippet:

- `hasRole()` is a built-in `SpEL` expression. We need to pass a valid `SpEL` expression, and it should be a string. Any variable used to form this `SpEL` expression should be final. Therefore, we have declared the final string constants in the `RoleEnum` enum (for example, `Const.ADMIN`).

- Now, the `DELETE /api/v1/addresses/{id}` REST API can only be invoked if the user has the `ADMIN` role.

- Spring Security provides various built-in `SpEL` expressions, such as `hasRole()`. Here are some others:

  - `hasAnyRole(String... roles)`: This returns `true` if the principal's role matches any of the given roles.

  - `hasAuthority(String authority)`: This returns `true` if the principal has given authority. Similarly, you can also use `hasAnyAuthority(String... authorities)`.

  - `permitAll`: This returns `true`.

  - `denyAll`: This returns `false`.

- `isAnonymous()`: This returns `true` if the current user is anonymous.

- `isAuthenticated()`: This returns `true` if the current user is not anonymous.

A full list of these expressions is available at `https://docs.spring.io/spring-security/site/docs/3.0.x/reference/el-access.html`.

Similarly, you can apply access restrictions for other APIs. Let's test security in the next section.

Testing security

By now, you must be looking forward to testing. You can find the API client collection at the following location. You can import it and then test the APIs, using any API client that supports the HAR type file import: `https://github.com/PacktPublishing/Modern-API-Development-with-Spring-6-and-Spring-Boot-3/blob/main/Chapter06/Chapter06-API-Collection.har`.

> **Important note**
>
> Make sure to generate the keys again, as keys generated by the JDK keytool are only valid for 90 days.

> **Building and running the Chapter 06 code**
>
> You can build the code by running `gradlew clean build` from the root of the project, and you can run the service using `java -jar build/libs/Chapter06-0.0.1-SNAPSHOT.jar`. Make sure to use Java 17 in the path.

Now, let's test our first use case.

Let's fire the `GET /api/vi/addresses` API without the `Authorization` header, as shown in the following command:

```
$ curl -v 'http://localhost:8080/api/v1/addresses' -H 'Content- Type:
application/json' -H 'Accept: application/json'
< HTTP/1.1 401
< Vary: Origin
< Vary: Access-Control-Request-Method
< Vary: Access-Control-Request-Headers
< WWW-Authenticate: Bearer
< X-Content-Type-Options: nosniff
< X-XSS-Protection: 1; mode=block
< Cache-Control: no-cache, no-store, max-age=0, must
revalidate
< Other information is removed for brevity
```

This returns the HTTP `401` status (`unauthorized`) and a `WWW-Authenticate: Bearer` response header, which suggests the request should be sent with an `Authorization` header.

Let's send the request again with an invalid token, as shown in the following command:

```
$ curl -v 'http://localhost:8080/api/v1/addresses' -H 'Content-Type:
application/json' -H 'Accept: application/json' -H 'Authorization:
Bearer eyJ0eXAiOiJKV1QiLCJhbGciOiJSUzI1NiJ9…
rest of the JWT string removed for brevity'
< HTTP/1.1 401
< Vary: Origin
< Vary: Access-Control-Request-Method
< Vary: Access-Control-Request-Headers
< WWW-Authenticate: Bearer
< Other information is removed for brevity
```

Again, it returns the `401` response.

We have created two users using a Flyway database migration script – `scott`/`tiger` and `scott2`/`tiger`. Now, let's perform a sign-in with the username `scott` to get the valid JWT, as follows:

```
$ curl -X POST http://localhost:8080/api/v1/auth/token -H  'Accept:
application/json' -H  'Content-Type: application/json' -d '{
```

```
    "username": "scott",
    "password": "tiger"
  }'
```

{"refreshToken":"9rdk5b35faafkneqg9519s6p4tbbqcdt412t7h5st9savkonqnnd
5e8ntr334m8rffgo6bio1vglng1hqqse3igdonoabc9711pdgt3bjoc3je3m811dp2au
vimts8p6","accessToken":"eyJ0eXAiOiJKV1QiLCJhbGciOiJSUzI1NiJ9.eyJzdWI
iOiJzY290dCIsInJvbGVzIjpbIlVTRVIiXSwiaXNzIjoiTW9kZXJuIEFQSSBEZXZlbG9
wbWVudCB3aXRoIFNwcmluZyBhbmQgU3ByaW5nIEJvb3QiLCJleHAiOjE2NzceOOTYzMzk
sImlhdCI6MTY3NzQ5NTQzQzOX0.a77O7ZbSAOw5v6Tb3w-MtBwotMEUvc1H1y2W0IU2QJh0m
1SJxSBCfdrNB1omVk1HnwX4kOpj4grbNasBjpIpHtyOLXdp-gngxdvVfaKSPuptrW4YzA3
ikxbUMWDdEtij_y2DRxJXQ6CrPTjA40L7yB_SXswnHT988Qq6ZALeGW-Lmz-vzAZiRcZUe
6bPPn7F-p41K_qi1nsUJ1rnWmmffLWCH37zt11cgh6bB1UJuOn9Hw2A1nQExfUutRKgFK0
-LxBUOKOKdRESOnJR9hwOL6v10IF19xNm53LVMIcuJrndCxvmv7mv0fUOxY63UwhO9kOT
RCXViGKCa3H8RxUFwG52q2nZelle_4I8CUSeDDdmD2R1ax2NyQNe-HHEJb9c91JSzhFm0
K0-c34-kiNGqaB3j1jndHoGXCBLM5prph1Sd1V4U9PYhmL8ZCaDv8q6rCPSAEcRoiOBPPn
dxEApHKulj9vrO_p7K1T9dLamJSFJKw9Yz8M3_ngiE3qtEBQ3tEUFkZsJpGop5HIxrkB0O
e7L_oETir_wUe1qs8AIZcKSwP9X6fpUuO1ONKDpDc-f-n5PjEAvts3BcxuM8Jrw80F6z6T
OJrcikrMt8DGaIXs2WHNP7C6051-JgwCVuZz_8S4LLtaCFnqq4xLU1Gy2qj5CBbALVoFcB
fjoVLN2fq4","username":"scott","userId":"a1b9b31d-e73c-4112-af7c-
b68530f38222"}

This returns with both a refresh and an access token. Let's use this access token to call the GET /api/
v1/addresses API again (please note that the Bearer token value in the Authorization
header is taken from the response of the previous POST /api/v1/ auth/token API call). The
command is shown in the following block:

```
$ curl -v 'http://localhost:8080/api/v1/addresses' -H 'Content-Type:
application/json' -H 'Accept: application/json' -H 'Authorization:
Bearer eyJ0eXAiOiJKV1QiLCJhbGci…
rest of the JWT string removed for brevity'
< HTTP/1.1 403
< Vary: Origin
< Vary: Access-Control-Request-Method
< Vary: Access-Control-Request-Headers
< WWW-Authenticate: Bearer error="insufficient_scope", error_
description="The request requires higher privileges than provided
by the access token.", error_uri="https://tools.ietf.org/html/
rfc6750#section-3.1"
```

This command execution returns 403. It means that the user was authenticated successfully. However,
the user doesn't contain the required role to access the endpoint.

Let's try again, with user scott2 this time, who has an ADMIN role:

```
$ curl -X POST http://localhost:8080/api/v1/auth/token -H  'Accept:
application/json' -H  'Content-Type: application/json' -d '{
    "username": "scott2",
    "password": "tiger"
  }'
```
{"refreshToken":"a6hidhaeb8scj3p6kei61g4a649dghcf5jit1v6rba2mn92o0dm0g

```
6gs6qfh7suiv68p2em0t0nnue8unm10bg079f39590jig0sccisecim5ep3ipuiu29ceao
c793h","accessToken":"eyJ0eXAiOiJKV1QiLCJhbGciOiJSUzI1NiJ9…","usernam
e":"scott2","userId":"a1b9b31d-e73c-4112-af7c-b68530f38223"}
## Some of the output removed for brevity
$ curl -v 'http://localhost:8080/api/v1/addresses' -H 'Content-Type:
application/json' -H 'Accept: application/json' -H 'Authorization:
Bearer eyJ0eXAiOiJKV1QiLCJhbGciOiJSUzI1NiJ9…'
[{"links":[{"rel":"self","href":"http://localhost:8080/
a731fda1-aaad-42ea-bdbc-a27eeebe2cc0"},{"rel":"self","hre
f":"http://localhost:8080/api/v1/addresses/a731fda1-aaad-
42ea-bdbc-a27eeebe2cc0"}],"id":"a731fda1-aaad-42ea-bdbc-
a27eeebe2cc0","number":"9I-999","residency":"Fraser Suites Le
Claridge","street":"Champs-Elysees","city":"Paris","state":"Île-de-
France","country":"France","pincode":"75008"}]
```

This time, the call is successful. Now, let's use the refresh token to get a new access token, as follows:

```
$ curl -X POST 'http://localhost:8080/api/v1/auth/token/refresh' -H
'Content-Type: application/json' -H 'Accept: application/json' -d '{
"refreshToken": "a6hidhaeb8scj3p6kei61g4a649dghcf5jit1v6rba2mn92o0dm0
g6gs6qfh7suiv68p2em0t0nnue8unm10bg079f39590jig0sccisecim5ep3ipuiu29ce
aoc793h"
}'
{"refreshToken":"a6hidhaeb8scj3p6kei61g4a649dghcf5jit1v6rba2mn92o0dm0g
6gs6qfh7suiv68p2em0t0nnue8unm10bg079f39590jig0sccisecim5ep3ipuiu29ceao
c793h","accessToken":"eyJ0eXAiOiJKV1QiLCJhbGciOiJSUzI1NiJ9… rest of
the token truncated for brevity","username":"scott2","userId":"a1b9b
31d-e73c-4112-af7c-b68530f38223"}
```

This time, it returns a new access token with the same refresh token given in the payload.

If you pass an invalid refresh token while calling the refresh token API, it will provide the following response:

```
{
    "errorCode":"PACKT-0010",
    "message":"Requested resource not found. Invalidtoken.",
    "status":404,
    "url":"http://localhost:8080/api/v1/auth/token/refresh",
    "reqMethod":"POST","timestamp":"2023-02-27T11:13:27.183172Z"
}
```

Similarly, you can call other API endpoints. Alternatively, you can import the following HAR file in an API client, such as Insomnia, and then test the remaining APIs: https://github.com/PacktPublishing/Modern-API-Development-with-Spring-6-and-Spring-Boot-3/blob/main/Chapter06/Chapter06-API-Collection.har.

Summary

In this chapter, you learned about JWTs, Spring Security, authentication using filters, and JWT token validation, using filters and authentication with the Spring OAuth 2.0 resource server. You also learned how you can add CORS and CSRF protection and why these are necessary.

You also learned about access protection based on roles and authorities. You have now the skills to implement JWTs, Spring Security, and the Spring Security OAuth 2.0 resource server to protect your web resources.

In the next chapter, you will develop a sample e-commerce app's UI using the Spring Security framework and APIs used in this chapter. This integration will allow you to understand the UI flows and how to consume REST APIs using JavaScript.

Questions

1. What is a security context and a principal?

2. Which is the preferred way to secure a JWT – signing or encrypting a token?

3. What are the best practices to use a JWT?

Answers

1. The security context stores the principal using `SecurityContextHolder` and is always available in the same thread of execution. The security context allows you to extract the principal during the flow execution and use it wherever you want. This is where a security annotation such as `@PreAuthorize` makes use of it for validation. The principal is the currently logged-in user. It can either be an instance of `UserDetails` or a string carrying a username. You can use the following code to extract it:

    ```
    Object principal = SecurityContextHolder
                        .getContext().getAuthentication()
                        .getPrincipal();
    if (principal instanceof UserDetails) {
      String username =
                  ((UserDetails)principal).getUsername();
    } else {
      String username = principal.toString();
    }
    ```

2. This is a subjective question. However, it is recommended to use the signing of tokens (JWS) if a JWT doesn't contain sensitive and private information, such as date of birth or credit card information. In such cases, you should make use of JWE to encrypt the information. If you want to use both together, then the preferred way is to use encryption for information carried by the token and then sign it with keys.

3. You can follow the following guidelines and add to them if you discover any new ones:

    - Make sure that JWT always has issuer and audience validations.

    - Make sure that a JWT validation does not allow a `none` algorithm (i.e., no algorithm mentioned in JWT or when the `alg` field in the JWT header contains a `none` value). Instead, make sure that you have verification in place that checks the specific algorithm (whatever you configured) and a key.

    - Keep an eye on the **National Vulnerability Database** (**NVD**).

    - Don't use a weak key (secret). Instead, use the asymmetric private/public keys with SHA 256, SHA 384, and SHA 512.

    - Use a minimum key size of 2,048 for normal cases and 3,072 for business cases.

    - A private key should be used for authentication, and the verification server should use a public key.

    - Make sure clients use the security guidelines to store the tokens, and web applications should use HTTPS for communication with servers.

    - Make sure the web application is tested thoroughly for **cross-site scripting** (**XSS**) attacks. It is always best to use a **content security policy** (**CSP**).

    - Keep a short expiration time, and use a refresh token to refresh an access token.

    - Keep an eye on OWASP security guidelines and new threats.

Further reading

- *Hands-On Spring Security 5.x* (video course): `https://www.packtpub.com/product/hands-on-spring-security- 5-x-video/9781789802931`

- Spring Security documentation: `https://docs.spring.io/spring-security/site/docs/current/reference/html5/`

- JWT: `https://tools.ietf.org/html/rfc7519`

- JWS: `https://www.rfc-editor.org/info/rfc7515`

- JWE: `https://www.rfc-editor.org/info/rfc7516`

- Spring Security in-built `SpEL` expressions: `https://docs.spring.io/spring-security/site/docs/3.0.x/refsseerence/el-access.html`

7
Designing a User Interface

In the previous chapter, you implemented authentication and authorization using Spring Security; that chapter also included all the example e-commerce app APIs. In this chapter, you will develop the frontend of an example e-commerce app using the React library. This UI app will then consume the APIs developed in the previous chapter, *Chapter 6, Securing REST Endpoints Using Authorization and Authentication*. This UI app will be a **single-page application** (**SPA**) that consists of interactive components such as *Login*, *Product Listing*, *Product Detail*, *Cart*, and *Order Listing*. This chapter will conclude the end-to-end development and communication between different layers of an online shopping app. By the end of this chapter, you will have learned about SPAs, UI component development using React, and consuming the REST APIs using the browser's built-in **Fetch API**.

This chapter will cover the following topics:

- Learning React fundamentals
- Exploring React components and other features
- Designing e-commerce app components
- Consuming APIs using Fetch
- Implementing authentication

Technical requirements

You need the following prerequisites for developing and executing the code:

- You should be familiar with JavaScript: `data types`, `variables`, `functions`, `loops`, and `array` methods such as `map()`, `Promises`, and `async`, and so on.
- Node.js 18.x with **Node Package Manager** (**npm**) 9.x (and optionally, Yarn, which you can install using `npm install yarn -g`).
- **Visual Studio Code** (**VS Code**): This is a free source code editor. You can use any other source code editor of your choice.
- React 18 libraries that will be included when you use `create-react-app`.

Let's get the ball rolling!

Please visit the following link to check the code for this chapter:

`https://github.com/PacktPublishing/Modern-API-Development-with-Spring-6-and-Spring-Boot-3/tree/main/Chapter07`

Learning React fundamentals

React is a declarative library used to build interactive and dynamic UIs, including isolated small components. It is also sometimes referred to as a framework because it is as capable as and comparable with other JavaScript frameworks such as AngularJS. However, React is a library and works with other supported libraries, including React Router, React Redux, and so on. You normally use it to develop SPAs, but it can also be used to develop full stack applications.

React is used to build the view layer of the application per the MVC architecture. You can build reusable UI components with their own state. You can use either plain JavaScript with HTML or **JavaScript Syntax Extension** (**JSX**) for templating. We'll be using JSX in this chapter, which employs a **virtual Document Object Model** (**VDOM**) for dynamic changes and interactions.

Let's create a new React app using the `create-react-app` utility next. This utility scaffolds and provides the basic app structure that you'll use to develop the example e-commerce app frontend.

Creating a React app

You can configure and build a React UI app from scratch. However, as mentioned, React provides a `create-react-app` utility that bootstraps and builds a basic running app template. You can take it further to build a full-fleshed UI application.

Its syntax is shown here:

```
npx create-react-app <app name>
```

npm package executor (**NPX**) is a tool that allows you to use **command-line interface** (**CLI**) tools and other executables available in the npm registry. It is by default available with npm 5.2.0, or you can install it using `npm i npx`. It executes the `create-react-app` React package directly.

Now, let's create an `ecomm-ui` application using the following command:

```
$ npx create-react-app ecomm-ui
Creating a new React app in /Users/dev/Modern-API-Development-with-
Spring-6-and-Spring-Boot-3/Chapter07/ecomm-ui.
Installing packages. This might take a couple of minutes.
Installing react, react-dom, and react-scripts with cra-template...
//... stripped output for brevity
added 1418 packages in 50s
```

```
Success! Created ecomm-ui at /Users/sourabhsharma/dev/pws/java/Modern-
API-Development-with-Spring-6-and-Spring-Boot-3/Chapter07/ecomm-ui
//… stripped output for brevity
Inside that directory
We suggest that you begin by typing:
  cd ecomm-ui
  npm start
```

Once it has been installed successfully, you can go to the app directory and start the installed application using `create-react-app` by running the following command:

```
$ cd ecomm-ui
$ code .
```

The `code .` command opens the `ecomm-ui` app project in VS Code. You can then use the following command in the terminal in VS Code to start the development server:

```
$ npm start
```

Once the server has started successfully, it will open a new tab on your default browser at `localhost:3000`, as shown in the following screenshot:

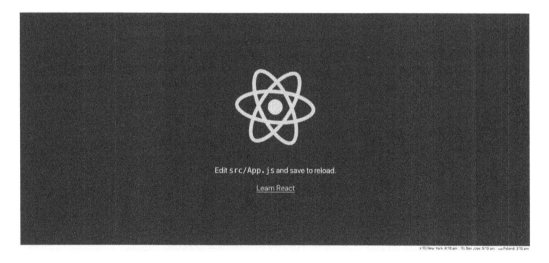

Figure 7.1 – Default UI app created by the create-react-app utility

Our bootstrapped React UI is up and running, but before you can build an e-commerce UI app on top of it, you need to understand the basic concepts and files generated by `create-react-app`.

Exploring the basic structures and files

A scaffolded React app contains the following directories and files inside the root project directory:

```
ecomm-ui
├── README.md
├── node_modules
├── package.json
├── package-lock.json
├── .gitignore
├── public
│   ├── favicon.ico
│   ├── index.html
│   ├── logo192.png
│   ├── logo512.png
│   ├── manifest.json
│   └── robots.txt
└── src
    ├── App.css
    ├── App.js
    ├── App.test.js
    ├── index.css
    ├── index.js
    ├── logo.svg
    ├── reportWebVitals.js
    └── setupTests.js
```

Let's understand the main parts, as follows:

- node_modules: You don't need to make any changes here. Node-based applications keep a local copy of all the dependent packages here.

- public: This directory contains all the static assets of an app, including index.html, images, favicon icon, and robots.txt.

- src: This directory contains all the dynamic code, including React code and **Cascading Style Sheets (CSS)** (including **Syntactically Awesome Style Sheets (Sass)**, **Leaner Style Sheets (Less)**, and so on). It also contains the test code.

- package.json: This **JavaScript Object Notation (JSON)** file contains all the metadata, commands (inside scripts), and dependent packages (inside dependencies and dev-dependencies).

You can remove the serviceWorker.js file (if generated), the logo.svg file, and test files from the src directory for now as we are not going to use them in this chapter.

Let's discuss the package.json file in the next subsection.

Understanding the package.json file

You can also view the `package.json` file that contains all the dependencies under the `dependencies` and `dev-dependencies` fields. It is similar in nature to the `build.gradle` file.

The main React libraries are `react` and `react-dom`, mentioned in the dependencies field; these are for React and the VDOM, respectively.

`package.json` also contains a script field that contains all the commands you can execute on this application. We have used the `yarn start` command to start the application in development mode. Similarly, you can execute other commands, as shown in the following code block, with `yarn` and npm:

```
"scripts": {
    "start": "react-scripts start",
    "build": "react-scripts build",
    "test": "react-scripts test",
    "eject": "react-scripts eject"
},
```

`react-scripts` is a CLI package installed by the `create-react-app` utility. It contains many dependencies, a few of the primary ones of which are listed here:

- **webpack** (`https://webpack.js.org`): This is a module bundler that bundles JavaScript, CSS, images, HTML, and so on. CSS and images may require extra loaders as dependencies. For example, it will pick all JavaScript files and bundle them into a single JavaScript file, though you can customize the way it bundles them by using a `webpack.config.js` configuration.

- **Jest** (`https://jestjs.io`): Jest is a JavaScript testing framework maintained by Facebook.

- **ESLint** (`https://eslint.org`): ESLint is a linter that allows you to maintain code quality. It is very similar to *Checkstyle* in the Java world.

- **Babel** (`https://babeljs.io`): Babel is a JavaScript transcompiler tool that converts JavaScript code to backward-compatible JavaScript code. The latest JavaScript draft version is **ECMAScript 2020**, also referred to as **ES10**. The latest JavaScript stable version is **ECMAScript 2018 (ES9)**. Babel allows you to generate optimized backward-compatible code from JavaScript code written using the latest versions.

You can find `react-scripts` under the `dependencies` field in `package.json`. Let's understand each of these commands, as follows:

- `start`: This command allows you to start the development server in a node environment. It also provides the hot reload feature, which means any changes to the React code would be reflected in the application, without a restart being required. Therefore, if there are any linting or code issues, they will show up accordingly in the console (terminal window) and web browser.

- `build`: This command packages the React application code for production deployment. It does the bundling of the JavaScript files in one, the CSS files into another, and then minifies and optimizes the code files. You can then deploy this bundle on any web server.

- `test`: This command executes a test using the test runner (the Jest tool). It executes all test files with extensions such as `.test.js` or `.spec.js`.

- `eject`: React comes with default build configurations such as webpack, Babel, and so on. The build configuration has the best practices already implemented to optimize the built app. This `eject` command allows you to eject the hidden configuration, after which you can override and customize the build configuration. However, you should do this with the utmost care because this is a one-way activity, and you can't reverse it.

Let's take a closer look at how React works in the next subsection.

Bootstrapping a React app

A web page is nothing but an HTML document. HTML documents contain the DOM, a tree-like structure of HTML elements. Any changes to the DOM are reflected in the rendering of the HTML document in the browser. Making changes in the actual DOM— and, specifically to the nth level—is a heavy operation in terms of traversal and rendering the DOM, because each change is done on the whole DOM, and this is a time- and memory-consuming operation.

React uses a VDOM to make these operations lightweight. A VDOM is an in-memory copy of the actual DOM. React maintains the VDOM using the `react-dom` package. Therefore, when you initialize the React app, you first pass the root HTML ID element to the `ReactDOM` object's `render` function. React writes the VDOM under this root element after its first render.

After the first render, only the necessary changes are written to the actual DOM based on the changes in React components and their state. The React components' `render` function returns the markup in JSX syntax. Then, React transforms it to HTML markup and compares the generated VDOM with the actual HTML DOM, and only makes the necessary changes to the actual DOM. This process then continues till the components get changed. Let's explore how the first render takes place.

The `pubic/index.html` file contains the main HTML file. This is an application skeleton that contains the web app's `title`, `meta` elements, and a `body` element. It also contains a `div` element (under `body`) with an `ID` as `root`. You pass this root element to the `render` function of `ReactDOM` in `index.js`, in the `src` directory. This is the entry point of the React app. Let's have a look at its code, as follows:

```
import React from 'react';
import ReactDOM from 'react-dom/client';
import './index.css';
import App from './App';
const root = ReactDOM.createRoot(document.getElementById('root'));
```

```
root.render(
  <React.StrictMode>
    <App />
  </React.StrictMode>
);
```

https://github.com/PacktPublishing/Modern-API-Development-with-Spring-6-and-Spring-Boot-3/tree/dev/Chapter07/ecomm-ui/src/index.js

Here, React uses the ReactDOM object from the react-dom package to render the page. First it creates the root object by calling the createRoot method and passing the root element. The document.getElementById('root') method fetches <div id= "root"> from inside the <body> element of index.html.

The render() function of the root object contains an argument: element of type ReactNode. You pass an <App /> tag component wrapped with React's strict mode component as an element argument.

App components can be a single component or a parent component with single or multilayer child components. A single component won't contain any other React component; it simply contains the JSX, and that's it. However, parent components may contain one or more child components, and those child components may contain one or more child components, and so on. For example, an App component may have header, footer, and content components. A content component may have a cart component, and then the cart component may have items inside it.

A <React.StrictMode> component is a special React component that gets rendered twice in development mode to check for best practices, deprecated methods, and potential risks in your React components, and prints warnings and suggestions in the console log. It has no impact on the production build because it only works in development mode.

The render() function transforms the JSX of the App component to HTML and adds it inside the <div id="root"> tag, then it compares the VDOM with the real DOM and makes the necessary changes in the real DOM. This is how React components get rendered on the browser.

You now understand that React components are key here. Let's deep dive into them in the next section.

Exploring React components and other features

Each page in an app is built up using React components — for example, the **Product Listing** page of Amazon can broadly be divided into *Header, Footer, Content, Product List, Filter and Sorting options*, and *Product Card* components. You can create components in React in two ways: by using JavaScript classes or functions.

Let's create an example header component in React with both a function and a class.

You can either write a plain old JavaScript function or an **ECMAScript 6 (ES6)** arrow function. Components created using either arrow functions or plain JavaScript functions are called React **functional components**. We'll mostly use functional components in our code. In the following code snippet, check out the `Header` component using a JavaScript arrow function:

```
export default const Header = (props) => {
  return (
    <div>
      <h1>{props.title}</h1>
    <div>
  )
}
```

Let's create the same `Header` component using a JavaScript class, as follows:

```
export default class Header extends React.Component {
  render() {
    return (
      <div>
        <h1>{this.props.title}</h1>
      <div>
      )
  }
}
```

Let's understand both components point by point, as follows:

- Both return JSX that looks like HTML, which gets rendered after transformation (from JSX to HTML).

- Both export the function and class respectively so that they can be imported by other components.

- Both have props—one as an argument and one bound with this scope, which is part of `React.Component`. Props represent the attributes and their values—for example, here, a `title` attribute is used. When it gets rendered, it is replaced by the `title` attribute's value.

- The class needs a `render()` function, whereas the function simply needs a `return` statement.

Let's see how the `Header` component could be used. You can use this `Header` component as you would use any other HTML tag in your JSX code, as shown next:

```
<Header title="Sample Ecommerce App" />
```

Here, `title` is the property of `Header` component. It describes how properties (props) of components are passed. When this `Header` component gets rendered, it will show the `title` value wrapped in an `<H1>` element.

Let's explore the JSX next. This is how you use the `props`: you add an attribute (such as `title`) to its value while using the component. Inside the component, you can access these attributes (properties) by using `props` directly or using the { `title` } de-structuring form in functional components and by using `this.props` in the class components.

Exploring JSX

React components would return the JSX. You can write HTML code to design the components because JSX is very similar to HTML, except for the HTML attributes. Therefore, you need to make sure to update attributes such as `class` to `className`, `for` to `htmlFor`, `fill-rule` to `fillRule`, and so on. The advantage of using the `React.StrictMode` component is that you get a warning and a suggestion to use the correct JSX attribute names if you use HTML attributes or have a typo.

You can also put any JavaScript expressions inside JSX or an element's attributes to make the component dynamic by using the expression wrapped in curly braces ({ }).

Let's have a look at some example code to understand both JSX and expressions. The following JSX code snippet has been taken from the `CartItem` component. Check out the highlighted code for expressions; the rest of the code is JSX, which is very similar to HTML:

```
<div className="w-32">
 <img className="h-24" src={item?.imageUrl} alt="" />
</div>
<div className="flex flex-col justify-between
   ml-4 flexgrow">
 <Link to={"/products/" + item.id} className="font-bold
     text-sm text-indigo-500 hover:text-indigo-700">
   {item?.name}
 </Link>
 <span className="text-xs">Author: {author}</span>
 <button className="font-semibold hover:text-red-500
   text-indigo-500 text-xs text-left"
     onClick={() => removeItem(item.id)}>
   Remove
 </button>
</div>
```

The preceding code fragment represents a cart item that shows the product image, product name, author, and **Remove** button. The product name is also a link that links to the product detail page. You can design these components using JSX (read HTML) as shown. Please also note that the `class` attribute name is changed to `className` because it is JSX. `Link` is a part of the `react-router-dom` library.

You are done with the cart item's design part. Now, you need a mechanism to populate the values and add the event handling in it. This is where a JSX expression helps you.

You use `item` — an object that represents the cart item, and `author` — a variable that contains the author's name. Both are part of the React component's state. You will learn more about the state in the next subsection, but for the time being you can think of them as variables defined in the `CartItem` component. Once you write the JSX (read HTML), dynamic values (from variables) and interaction (for events) can be defined using the expressions wrapped inside curly braces (`{ }`).

Let's understand each of the expressions as follows:

- `src={item?.imageUrl}`: You get the item (product) image URL as part of the API response. You simply assign it to the `src` attribute of the `img` tag. Note that the dot operator (`.`) allows you to access the property of an object. The code may throw an error if you try to read the property of any `null` or `undefined` object. You can avoid that by using the `?.` operator. Then, the property (in this case, `imageUrl`) will only be read if an object (in this case, `item`) is not `null` or `undefined`.

- `to={"/products/" + item.id}`: Here, links to an attribute are formed by using the object item's `id` property.

- `{item?.name}`: Here, the name of the product is displayed using the name property of the `item` object.

- `Author: {author}`: The author value is displayed using the `author` variable.

- `onClick={() => removeItem(item.id)}`: This is the way you associate a user-defined function with an event. Here, `removeItem()` will be called by passing the item object's `id` property on the click of a button. If you are not passing any argument or using multiple statements, then you can directly pass the function name instead of using the arrow function—for example, `onClick={removeItem}`.

Next, we will deep dive into the state of React components. Let's see how this works.

Understanding React hooks

Components are dynamic and contain a state. The state represents the data and metadata held by the component at a given point in time. There are two levels of state: a global (app-level) state and a local (component-level) state.

Earlier (prior to React version 16.8), the state was only supported in components defined using classes. Now, React supports the state in both functional and class components. React supports the state in functional components using **hooks** such as `useState()`, `useContext()`, and so on.

> **What are hooks?**
>
> Hooks are special React built-in functions or user-defined functions that can be stateful and are used to manage the side effects of React functional components. Popular and frequently used hooks are `useState()` and `useEffect()`.

React introduced hooks (a set of functions) in the 16.8 version, which introduced many features to functional components that were earlier not supported, including state and events such as `componentDidMount` (a lifecycle method in the class that indicates a component was mounted), and you can now perform certain operations such as loading data using APIs, among other things.

Let's discuss React hooks next.

Each hook in React represents a special feature that you can use in functional components. Let's discuss the most popular and common hooks one by one, as follows:

- useState: `useState()` allows you to define and maintain the state. Let's examine how you use this hook. First, you import the `useState()` hook at the top of the component code file, as follows:

    ```
    import {useState} from "react";
    ```

    Next, inside your component's arrow function code, define the state before the `return` statement, as shown next:

    ```
    const [total, setTotal] = useState(0);
    ```

    You need to define both state and state setter functions in an array while declaring the state. Here, the `total` state is defined with its setter function. You can use any type of state, such as an `object`, `array`, `string`, or `number`. The total state is of type `number`; therefore, it is initialized with `0`. `setTotal()` is a setter function. The setter function allows you to update the state (`total` here)—for example, you could update the total state by calling `setTotal(100)`, in which case the `total` state would be changed from `0` to `100`.

    React tracks the state's setter function and whenever it is called. React updates the state of the component and re-renders the component. The naming convention of the setter function is to prefix the state name with `set` and make the state's first letter a capital letter. Therefore, we have used the `setTotal()` name for the `total` state. You'll use `useState()` for local state management in most components.

- useEffect: You use a `useEffect()` hook when you want to do something after rendering a component. This gets called after each render. You can also use it when you want to load the initial data from an API or add an event listener. However, if an API call should be made once, then you can pass the empty array (`[]`) dependency while calling it. You'll find multiple instances of `useEffect` in the `ecomm-ui` code when an empty array is passed for a single call.

React recommends using multiple `useEffect` functions inside components for separating the concern. Also, make sure it returns an arrow function for cleanup. For example, when you add the event listener for any component, it should return an arrow function that removes the event listener.

- `useContext`: You can pass props from one component to another. Sometimes, you must use props drilling to the nth level. React also provides an alternative way to define these props so that they can be used in any component in a tree without using prop drilling. You would use it for props that are common across components, such as `theme` or `isUserLoggedIn`.

- React provides a `createContext()` function to create a context. It returns a provider and consumer to provide access to its values and changes respectively (see the next code block). However, `useContext` can easily make use of the context by removing the usage of the consumer, which is returned by `createContext()`. The following code snippet depicts `useContext` usage:

```
import {createContext} from "react";
import ReactDOM from "react-dom";
const LoggedInContext = createContext();
const App = () => {
  return (
    <LoggedInContext.Provider isUserLoggedIn=true>
      <ProductList/>
    <LoggedInContext.Provider/>
  );
}
const ProductList = () => {
  return (
    <LoggedInContext.Consumer> { isUserLoggedIn =>
    <div>Is user logged-in: {isUserLoggedIn}</div>
    } <LoggedInContext.Consumer>
  );
}
ReactDOM.render(<App/>,document.getElementById("root"));
```

You can simplify the `ProductList` component's `return` block in the previous code snippet (check the highlighted code) with `useContext`, as follows:

```
import {createContext, useContext} from "react";
import ReactDOM from "react-dom";
const LoggedInContext = createContext();
const App = () => {
  return (
    <LoggedInContext.Provider isUserLoggedIn=true>
      <ProductList/>
```

```
        <LoggedInContext.Provider/>
    );
}
const ProductList = () => {
  const isUserLoggedIn = useContext(LoggedInContext);
  return (
    <div>Is user logged-in: {isUserLoggedIn}</div>
    );
}
ReactDOM.render(<App/>, document.getElementById("root"));
```

This is how you can use `createContext` and `useContext` hooks.

- useReducer: This is an advanced version of the `useState` hook that not only allows you to use a component's state but also provides better controls to manage its state by taking the `reducer` function as a first argument. It takes the initial state as a second argument. Check out its syntax, as seen in the following code block:

```
const [state, dispatch] = useReducer(reducer, initialState);
```

The `reducer` function is a special function that takes state and action as arguments and returns a new state. We'll explore this more when we build the `CartContext` component later in this chapter.

Now that you have learned the basic concepts of React, let's add some styling to the `ecomm-ui` application using Tailwind CSS.

Styling components using Tailwind

Tailwind CSS is a utility CSS framework that helps you to design a responsive UI. It supports theming, animation, pre-defined padding and margins, flex, grids, and so on. You can install Tailwind and its peer packages using npm, as shown in the following command (executing it from the project root directory):

```
$ npm install -D tailwindcss
```

Here, version 3.2.7 of Tailwind CSS was installed.

Next, let's configure Tailwind CSS. The `npx tailwindcss init` command generates the Tailwind CSS configuration file with the default empty values:

```
$ npx tailwindcss init
Created Tailwind CSS config file: tailwind.config.js
```

The previous command generates the following file:

```
/** @type {import('tailwindcss').Config} */
module.exports = {
  content: [],
  theme: {
    extend: {},
  },
  plugins: [],
}
```

https://github.com/PacktPublishing/Modern-API-Development-with-Spring-6-and-Spring-Boot-3/blob/main/Chapter07/ecomm-ui/tailwind.config.js

Now, we can modify the Tailwind CSS configuration to purge unused styles in production.

Configuration to remove unused styles in production

You would like to keep the style sheet size down in a production environment because this improves the performance of the application. You can purge unnecessary styles by adding the following filters to the content block of the tailwind.config.js file. Then, Tailwind can tree-shake unused styles while building the production build. Generated CSS files only contain the used styles in files matching the given filters. The code is illustrated in the following snippet:

```
module.exports = {
  content: ["./src/**/*.{js,jsx,ts,tsx}",
            "./public/index.html"],
  theme: {
    extend: {},
  },
  plugins: [],
}
```

Next, we will add Tailwind to React.

Including Tailwind in React

Open the src/index.css file that create-react-app generates for you by default and import Tailwind's base, components, and utilities styles, replacing the original file contents as follows:

```
@tailwind base;
@tailwind components;
@tailwind utilities;
```

```
https://github.com/PacktPublishing/Modern-API-Development-with-
Spring-6-and-Spring-Boot-3/tree/dev/Chapter07/ecomm-ui/src/index.css
```

These statements import the styles generated by the build based on the Tailwind configuration when you execute the build.

Finally, make sure that the CSS file is imported in the `src/index.js` file, as shown in following code:

```
import React from 'react';
import ReactDOM from 'react-dom/client';
import './index.css';
import App from './App';
import reportWebVitals from './reportWebVitals';
const root = ReactDOM.createRoot(document.getElementById('root'));
// … rest of the code removed for brevity
```

Done! Next, execute `yarn start` to use Tailwind CSS in the `ecomm-ui` app.

However, before that, you are going to add the basic structure to your web app – the header, container (content), and footer.

Adding basic structural React components

Before adding the `Header`, `Footer`, and `Container` components, you need to remove the following files created by `create-react-app`:

- `App.css`
- `logo.svg`

Don't forget to remove these file references from `/src/App.js` too.

Then, create a `components` directory under `/src`. You will create all new components under this directory, as shown in *Figure 7.2*. Let's create three new components, as follows:

- `Header`: This will be displayed at the top and contains header items such as the app name and **Login/Logout** button
- `Container`: This will contain the main content, such as the product list
- `Footer`: This will be displayed at the bottom and contains footer items such as the copyright information

The basic structure can be seen in the following screenshot:

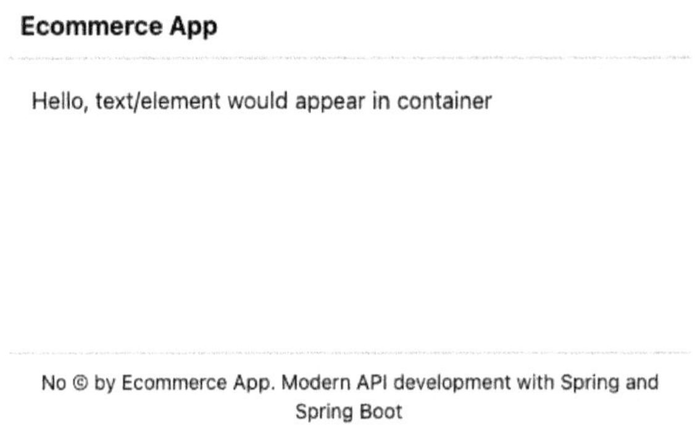

Ecommerce App

Hello, text/element would appear in container

No © by Ecommerce App. Modern API development with Spring and
Spring Boot

Figure 7.2 – Basic structure of the app containing the Header, Footer, and Container components

Let's add these basic components. First, let's create a `Header` component, as shown in the following code snippet:

```
const Header = () => {
  return (
    <div>
      <header className="p-2 border-b-2
        border-gray-300 bggray-200">
        <h1 className="text-lg font-bold">
          Ecommerce App</h1>
      </header>
    </div>
  );
};
export default Header;
```

Similarly, create a `Footer` component, as shown in the following code snippet:

```
const Footer = () => {
  return (
    <div>
      <footer
        className="text-center p-2 border-t-2 bggray-
          200 border-gray-300 text-sm">
        No &copy; by Ecommerce App.{" "}
        <a href=https://github.com/PacktPublishing/Modern-
          API-Development-with-Spring-and-Spring-Boot>
          Modern API development with Spring and
            Spring Boot
```

```
        </a>
      </footer>
    </div>
  );
};
export default Footer;
```

Next, create a `Container` component, as shown in the following code snippet:

```
const Container = () => {
  return (
    <div className="flex-grow flex-shrink-0 p-4">
      <p>Hello, text/element would appear in container</p>
    </div>
  );
};
export default Container;
```

And finally, modify the `/src/App.js` file as shown in the following code snippet:

```
import Header from "./components/Header";
import Footer from "./components/Footer";
import Container from "./components/Container";
function App() {
  return (
    <div className="flex flex-col min-h-screen h-full">
      <Header />
      <Container />
      <Footer />
    </div>
  );
}
export default App;
```

This is how you can create and use new components. These components are in their simplest form and are kept as such to be understood more easily. However, you can find refined and improved versions of these components on GitHub, as follows:

- *Header component source*: https://github.com/PacktPublishing/Modern-API-Development-with-Spring-6-and-Spring-Boot-3/tree/dev/Chapter07/ecomm-ui/src/components/Header.js

- *Footer component source*: https://github.com/PacktPublishing/Modern-API-Development-with-Spring-6-and-Spring-Boot-3/tree/dev/Chapter07/ecomm-ui/src/components/Footer.js

- The `Container` component (which contains the actual content in the center) could be replaced with the `switch` component from `react-router-dom`, which would display the components based on a given `route`, such as `cart`, `orders`, and `login`.

Now, you can start writing the actual `ecomm-ui` components next.

Designing the e-commerce app components

Design is not only a key part of UX and UI work, but is also important for frontend developers. Based on the design, you can create reusable and maintainable components. Our example e-commerce app is a simple application that does not need much attention. You will create the following components in this application:

- **Product listing component**: A component that displays all the products and acts as a home page. Each product in the list will be displayed as a card with the product name, price, and two buttons—**Buy now** and **Add to bag**. The following screenshot displays the **Product listing** page, which shows the product information along with an image of each product:

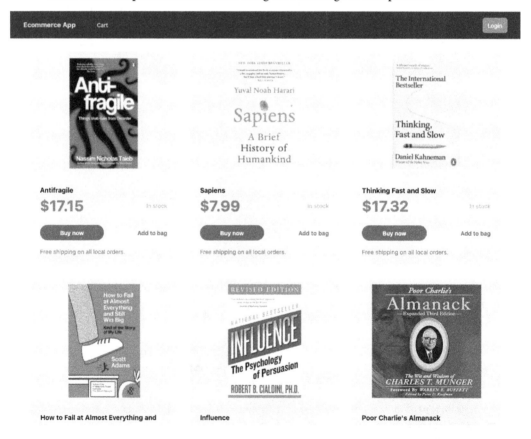

Figure 7.3 – Product listing page (home page)

- **Product detail component**: This component displays the details of a product when clicked. It displays the product image, product name, product description, tags, and the **Buy now** and **Add to bag** buttons, as shown next:

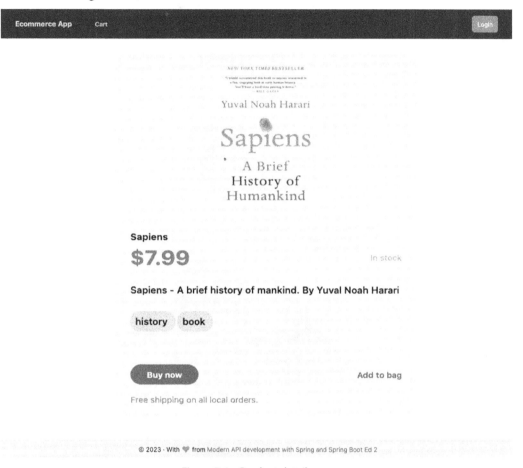

Figure 7.4 – Product detail page

- **Login component**: Login components allow a user to log in to an app by using their username and password, as illustrated in the following screenshot. This component displays an error message when a login attempt fails. Click on **Cancel** to go back to the **Product listing** page. The **Product listing** page shows a list of products a customer can buy.

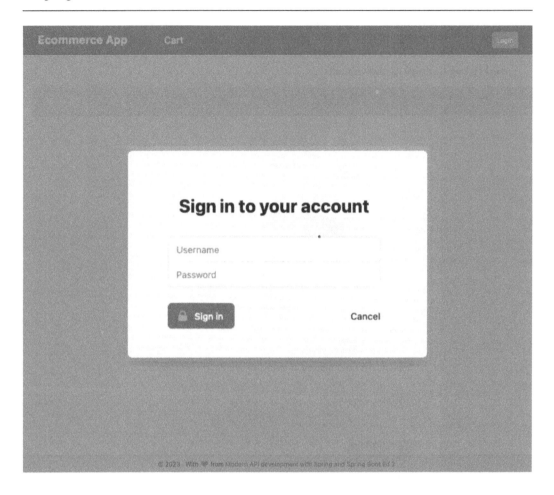

Figure 7.5 – Login page

- **Cart component**: The Cart component lists all the items that have been added to the cart. Each item displays the given product image, name, description, price, quantity, and total. It also provides a button to decrease or increase the quantity, and a button to remove the item from the cart.

  Product name is a link that takes the user back to the **Product detail** page. The **Continue shopping** button takes the user to the **Product listing** page. The **CHECKOUT** button performs the checkout process. On successful checkout, an order is generated and the user is redirected to the **Orders** page.

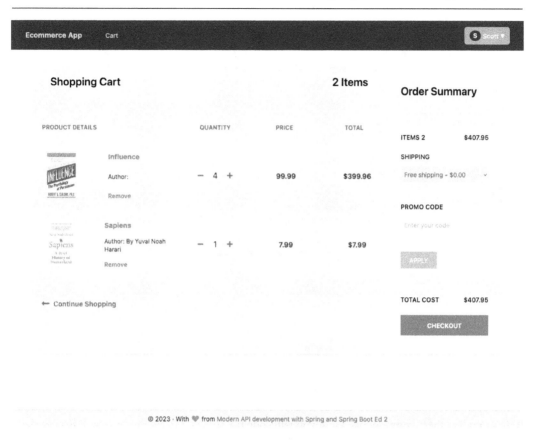

Figure 7.6 – Cart page

- **Orders component**: The **Orders** page shows all orders placed by the user in a tabular form. The **Orders** table displays the order date, ordered items, order status, and order amount for each order.

The order date will be displayed in the user's local time, but on the server it will be in **Universal Coordinated Time** (**UTC**) format. Order items will be displayed in an order list, with their quantity and unit price in brackets, as illustrated in the following screenshot:

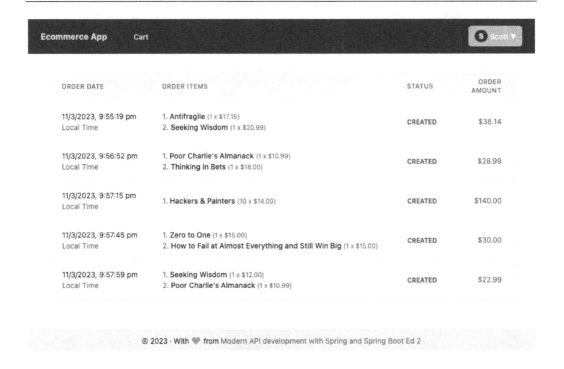

Figure 7.7 – Order page

Let's start coding these components. First, you will code the **Product listing** page, which fetches the products from the backend server using the REST API.

Consuming APIs using Fetch

Let's create the first component — that is, the **Product Listing** page. Create a new file in the `src/components` directory with the name `ProductList.js`. This is the parent component of the **Product Listing** page.

This component fetches the products from the backend server and passes them to the child component, `Products` (it creates a new `Products.js` file under the `src/components` directory).

`Products` contains the logic responsible for looping through the fetched product list. Each iteration renders the card UI of each product. This product card component is represented using `ProductCard`, another component. Therefore, let's create a `ProductCard.js` file, under `src/components`.

You can write the product card code inside `products` (`product list` component), but to single out the responsibilities, it's better to create a new component.

The `ProductCard` component has a **Buy now** button and an **Add to bag** link. These links should only work if the user is logged in, else it should redirect the user to the login page.

You now have an idea about the **Product Listing** page component tree structure. Now, our first task is to have an API client that fetches products we can render in these components.

Writing the product API client

You are going to use the `Fetch` browser built-in library as a REST API client. You can also use a third-party library such as `axios`. `Fetch` can do the job for this example app and reduce the number of third-party dependencies.

You can create a configuration file for API clients settings. Let's name it `Config.js` and place it in the `src/api` directory.

`Config` is a JavaScript class that contains constants such as URLs and common methods such as `DefaultHeaders()` and `tokenExpired()`. Check out its code in the following snippet:

```
class Config {
  SCHEME = process.env.SCHEME ? process.env.
    SCHEME : "http";
  HOST = process.env.HOST ? process.env.HOST : "localhost";
  PORT = process.env.PORT ? process.env.PORT : "8080";
  CART_URL = `${this.SCHEME}://${this.HOST}:$
    {this.PORT}/api/v1/carts`;
  // truncated code for brevity
  defaultHeaders() {
    return {
      "Content-Type": "application/json",
      Accept: "application/json",
    };
  }
  …
}
```

https://github.com/PacktPublishing/Modern-API-Development-with-Spring-6-and-Spring-Boot-3/tree/dev/Chapter07/ecomm-ui/src/api/Config.js

Here, you can see that we have created constants that are formed using environment variables. The `defaultHeaders()` function returns the common headers used in all API calls, and `headersWithAuthorization()` returns common headers with the `Authorization` header. `headersWithAuthorization()` uses object destruction to retrieve the default headers. The `Authorization` header is fetched from the local storage, which is set when a user is logged in successfully and is removed once the user logs out.

It also has a `tokenExpired()` function that simply checks the expiration time for a token stored in local storage. This expiration time is extracted from the access token (**JSON Web Token**, or **JWT**). It returns true if the expiration time is past the current time. Check out this function's code in the following code snippet:

```
// src/api/Config.js
tokenExpired() {
  const expDate = Number(localStorage.getItem
    (this.EXPIRATION));
  if (expDate > Date.now()) {
      return false;
  }
  return true;
}
```

The `Config` class also contains a `storeAccessToken()` function that simply stores the access token and expiration time in local storage. It uses a `getExpiration()` function to extract the expiration time from the access token. Put simply, this function first extracts the payload from the token string and then decodes the payload and converts it to JSON. Finally, it returns the expiration time if a payload is a valid object, else it returns 0. You can find these functions in the following code block:

Relevant code of src/api/Config.js

```
storeAccessToken(token) {
  localStorage.setItem(this.ACCESS_TOKEN, `Bearer ${token}`);
  localStorage.setItem(this.EXPIRATION,
                              this.getExpiration(token));
}
getExpiration(token) {
  let encodedPayload = token ? token.split(".")[1] : null;
    if (encodedPayload) {
        encodedPayload = encodedPayload.replace
          (/-/g, "+").replace(/_/g, "/");
        const payload = JSON.parse(window.atob
          (encodedPayload));
        return payload?.exp ? payload?.exp * 1000 : 0;
    }
  return 0;
}
```

Next, let's make use of this `Config` class in the `src/api/ProductClient.js` file, as shown in the following code block. This file will act as the API client for product-related APIs:

```
import Config from "./Config";
class ProductClient {
```

```
constructor() { this.config = new Config(); }
async fetchList() {
  return fetch(this.config.PRODUCT_URL, {
    method: "GET",
    mode: "cors",
    headers: { ...this.config.defaultHeaders(),},
  })
  .then((res) => Promise.all([res, res.json()]))
  .then(([res, json]) => {
    if (!res.ok) { return { success: false, error: json };}
    return { success: true, data: json };
  }).catch((e) => {
    return this.handleError(e);
  });
}
...
...
```

https://github.com/PacktPublishing/Modern-API-Development-with-Spring-6-and-Spring-Boot-3/tree/dev/Chapter07/ecomm-ui/src/api/ProductClient.js

ProductClient is a class, and a config instance is instantiated in its constructor. This class contains two asynchronous functions for fetching the products: fetchList() and fetch(). The former fetches all products, and the latter is for fetching a single product based on its ID. fetchList() makes use of the fetch browser function to fetch the product list. You pass the URL as the first argument input and request an initialization object that contains the HTTP method, mode, and headers as the second argument. The fetch browser call returns a promise that you use to handle the request. First, you resolve both the promises – the response and response JSON and then check response. ok. response.ok returns true for a status in the 200 to 299 range. Upon a successful response, the fetchList() method returns an object with data and success fields as true. Upon an unsuccessful response, it returns success as false and shows an error response in the data field.

Similarly, you can write a fetch function to retrieve the product by ID. Everything will be the same except the URL, as you can see in the following code block:

Remaining code of src/api/ProductClient.js

```
async fetch(prodId) {
  return fetch(this.config.PRODUCT_URL + "/" + prodId, {
    method: "GET",
    mode: "cors",
    headers: { ...this.config.defaultHeaders(),},
  })
```

```
      .then((res) => Promise.all([res,  res.json()]))
      .then(([res, json]) => {
        if (!res.ok) { return { success: false, error: json }; }
        return { success: true, data: json };
      }).catch((e) => {
        this.handleError(e);
      });
  }
  handleError(error) {
    const err = new Map([
      [TypeError, "Problem fetching the response."],
      [SyntaxError, "Problem parsing the response."],
      [Error, error.message],
    ]).get(error.constructor);
    return err;
  }
}
export default ProductClient;
```

Here, the handleError() function checks the type of the error (using error. constructor) and, based on that, returns the appropriate error message.

Please note that other API clients such as CartClient, CustomerClient, and OrderClient are developed in a similar fashion. The code is available at the following locations:

- CartClient: https://github.com/PacktPublishing/Modern-API-Development-with-Spring-6-and-Spring-Boot-3/tree/dev/Chapter07/ecomm-ui/src/api/CartClient.js

- CustomerClient: https://github.com/PacktPublishing/Modern-API-Development-with-Spring-and-Spring-Boot/blob/main/ Chapter07/ecomm-ui/src/api/CustomerClient.js

- OrderClient: https://github.com/PacktPublishing/Modern-API-Development-with-Spring-and-Spring-Boot/blob/main/ Chapter07/ecomm-ui/src/api/OrderClient.js

Now, we can use ProductClient to fetch the products from the backend using REST APIs. Let's code the ProductList component and its child components.

Coding the Product Listing page

`ProductList` is a straightforward component that loads the products after their first render using `ProductClient`. You know that for this purpose, `useEffect` hooks should be used. Let's code it as follows:

```
// other imports
import Products from "./Products";
const ProductList = ({ auth }) => {
  const [productList, setProductList] = useState();
  const [noRecMsg, setNoRecMsg] = useState("Loading...");
  const { dispatch } = useCartContext();
  useEffect(() => {
    async function fetchProducts() {
      const res = await new ProductClient().fetchList();
      if (res && res.success) { setProductList(res.data); }
      else { setNoRecMsg(res); }
    }
    async function fetchCart(auth) {
      const res = await new CartClient(auth).fetch();
      if (res && res.success) {
        dispatch(updateCart(res.data.items));
        if (res.data?.items && res.data.items?.length < 1)
        {
          setNoRecMsg("Cart is empty.");
        }
      } else {
        setNoRecMsg(res && typeof res === "string"?
          res : res?.error?.message);
      }
    }
    if (auth?.token) fetchCart(auth);
    fetchProducts();
  }, []);
  // rest of the code …
```

https://github.com/PacktPublishing/Modern-API-Development-with-Spring-6-and-Spring-Boot-3/tree/dev/Chapter07/ecomm-ui/src/components/ProductList.js

The `ProductList` component uses `auth` as a prop. It contains authentication information such as a token. The `ProductList` component is used as the main `App` component, and `auth` is passed to the `ProductList` component by it.

Please note that you have passed an empty array ([]) as a dependency to make sure that the API is called only once. You are using a useState hook to store the product list (productList) and message states (noRecMsg—no record) via setter methods.

Why does the cart need to be fetched in ProductList?

The ProductList component and its child components are available for non-authenticated users. Once the user clicks on the **Buy now** button or the **Add to bag** link, it will ask the user to log in. Once logged in, the user can add items to the cart. It is quite possible that the user might already have some items in the cart. Therefore, when you add an item to the cart, the quantity of existing products should be increased, and if a clicked item does not exist in the cart, then it should be added to the cart.

Cart is a separate component altogether; this means you can't access the cart unless you do cart prop drilling from the App component to both the Cart and ProductCard components or have a useContext hook for the cart. We have built a custom store to maintain the cart state, very similar to **Redux** (a library used to maintain the state in React app). We'll learn more about this library later in this chapter. Dispatch is an action that updates the cart items received from the backend server to the cart context.

Next, create a JSX template and pass the fetched productList component to the child component, Products, for further rendering, as illustrated in the following code snippet:

Remaining code of src/components/ProductList.js

```
  return (
    <div className="max-w-7xl mx-auto px-4 sm:px-6 lg:px-8">
      {productList ? (
        <div className="flex flex-wrap -mx-1 lg:-mx-4">
          <Products auth={auth} productList={productList ?
            productList : []} />
        </div>
      ) : (
        <div className="text-lg
          font-semibold">{noRecMsg}</div>
      )}
    </div>
  );
};
export default ProductList;
```

Here, it also passes the auth object as a prop to Products.

Let's have a look at the `Products` code, as follows:

```
import ProductCard from "./ProductCard";
const Products = ({ auth, productList }) => {
 return ( <> {productList.map((item) => (
  <ProductCard key={item.id} product={item} auth={auth} /> ))}
  </>
 );
};
export default Products;
```

https://github.com/PacktPublishing/Modern-API-Development-with-Spring-6-and-Spring-Boot-3/tree/dev/Chapter07/ecomm-ui/src/components/Products.js

Put simply, the code does the job of iterating the product list passed by the `ProductList` component and passes each item with the `product` props to the `ProductCard` component along with the `auth` object.

You can observe the usage of two React concepts here, as follows:

- The code uses a `<></>` fragment, which is an empty tag. Ideally, this is used when a component returns more than one top-level tag because React needs only one top-level tag in each component. Then, you can wrap those tags with a fragment. You can also use `<React.Fragment>` in place of an empty tag after importing `React` from the `react` package.

- Another usage is for the `key` props in the `ProductCard` component. When you generate components based on a collection, React requires the `key` props to uniquely identify them. This will allow React to identify which item is changed, removed, or added. We have used the item ID here. If you don't have an ID in your collection, you can also use the `index`, as shown in the following code example:

```
{productList.map((item, index) => (
  <ProductCard key={index} product={item} auth={auth} />
))}
```

Now, let's have a look at the last child component of the `ProductList` component: `ProductCard`. The `ProductCard` component simply passes `Product` values to JSX template expressions for rendering.

We have added some extra code to add the functionality associated with the **Add to bag** and **Buy now** click events.

Configuring routing

You are creating an SPA. Here, routing is not available by default. Routing is the mechanism that provides the routing to an SPA, which means that with each new page, the browser URL will reflect the change and allow you to bookmark the page. SPA routing also maintains the browser history. You are going to use the react-router-dom package for routing management. You need to add the react-router-dom package to use routing, as shown in the following code snippet. Make sure to execute it from the project root directory:

```
$ npm install react-router-dom
```

The routing will be configured in the App component because it is the root component of the ecomm-ui application. In the ProductList component, you are going to use the Link component and the useNavigate() hook from the react-router-dom package. Let's examine them, as follows:

- Link: This is like a <a> HTML anchor tag. Instead of a href attribute, it uses a to attribute to link the URL. The route library maintains the links; therefore, it knows which component to render when a link is passed in with a to attribute when Link is clicked.

- useNavigate(): This allows navigation inside the component and accesses the state of the router. You would use the navigate("/path") function if navigate is declared as const navigate = useNavigate() to navigate from one component to another, as shown in the checkLogin() function of the ProductCard component.

Let's continue with the development of the next product-based component: ProductCard.

Developing the ProductCard component

First, let's import the required packages. Then, declare the state (using useCartContext and useState) and variables. Please note that the following code snippet has auth and product as props:

```
import { useState } from "react";
import { Link, useNavigate } from "react-router-dom";
import CartClient from "../api/CartClient";
import { updateCart, useCartContext } from "../hooks/CartContext";
const ProductCard = ({ auth, product }) => {
  const navigate = new useNavigate();
  const cartClient = new CartClient(auth);
  const { cartItems, dispatch } = useCartContext();
  const [msg, setMsg] = new useState("");
  // continue …
```

```
https://github.com/PacktPublishing/Modern-API-Development-with-
Spring-6-and-Spring-Boot-3/tree/dev/Chapter07/ecomm-ui/src/components/
ProductCard.js
```

To begin, you write the add() asynchronous function that adds the product to the cart. It first checks whether the user is logged in or not. If not, it redirects the user to the login page. checkLogin() uses the useNavigate hook's push method to redirect. The token property of auth is used to identify whether the user is logged in or not.

Once it has identified that the user is logged in, it calls the callAddItemApi function to add a product to the cart. The callAddItemApi function first finds out whether the product exists in the cart or not. If it exists, it finds out the quantity and adds one more to it. The callAddItemApi function then calls the REST API using the CartClient to add a new item or update the quantity in the existing cart item.

Finally, the add function calls dispatch to update the state of cartItems in the cart context.

The following code snippet contains the same logic:

// Continue code of src/components/ProductCard.js

```javascript
const add = async () => {
  const isLoggedIn = checkLogin();
  if (isLoggedIn && product?.id) {
    const res = await callAddItemApi();
    if (res && res.success) {
      if (res.data?.length > 0) {
        setMsg("Product added to bag.");
        dispatch(updateCart(res.data));
      }
    } else { setMsg(res && typeof res === "string" ? res :
      res.error.message); }
  }
};
const checkLogin = () => {
  if (!auth.token) {
    navigate("/login");
    return false;
  }
  return true;
};
const callAddItemApi = async () => {
  const qty = findQty(product.id);
  return cartClient.addOrUpdate({
    id: product.id,
```

```
    quantity: qty + 1,
    unitPrice: product.price,
  });
};
const findQty = (id) => {
  const idx = cartItems.findIndex((i) => i.id === id);
  if (~idx) { return cartItems[idx].quantity; }
  return 0;
};
```

Here, the add function is called when the **Add to bag** link is clicked. Similarly, the buy function shown in the following code snippet will be called when the user clicks on the **Buy now** button:

```
// ProductCard.js
const buy = async () => {
  const isLoggedIn = checkLogin();
  if (isLoggedIn && product?.id) {
    const res = await callAddItemApi();
    if (res && res.success) { navigate("/cart"); }
    else { setMsg(res && typeof res === "string" ? res :
        res.error.message); }
  }
};
```

Here, the buy function is very similar to the add function. Contrary to the add function, on a successful response from callAddItemApi, the buy function redirects the user to the cart page using a useNavigate hook instance.

Let's have a look at a JSX template. In the following code snippet, the className attribute values have been stripped for better readability:

```
return (
 <div id={product.id} className="my-1 px-1 w-full...">
  <figure className="bg-gray-100 rounded-xl p-8 ...">
   <img className="w-72 h-72 mx-auto"
     src={product.imageUrl}alt={product.name} />
   <div className="pt-4 md:p-6 text-center xs:pl-2 ...">
    <form className="flex-auto">
     <div className="flex flex-wrap items-center ...">
      <h1 className="w-full flex-none font-bold mb-2.5 ...">
       <Link to={`/products/${product.id}`}>
         {product.name}</Link>
      </h1>
      <div className="text-4xl leading-7 font-bold ...">
        {"$"}
```

```
      {product.price.toFixed(2)}
    </div>
    <div className="text-sm font-medium text-gray-400 …">
      In stock
    </div>
  </div>
  <div className="flex space-x-3 mt-8 mb-4 text-sm …">
    <div className="flex-auto flex justify-between">
      <button className="w-1/2…"
        type="button" onClick={buy}>
      Buy now
      </button>
      <button className="flex…"
        type="button" onClick={add}>
      Add to bag
      </button>
    </div>
  </div>
  <p className="text-sm text-gray-500 text-left">
    Free shipping on all local orders.
  </p>
  </form>
  </div>
  </figure>
  </div>
  );
};
export default ProductCard;
```

https://github.com/PacktPublishing/Modern-API-Development-with-Spring-6-and-Spring-Boot-3/tree/dev/Chapter07/ecomm-ui/src/components/ProductCard.js

The onClick event has been bound to buy and add for the **Buy now** button and the **Add to bag** link, respectively. Also, the product name is a link created using Link. The to attribute of Link contains the path that points to the ProductDetail component. This path also contains the path parameter ID. You can use this parameter to perform certain operations on it. Similarly, you can also pass the query parameters the way you do in the browser URL.

When the user clicks on the product name, the user is redirected to the ProductDetail component (ProductDetail.js). Let's develop this next.

Developing the ProductDetail component

The `ProductDetail` component is like the `ProductCard` component, except that it loads the product details from the backend by using the ID from the path.

Let's see how this is done. Only code related to the `Fetch` product has been shown in the following snippet. The rest of the code is the same as for the `ProductCard` component. However, you can refer to the full code in the GitHub repository:

```
import {Link, useParams, useNavigate} from "react-router-dom";
import ProductClient from "../api/ProductClient";
// Other imports removed for brevity
const ProductDetail = ({ auth }) => {
  const { id } = useParams();
  // Other declaration removed for brevity
  // Other functions removed for brevity
  useEffect(() => {
    async function getProduct(id) {
      const client = new ProductClient();
      const res = await client.fetch(id);
      if (res && res.success) { setProduct(res.data); }
    }
    async function fetchCart(auth) {
      const res = await new CartClient(auth).fetch();
      if (res && res.success) {
        console.log(res.data);
        dispatch(updateCart(res.data.items));
      }
    }
    if (auth?.token) fetchCart(auth);
    getProduct(id);
  }, [id]);
  return ( /* JSX Template */ );
};
export default ProductDetail;
```

https://github.com/PacktPublishing/Modern-API-Development-with-Spring-6-and-Spring-Boot-3/tree/dev/Chapter07/ecomm-ui/src/components/ProductDetail.js

You have used `useParams()` from the `react-router-dom` package to retrieve the product ID passed from the `ProductCard` component. This `id` property is then used to fetch the product from the backend server using the `ProductClient` component. Upon a successful response, the retrieved product detail is set in the state product using the `setProduct` state function.

We are done with the development of product-based components such as `ProductList`, `Products`, `ProductCard`, and `ProductDetail`. We will now focus on authentication functionality so that we can later work on the `cart` and `orders` components, which require an authenticated user.

Implementing authentication

Before you jump into the `Login` component development, you will want to figure out how to manage a token received from a successful login response and how to make sure that if the access token has expired, then a refresh token request should be fired before making any call that requires authentication.

The browser allows you to store tokens or any other information in cookies, session storage, and local storage. From the server side, we haven't opted for cookie or stateful communication, therefore we are left with the remaining two options. Session storage is preferable for more secure applications because it is specific to a given tab, and it gets cleared as soon as you click on the **Refresh** button or close the tab. We want to manage login persistence between different tabs and page refresh; therefore, we'll opt for local storage of the browser.

On top of that, you can also store them in the state in the same way you will manage the cart state. However, this will be very similar to session storage. Let's leave that option for now.

Creating a custom useToken hook

You have now used different React hooks. Let's move a step forward and create a custom hook. First, create a new `hooks` directory under the `src` directory, and create a `useToken.js` file in it.

Then, add the following code to it:

```
import { useState } from "react";
export default function useToken() {
  const getToken = () => {
    const tokenResponse = localStorage.getItem
      ("tokenResponse");
    const info = tokenResponse ? JSON.parse
      (tokenResponse) : "";
    return info;
  };
  const [token, setToken] = useState(getToken());
  const saveToken = (tokenResponse) => {
    localStorage.setItem("tokenResponse", JSON.stringify
      (tokenResponse));
    setToken(tokenResponse);
  };
  return { setToken: saveToken, token, };
}
```

```
https://github.com/PacktPublishing/Modern-API-Development-with-
Spring-6-and-Spring-Boot-3/tree/dev/Chapter07/ecomm-ui/src/hooks/
useToken.js
```

Here, you are using a `useState` hook to maintain the token state. The token state is initialized while declaring the token state by calling the `getToken` function in the constructor of `useState`. Now, you need to provide a mechanism that should update the initial token state whenever there is a change in action, such as login or logout. You can create a new function, `saveToken`, for this purpose.

Both the `getToken` and `saveToken` functions use `localStorage` to retrieve and update the token respectively. Finally, both the token state and the `saveToken` function (in the form of `setToken`) are returned for their usage.

Next, you create another REST API client for authentication. Let's add another client, `Auth.js` (`https://github.com/PacktPublishing/Modern-API-Development-with-Spring-6-and-Spring-Boot-3/tree/dev/Chapter07/ecomm-ui/src/api/Auth.js`), in the `src/api` directory.

This `Auth.js` client is very similar to other API clients. It has three functions that perform the `login`, `logout`, and `refresh` access token operations by using the backend server REST APIs, outlined as follows:

- The login operation sets the access token, refresh token, user ID, and username in the `responseToken` key of the local storage by using the state arguments passed by the `App` component. The `App` component, as usual, uses the `useToken` custom hook. The login operation also sets the access token's expiration time.

- The refresh access token operation updates the access token and its expiration time.

- The logout operation removes the tokens and sets the expiration time to zero.

You are done with the prerequisite work for implementing the login functionality and can now move on to creating the `Login` component.

Writing the Login component

Let's create a new `Login.js` file under the `src/components` directory and then add the following code:

```
import { useNavigate } from "react-router-dom";
import { useState } from "react";
import PropTypes from "prop-types";
Login.propTypes = { auth: PropTypes.object.isRequired, };
const Login = ({ uri, auth }) => {
  const [username, setUserName] = useState();
  const [password, setPassword] = useState();
```

```
const [errMsg, setErrMsg] = useState();
const navigate = useNavigate();
const cancel = () => {
  const l = navigate.length;
  l > 2 ? navigate.goBack() : navigate("/");
};
const handleSubmit = async € => {
  e.preventDefault();
  const res = await auth.loginUser({ username, password });
  if (res && res.success) {
    setErrMsg(null);
    navigate(uri ? uri":""/");
  } else { setErrMsg(
   res && typeof res ="= "str"ng" ?
     res : "Unsuccessful");
  }
};
return (/* JSX Template */ );
}
```

https://github.com/PacktPublishing/Modern-API-Development-with-Spring-6-and-Spring-Boot-3/tree/dev/Chapter07/ecomm-ui/src/components/Login.js

Before you start understanding the code, it's useful to know that PropTypes provides a way to check the type of passed props. Here, we have made sure that the auth prop is an object and a required prop. You may see messages in the console if typing fails during argument passing or assignments. Normally, you add this props check at the end of a file (see bottom of the source code), but here it has been added at the top for better readability.

This component contains two props: auth and uri. The auth prop represents the authentication client, and uri is a string that sends the user back to the appropriate page after a successful login.

Login.js has two functions: handleSubmit and cancel. The cancel function just sends the user back to the previous page or the home page. The handleSubmit function makes use of the authentication client and calls the login API with the username and password.

The handleSubmit function is called when a form is submitted (i.e., when the user clicks on the **Sign in** button). The cancel function is called when the user clicks on the **Cancel** button. Another noteworthy point relates to setting the username and password states. These are set on onChange events respectively (refer to the Login.js code file on GitHub for truncated JSX content). The e.target.value argument represents the user input value in the input field. The e instance represents the event and target represents the target input field for the respective event.

So, now you know the complete flow: the user logs in and the app sets the required token and information in local storage. The API client uses this information to call the authenticated APIs. The `logout` operation, which is a part of the `Header` component (`https://github.com/PacktPublishing/Modern-API-Development-with-Spring-6-and-Spring-Boot-3/tree/dev/Chapter07/ecomm-ui/src/components/Header.js`), calls the Auth client's `logout` function, which calls the remove refresh token backend server's REST API and removes the authentication information from the local storage.

After authentication implementation, you need to write one more piece of code before you jump to writing the `Cart` component: cart context. Let's do that now.

Writing the custom cart context

You can use the Redux library for centralizing and maintaining an application's global state. However, you are going to use a Redux-like custom hook to maintain the state for the cart. This uses the `createContext`, `useReducer`, and `useContext` hooks from the React library.

You already know that `createContext` returns `Provider` and `Consumer`. Therefore, when you create a `CartContext` using `createContext`, it will provide `CartContext.Provider`. You won't use `Consumer`, as you are going to use a `useContext` hook.

Next, you need a cart state (`cartItems`) that you pass to the value in `CartContext.Provider` so that it will be available in the component that uses `CartContext`. Now, we just need a `reducer` function. A `reducer` function accepts two arguments: `state` and `action`. Based on the provided action, it updates (mutates) the state and returns the updated state.

Now, let's jump into the code and see how it turns out. Have a look at the following snippet:

```
import React, { createContext, useReducer, useContext }
  from "react";
export const CartContext = createContext();
function useCartContext() { return useContext(CartContext); }
export const UPDATE_CART = "UPDATE_CART";
export const ADD_ITEM = "ADD_ITEM";
export const REMOVE_ITEM = "REMOVE_ITEM";
export function updateCart(items) {
  return { type: UPDATE_CART, items };
}
export function addItem(item) {
  return { type: ADD_ITEM, item };
}
export function removeItem(index) {
  return { type: REMOVE_ITEM, index };
}
```

```
export function cartReducer(state, action) {
  switch (action.type) {
    case UPDATE_CART:
      return [...action?.items];
    case ADD_ITEM:
      return [...state, action.item];
    case REMOVE_ITEM:
      const list = [...state];
      list.splice(action.index, 1);
      return list;
    default:
      return state;
  }
}
const CartContextProvider = (props) => {
 const [cartItems, dispatch] = useReducer(cartReducer, []);
 const cartData = { cartItems, dispatch };
 return <CartContext.Provider value={cartData} {...props} />;
};
export { CartContextProvider, useCartContext };
```

https://github.com/PacktPublishing/Modern-API-Development-with-Spring-6-and-Spring-Boot-3/tree/dev/Chapter07/ecomm-ui/src/hooks/CartContext.js

First, you have created a CartContext with a createContext hook. Then, you have declared a function that uses a useContext hook and returns the value field's value declared in the CartContext.Provider tag.

Next, you need a reducer function that uses the action and state. Therefore, we first define action types such as UPDATE_CART and then write functions that return an action object that contains both the action type and argument value, such as updateCart. Finally, you can write a reducer function that takes state and action as arguments and, based on the passed action type, will mutate the state and return the updated state.

Next, you define a CartContextProvider function that returns the CartContext.Provider component. Here, you use the reducer function in useReducer hook, and in its second argument, you pass the empty array as an initial state. The useReducer hook returns to the state and dispatch functions. The dispatch function takes the action object as an argument. You can use the function that returns the action object, such updateCart and addItem. You wrap the state (cartItems) and dispatcher function (dispatch) in the cartData object and pass it to the value attribute in the CartContext.Provider component. Finally, it exports both the CartContextProvider and useCartContext functions.

You are going to use CartContextProvider as a component wrapper in the App component. This makes cartData (cartItems and dispatch) available to all components inside CartContextProvider, which can be accessed and used via useCartContext.

Now, finally, you can write the Cart component in the next subsection.

Writing the Cart component

The Cart component is a parent component because it can have multiple items (CartItem components) in it. Let's create a new cart.js file in the src/components directory and add the following code to it:

```
// other imports
import { removeItem, updateCart, useCartContext }
        from "../hooks/CartContext";
import CartItem from "./CartItem";
const Cart = ({ auth }) => {
  const [grandTotal, setGrandTotal] = useState(0);
  const [noRecMsg, setNoRecMsg] = useState("Loading...");
  const navigate = useNavigate();
  const cartClient = new CartClient(auth);
  const orderClient = new OrderClient(auth);
  const customerClient = new CustomerClient(auth);
  const { cartItems, dispatch } = useCartContext();
  // continue …
```

https://github.com/PacktPublishing/Modern-API-Development-with-Spring-6-and-Spring-Boot-3/tree/dev/Chapter07/ecomm-ui/src/components/Cart.js

Here, you used the useCartContext that was created in the previous subsection. You also use import functions such as updateCart that return the action object (consumed by the dispatch function). Apart from the CartClient Fetch-based API client, you also used OrderClient and CustomerClient here for cart checkout operations.

Let's add functions for calculating the total (calTotal) and increasing the quantity (increaseQty) of a given product ID, as shown next:

```
// continue src/components/Cart.js
const calTotal = (items) => {
  let total = 0;
  items?.forEach((i) => (
    total = total + i?.unitPrice * i?.quantity));
  return total.toFixed(2);
};
const increaseQty = async (id) => {
```

```
  const idx = cartItems.findIndex((i) => i.id === id);
  if (~idx) {
    cartItems[idx].quantity = cartItems[idx].quantity + 1;
    const res = await cartClient.addOrUpdate
        (cartItems[idx]);
    if (res && res.success) {
      refreshCart(res.data);
      if (res.data?.length < 1) { setNoRecMsg
          ("Cart empty"); }
    } else {
      setNoRecMsg(res && typeof res === "string" ?
          res : res.error.message);
    }
  }
}; // continue …
```

Here, the `increaseQty` function first finds whether the given ID exists among the cart items or not. If it exists, then it increases the quantity of a product by 1. Finally, it calls the REST API to update the cart items and uses the response to update the cart by calling the `refreshCart` function.

Let's add a `decreaseQty` function, which is like `increaseQty` but rather decreases the quantity by one. Also, the `deleteItem` function will remove a given cart item from the cart. The code is shown in the following snippet:

```
// continue src/components/Cart.js
const decreaseQty = async (id) => {
  const idx = cartItems.findIndex((i) => i.id === id);
  if (~idx && cartItems[idx].quantity <= 1) {
    return deleteItem(id);
  } else if (cartItems[idx]?.quantity > 1) {
    cartItems[idx].quantity = cartItems[idx].quantity - 1;
    const res = await cartClient.addOrUpdate
        (cartItems[idx]);
    if (res && res.success) {
      refreshCart(res.data);
      if (res.data?.length < 1) { setNoRecMsg
          ("Empty cart"); }
      return;
    } else {
      setNoRecMsg(res && typeof res === "string" ?
          res : res?.error?.message);
    }
  }
};
```

```
const deleteItem = async (id) => {
  const idx = cartItems.findIndex((i) => i.id === id);
  if (~idx) {
    const res = await cartClient.remove(cartItems[idx].id);
    if (res && res.success) {
      dispatch(removeItem(idx));
      if (res.data?.length < 1) { setNoRecMsg
          ("Item removed");}
    } else {
      setNoRecMsg(
        res && typeof res === "string" ? res:
          "There is an error performing the remove.");
    }
  }
}; // continue …
```

Here, the decreaseQty function does one extra step in comparison to increaseQty — it removes the item if the existing quantity is 1 by calling the deleteItem function.

The deleteItem function first finds the product based on a given ID. If it exists, then it calls the REST API to remove the product from the cart and updates the cart item state by calling the dispatch function with the action object returned by the removeItem function.

Let's define the refreshCart and useEffect functions, as shown in the following code snippet:

```
// continue src/components/Cart.js
const refreshCart = (items) => {
  setGrandTotal(calTotal(items));
  dispatch(updateCart(items));
};
useEffect(() => {
  async function fetch() {
    const res = await cartClient.fetch();
    if (res && res.success) {
      refreshCart(res.data.items);
      if (res.data?.items && res.data.items?.length < 1) {
        setNoRecMsg("Cart is empty.");
      }
    } else {
      setNoRecMsg(res && typeof res === "string" ?
          res : res.error.message);
    }
  }
  fetch();
}, []); // continue …
```

The `refreshCart` function updates the total and dispatches the `updateCart` action. `useEffect` loads the cart items from the backend server and calls `refreshCart` to update the `cartItems` global state.

Let's add the last function of the `Cart` component to perform the checkout operation, as shown in the following code snippet:

```
// continue src/components/Cart.js
  const checkout = async () => {
    const res = await customerClient.fetch();
    if (res && res.success) {
      const payload = {
        address: { id: res.data.addressId },
        card: { id: res.data.cardId },
      };
      const orderRes = await orderClient.add(payload);
      if (orderRes && orderRes.success) {
        navigate("/orders");
      } else {
        setNoRecMsg(orderRes && typeof
            orderRes === "string"
          ? orderRes : "Couldn't process checkout.");
      }
    } else {
      setNoRecMsg(res && typeof res === "string" ?
        res : "error retreiving customer");
    }
  };
  return (/* JSX Template */ );
}
```

https://github.com/PacktPublishing/Modern-API-Development-with-Spring-6-and-Spring-Boot-3/tree/dev/Chapter07/ecomm-ui/src/components/Cart.js

Here, the `checkout` function first fetches the customer information and forms a payload for placing the order. On a successful POST order API response, the user is redirected to the `Orders` component.

Finally, the `Cart` component returns a JSX template, which is taken from CodePen user `abdelrhman`.

In the JSX template, you will find that when the **Checkout** button is clicked, the checkout function to place the user order is called. Cart items are rendered using the `CartItem` component that you create next. You pass the `removeItem`, `increaseQty`, and `decreaseQty` functions as props to it.

Let's write the `CartItem` component by creating a new file (`src/components/ CartItem. js`) and adding the following code:

```
// imports
const CartItem=({item, increaseQty, decreaseQty, removeItem }) => {
  const d = item ? item.description?.split(".") : [];
  const author = d && des.length > 0 ? d
      [d.length - 1] : "";
  const [total, setTotal] = useState();
  const calTotal = (item) => {
    setTotal((item?.unitPrice * item?.quantity)
        ?.toFixed(2));
  };
  const updateQty = (qty) => {
    if (qty === -1) { decreaseQty(item?.id); }
    else if (qty === 1) { increaseQty(item?.id); }
    else { return false; }
    calTotal(item);
  };
  useEffect(() => { calTotal(item); }, []);
  return (/* JSX Template */ );
}
```

https://github.com/PacktPublishing/Modern-API-Development-with-Spring-6-and-Spring-Boot-3/tree/dev/Chapter07/ecomm-ui/src/components/CartItem.js

Here, you maintain the state of the total that is a product of the quantity and the unit price (`calTotal` function), and use the `updateQty` helper function to perform the increase or decrease quantity operations. The `useEffect` hook also calls `calTotal` to update the total on the **Cart** page.

Now, you can write the last component (page) of this application in the next subsection: the `Order` component.

Writing the Order component

The `Order` component contains the order details fetched from the backend server. It shows the date, status, amount, and items in a tabular format. It loads the order details on the first render with the `useEffect` hook and then the `orders` state is used in the JSX expression to display it.

Let's create a new file, `Orders.js`, in the `src/components` directory and add the following code to it:

```
// imports
const Orders = ({ auth }) => {
```

```
const [orders, setOrders] = useState([]);
const formatDate = (dt) => {
  return dt && new Date(dt).toLocaleString();
};
useEffect(() => {
  async function fetchOrders() {
    const client = new OrderClient(auth);
    const res = await client.fetch();
    if (res && res.success) { setOrders(res.data); }
  }
  fetchOrders();
}, []);
return (/* JSX Template */ );
}
```

https://github.com/PacktPublishing/Modern-API-Development-with-Spring-6-and-Spring-Boot-3/tree/dev/Chapter07/ecomm-ui/src/components/Orders.js

Here, the code is straightforward. It simply displays the information fetched from the `orders` state.

Now, we can update the root component to complete the flow and test the application after starting it again with the `yarn start` command.

Writing the root (App) component

The App component is a root component of the React application. It contains routing information and the application layout with all the parent components, such as the product list and orders components.

Update the App.js file available in the project `src` directory with the following code:

```
import { BrowserRouter as Router, Route, Routes }
  from "react-router-dom";
// other imports
function App() {
  const { token, setToken } = useToken();
  const auth = new Auth(token, setToken);
  const LoginComponent = (props) => (
    <Login {...props} uri="/login" auth={auth} />
  );
  const ProductListComponent = (props) =>
      <ProductList auth={auth} />;
  // continue …
```

https://github.com/PacktPublishing/Modern-API-Development-with-Spring-6-and-Spring-Boot-3/tree/dev/Chapter07/ecomm-ui/src/App.js

Here, the first `import` statement imports the required components from `react-router-dom`. There are other imports that you can check on the GitHub repository (the link is given at the bottom of the preceding code block). Then, the `useToken()` hook and the `Auth` authentication REST API client are used for authentication purposes. You create functions that return `loginComponent` and `productListComponent`.

The `App.js` JSX template is different from what we have used till now. It uses the `BrowserRouter` (`Router`), `Route`, and `Routes` components from the `react-router-dom` package. You define all the `Route` components inside the `BrowserRouter` component. Here, we are also using the `Routes` component because we want to render components exclusively. `Route` also allows you to render the `NotFound` component (the typical *404 – not found* page) if no path matches. The `Route` component allows you to define the path and component to be rendered. You used the `element` property to represent the component that you want to render based on the given `path`. The following code snippet contains the logic explained here:

```
// App.js continue
return (
 <div className="flex flex-col min-h-screen h-full ">
  <Router>
   <Header userInfo={token} auth={auth} />
    <div className="flex-grow flex-shrink-0 p-4">
     <CartContextProvider>
      <Routes>
       <Route path="/" exact
          element={<ProductListComponent />} />
       <Route
         path="/login"
         element={token ? <ProductListComponent /> :
            <LoginComponent />} />
       <Route
         path="/cart"
         element={token ? <Cart auth={auth} /> :
            <LoginComponent />} />
       <Route
         path="/orders"
         element={token ? <Orders auth={auth} /> :
         <LoginComponent />} />
       <Route
         path="/products/:id"
         element={<ProductDetail auth={auth} />} />
       <Route path="*" exact element={<NotFound />} />
```

```
      </Routes>
    </CartContextProvider>
  </div>
  <Footer />
 </Router>
</div>
);
}
export default App;
```

All components are wrapped inside `CartContextProvider` to allow `cartItems` and `dispatch` to be accessible in all components provided they use the `useCartContext` custom hook.

You are done with the major development work. Let's run the code using the instructions given in the next section.

Running the application

You need a backend server for testing the UI because the UI fires REST APIs to get the data. You are going to use code from *Chapter 6*.

Go to the home directory of the *Chapter 6* code. You can build the code by running `gradlew clean build` from the root of the `Chapter06` project and run the backend using the following command:

```
$ java -jar build/libs/Chapter06-0.0.1-SNAPSHOT.jar.
```

Make sure to use Java 17 in the path.

Once the backend is up and running, you can open another terminal and start the `ecomm-ui` app by executing the following command from the `Chapter07/ecomm` project root directory:

```
$ yarn start
```

If the application starts successfully, the UI will be accessible at `http://localhost:3000`. You can open `http://localhost:3000` in your favorite browser.

Once the product listing page loads, you can log in to the example e-commerce UI app with the username/password (`scott/tiger`) and perform all the operations such as checkout, orders, and so on.

Let's review what you have learned and summarize this chapter in the next section.

Summary

In this chapter, you have learned some basic concepts of React and created different types of components using them. You have also learned how to use the browser's built-in Fetch API to consume the REST APIs. You acquired the following skills in React: developing a component-based UI, implementing routing, consuming REST APIs, implementing functional components with hooks, writing custom hooks, and building a global state store with a React context API and a `useReducer` hook.

The concepts and skills you acquired in this chapter have laid a solid foundation for modern frontend development and advance you toward gaining a 360-degree perspective of application development.

In the next chapter, you will learn about writing automated tests for REST-based web services.

Questions

1.   What is the difference between props and state?

2.   What is an event and how can you bind events in a React component?

3.   What is a higher-order component?

Answers

1.   Props are special objects that you use to pass the values/objects/functions from the parent component to a child component, whereas state belongs to a component – it could be global or local to the component. From a functional component perspective, you use the `useState` hook for local state and `useContext` for global state.

2.   In general, events are objects generated by the browser on input such as `keydown` or `onclick`. React uses `SyntheticEvent` to ensure that the browser's native events work identically across all browsers. `SyntheticEvent` wraps on top of the native event. You used the `onChange={(e) => setUserName(e.target.value)}` code in the login component. Here, e is `SyntheticEvent` and `target` is one of its attributes. The onChange event is bound in JSX that calls `setUserName` when the input value is changed. You can also use the same JavaScript technique to bind events such as `window. addEventListener("click", handleClick)`.

     Ideally, you would do this in the `useEffect` hook; however, the event should be removed as a part of the cleanup. That can also be done in `useEffect` when you return the arrow function that removes the binding, for example, `return () => { window. removeEventListener("click", handleClick); }`. You can find this example in the `Header.js` file in the `src/components` directory.

3. In JavaScript, higher-order functions take a function as an argument and/or return a function, such as that of an array (map, filter, and so on). Similarly, in React, **higher-order components** (**HOCs**) are a pattern that involves the use of composition with an existing component and returns a new component. Basically, you write a new function that takes a component as an argument and returns it. An HOC allows you to reuse the existing component and its logic.

    In the `ecomm-ui` application, the `ProductCard` and `ProductDetail` components are similar in nature, and you can use an HOC to reuse the logic.

Further reading

- *React 18 Design Patterns and Best Practices - Fourth Edition*: `https://www.packtpub.com/product/react-18-design-patterns-and-best-practices-fourth-edition/9781803233109`

- React documentation: `https://reactjs.org/docs/getting-started.html`

- React Router guide: `https://reactrouter.com/en/main`

8
Testing APIs

Proper automated testing helps you to reduce regression bugs and keeps your application stable. It makes sure that every change you make will fail during the build or testing phase if the change has any side effects on existing code. Investing in a test automation suite can give you peace of mind and will prevent any surprises in production.

This chapter will help you learn about test automation by showing you how to implement unit and integration test automation. You will learn how to test APIs manually and automatically. First, you will learn about automating unit and integration tests. After learning about these forms of automation, you will be able to make both types of testing an integral part of any build. You will also learn how to set up the **Java Code Coverage** (**JaCoCo**) tool to calculate different code coverage metrics.

In this chapter, we will cover the following topics:

- Testing APIs and code manually
- Testing automation

Let's get started!

Technical requirements

The code for this chapter is available at `https://github.com/PacktPublishing/Modern-API-Development-with-Spring-6-and-Spring-Boot-3/tree/dev/Chapter08`.

Testing APIs and code manually

Testing is a continuous process in software development and maintenance cycles. You need to do full testing that covers all possible use cases and the respective code for each change. Different types of testing can be performed for APIs, including the following:

- **Unit testing**: Unit testing is performed by developers to test the smallest unit (such as a class method) of code.

- **Integration testing**: Integration testing is performed by developers to test the integration of different layers of components.

- **Contract testing**: Contract testing is performed by developers to make sure any changes that are made to the API won't break the consumer code. The consumer code should always comply with the producer's contract (API). It is primarily required in microservices-based development.

- **End-to-end (E2E) testing**: E2E testing is performed by the **quality assurance** (**QA**) team to test end-to-end scenarios, such as from the UI (consumer) to the backend.

- **User acceptance testing** (UAT): UAT is performed by business users from a business perspective and may overlap with E2E testing.

You performed manual API testing by using the cURL and Postman tools earlier in this book. Every change requires the APIs to be completely tested – not only the impacted APIs. There is a reason for this. You may assume that it only impacts certain APIs, but what if your underlying assumptions are wrong? It may impact the other APIs that you skipped, which would lead to production issues. This can create panic and may require a release to be rolled over or a patch to be released with a fix.

You don't want to be in such situations, so products have a separate QA team that ensures releases are delivered with the best possible quality. QA teams do the separate E2E and acceptance testing (along with business/domain users), apart from the testing that's done by the development team.

This extra assurance for high-quality deliverables needs more time and effort. The time taken now is much shorter because of automated testing. It was longer previously because we performed manual testing; therefore, software development cycles used to be huge in comparison to today. **Time to market** (**TTM**) is a huge factor in today's competitive software industry. Today, you need faster release cycles. Moreover, quality checks, also known as testing, are an important and major part of release cycles.

You can reduce the testing time by automating the testing process and making it an integral part of the CI/CD pipeline. **CI** stands for **continuous integration**, which means *build > test > merge* in a code repository. **CD** stands for **continuous delivery** and/or **continuous deployment**, both of which may be used interchangeably. Continuous delivery is a process where code is automatically tested and released (read and uploaded) to an artifact repository or container registry. Then, it can be picked and deployed to a production environment after manual approval. Continuous deployment is one step ahead of continuous delivery and automates all the steps. Continuous deployment also performs the automatic deployment to production once all tests are passed. Products that don't release their code for public access use this approach, such as Facebook and Twitter. On the other hand, products/services that are available publicly, such as the Spring Framework and Java, use continuous delivery pipelines.

We'll automate the manual testing we have done so far in the next section.

Testing automation

Whatever testing you are doing manually can be automated and made part of the build. This means that any change or code commit will run the test suite as a part of the build. A build will only be successful if all the tests are passed.

You can add automated integration tests for all the APIs. So, instead of firing each API manually using cURL or Insomnia, the build will fire them, and the test result will be available at the end of the build.

In this section, you are going to write an integration test that will replicate the REST client call and test all the application layers, starting from the controller, all the way down to the persistence layer, including the database (H2).

But before that, you will add the necessary unit tests. Ideally, these unit tests should have been added alongside the development process, or before the development process in the case of **test-driven development** (TDD).

Unit tests are tests that validate the expected results of small units of code, such as a class's methods. You can avoid most bugs if you have proper tests in place with good code (90% or above) and branch coverage (80% and above). Code coverage refers to metrics such as the number of lines and branches (such as `if-else`), which are validated when the tests are executed.

Some classes or methods have dependencies on other classes or infrastructure services. For example, controller classes have dependencies on service and assembler classes, while repository classes have dependencies on Hibernate APIs. You can create mocks to replicate dependency behaviors and assume these are working as expected or behave as per the defined tests. This approach will allow you to test the actual code unit (such as a method) and validate its behavior.

In the next section, we'll explore how to add unit tests before writing the integration tests.

Unit testing

I advise you to go back to *Chapter 6* as a base for this chapter's code. You don't have to add any additional dependencies for unit tests. You already have the following dependency in `build.gradle` (https://github.com/PacktPublishing/Modern-API-Development-with-Spring-6-and-Spring-Boot-3/tree/dev/Chapter08/build.gradle):

```
testImplementation('org.springframework.boot:spring-boot-starter-test')
```

Here, `spring-boot-starter-test` adds all the required test dependencies, not only for the unit tests but also for the integration tests. You are going to primarily use the following libraries for testing:

- **JUnit 5**: JUnit 5 is a bundle of modules, including the JUnit Platform, JUnit Jupiter, and JUnit Vintage:

  - **The JUnit Platform** allows you to launch tests on JVM and its engine provides APIs for writing testing frameworks that run on the platform. The JUnit Platform consists of `junit-platform-commons`.

  - **JUnit Jupiter** provides the programming and extension models for writing tests and extensions. It has a separate library called `junit-jupiter-engine` that allows you to run Jupiter-based tests on the JUnit Platform. It also provides the `junit-jupiter`, `junit-jupiter-api`, and `junit-jupiter-params` libraries.

  - **JUnit Vintage** supports older versions of JUnit, such as versions 3 and 4. You are going to use the latest version in this book, which is 5, so you don't need this bundle.

  You can find out more about JUnit at `https://junit.org/`.

- **AssertJ**: AssertJ is a test assertion library that simplifies assertion writing by providing fluent APIs. It is also extendable. You can write custom assertions for your domain objects. You can find more about it at `https://assertj.github.io/doc/`.

- **Hamcrest**: Hamcrest is another assertion library that provides assertions based on matchers. It also allows you to write custom matchers. You'll find an example of both in this chapter, though AssertJ is preferable because it has fluent APIs. Chained methods help IDEs to suggest appropriate assertions based on a given object. You can choose one of the assertion libraries or both based on your use cases and liking. You can find out more about it at `http://hamcrest.org/`.

- **Mockito**: Mockito is a mocking framework that allows you to mock objects (read dependencies) and to stub method calls. You can find out more about it at `https://site.mockito.org/`.

You already know that unit tests test the smallest testable code unit. But how can we write a unit test for controller methods? The controller runs on web servers and has the Spring web application context. If you write a test that uses `WebApplicationContext` and is running on top of a web server, then you can call it an integration test rather than a unit test.

Unit tests should be lightweight and must be executed quickly. Therefore, you must use `MockMvc`, a special class provided by the Spring test library, to test the controllers. You can use the standalone setup for `MockMvc` for unit testing. You can also use `MockitoExtension` to run the unit test on the JUnit Platform (JUnit 5 provides an extension for runners), which supports object mocking and method stubbing. You will also use the Mockito library to mock the required dependencies. These tests are fast and help developers build faster.

Let's write our test using AssertJ assertions.

Testing using AssertJ assertions

Let's write our first unit test for `ShipmentController`. The following code can be found in `src/test/java/com/packt/modern/api/controller/ ShipmentControllerTest.java`:

```
@ExtendWith(MockitoExtension.class)
public class ShipmentControllerTest {
    private static final String id =
        "a1b9b31d-e73c-4112-af7c-b68530f38222";
    private MockMvc mockMvc;
    @Mock
    private ShipmentService service;
    @Mock
    private ShipmentRepresentationModelAssembler assembler;
    @Mock
    private MessageSource msgSource;
    @InjectMocks
    private ShipmentController controller;
    private ShipmentEntity entity;
    private Shipment model = new Shipment();
    private JacksonTester<List<Shipment>> shipmentTester;
    // continue …
```

https://github.com/PacktPublishing/Modern-API-Development-with-Spring-6-and-Spring-Boot-3/tree/dev/Chapter08/src/test/java/com/packt/modern/api/controller/ShipmentControllerTest.java

Here, our test is using a Jupiter-based annotation (`ExtendWith`) that registers the extension (`MockitoExtension`) for running tests and supporting Mockito-based mocks and stubbing.

The Spring test library provides the `MockMvc` class, which allows you to mock the Spring MVC. As a result, you can execute the controller methods by calling the associated API endpoints' URI. The dependencies of the `ShipmentController` controller class, such as the service and assembler, are marked with @Mock annotations to create the mock instances of its dependencies. You can also use `Mockito.mock(classOrInterface)` to create the mock objects.

Another noticeable annotation is `@InjectMocks` on the controller declaration. It finds out all the declared mocks that are required for a testing class and injects them automatically. `ShipmentController` uses the `ShipmentService` and `ShipmentRepresentation ModelAssembler` instances, which are injected using its constructor. The Mockito-based `InjectMocks` annotation finds the dependencies in the `ShipmentController` class (service and assembler). Then, it looks for mocks of the service and assembler in the test class. Once it finds them, it injects these mock objects into the `ShipmentController` class. If required, you can also create an instance of the testing class using a constructor instead of using @InjectsMocks, as shown here:

```
controller = new ShipmentController(service, assembler);
```

A mock of MessageSource is created for RestApiHandler, which is being used in the setup method. You'll explore it further in the following code block.

The last part of the declaration is JacksonTester, which is part of the Spring testing library. JacksonTester is a custom JSON assertion class that's created using the AssertJ and Jackson libraries.

The JUnit Jupiter API provides the @BeforeAll and @BeforeEach method annotations, which can be used to set up the prerequisites. As their names suggest, @BeforeAll is run once per test class, while @BeforeEach gets executed before each test execution. @BeforeEach can be placed on public non-static methods, whereas @BeforeAll should be used to annotate public static methods.

Similarly, JUnit provides the @AfterAll and @AfterEach annotations, which execute the associated methods after each test is executed and after each test is executed, respectively.

Let's use the @BeforeEach annotation to set up the prerequisites for the ShipmentControll erTest class, as shown here:

```
// continue ShipmentControllerTest.java
@BeforeEach
public void setup() {
  ObjectMapper mapper = new AppConfig().objectMapper();
  JacksonTester.initFields(this, mapper);
  MappingJackson2HttpMessageConverter mappingConverter =
    new MappingJackson2HttpMessageConverter();
  mappingConverter.setObjectMapper(mapper);
  mockMvc = MockMvcBuilders.standaloneSetup(controller)
      .setControllerAdvice(new RestApiErrorHandler
        (msgSource))
      .setMessageConverters(mappingConverter).build();
  final Instant now = Instant.now();
  entity = // entity initialization code
  BeanUtils.copyProperties(entity, model);
  // extra model property initialization
}
```

First, we initialize the JacksonTester fields with the object mapper instance received from AppConfig. This creates a custom message converter instance (MappingJackson2HttpMes sageConverter).

Next, you can create a `mockMvc` instance using the standalone setup and initialize the controller advice using its setter method. The `RestApiErrorHandler` instance uses the mock object of the `MessageResource` class. You can also set the message converter to `mockMvc` before building it.

Finally, you initialize the instances of `ShipmentEntity` and `Shipment` (model).

Next, you are going to write the test for the GET `/api/v1/shipping/{id}` call, which uses the `getShipmentByOrderId()` method of the `ShipmentController` class. Tests are marked with `@Test`. You can also use `@DisplayName` to customize a test's name in the test reports:

```
@Test
@DisplayName("returns shipments by given order ID")
public void testGetShipmentByOrderId() throws Exception {
  // given
  given(service.getShipmentByOrderId(id))
      .willReturn(List.of(entity));
  given(assembler.toListModel(List.of(entity)))
      .willReturn(List.of(model));
  // when
  MockHttpServletResponse response = mockMvc.perform(
      get("/api/v1/shipping/" + id)
          .contentType(MediaType.APPLICATION_JSON)
          .accept(MediaType.APPLICATION_JSON))
      .andDo(print())
      .andReturn().getResponse();
  // then
  assertThat(response.getStatus())
      .isEqualTo(HttpStatus.OK.value());
  assertThat(response.getContentAsString())
      .isEqualTo(shipmentTester.write(
          List.of(model)).getJson());
}
```

Here, you are using the **behavior-driven development** (**BDD**) test style. You can find out more about BDD at `https://cucumber.io/docs/bdd/`. BDD tests are written using the Gherkin `Given > When > Then` language (`https://cucumber.io/docs/gherkin/`), which can be defined as follows:

- `Given`: Context of the test

- `When`: Test action

- `Then`: Test result, followed by validation

Let's read this test from a BDD perspective:

- `Given`: The service is available and returns the list of shipments based on the given order ID and an assembler, which converts the list of entities into a list of models. It also adds HATEOAS links.

- `When`: The user calls the API via `GET /api/shipping/a1b9b31d-e73c- 4112-af7c-b68530f38222`.

- `Then`: The test validates the received shipments associated with the given order ID.

Mockito's `MockitoBDD` class provides the `given()` fluent API to stub the mock objects methods. When `mockMvc.perform()` is called, internally, it calls the respective service and assembler mocks, which, in turn, call the stubbed methods and return the values defined in the stub (using `given()`).

The `andDo(MockMvcResultHandlers.print())` method logs the request and response trace, including the payload and response body. If you want to trace all the `mockMvc` logs inside a test class, then you can configure them directly while initializing `mockMvc` instead of defining them individually in `mockMvc.perform()` calls, as shown here (the highlighted code):

```
mockMvc = MockMvcBuilders.standaloneSetup(controller)
    .setControllerAdvice(new RestApiErrorHandler
      (msgSource))
    .setMessageConverters(mappingJackson2HttpMessageConverter)
    .alwaysDo(print())
    .build();
```

At the end, you perform assertions (whether the status is `200 OK` or not and whether the returned JSON object matches the expected object or not) using AssertJ fluent APIs. First, you use the `Asserts.assertThat()` function, which takes the actual object and compares it with the expected object using the `isEqualTo()` method.

So far, you have used AssertJ assertions. Similarly, you can also use Spring and Hamcrest assertions.

Testing using Spring and Hamcrest assertions

At this point, you know how to write JUnit 5 tests using `MockitoExtension`. You'll use the same approach to write a unit test, except with assertions. This time, you will write an assertion using Hamcrest assertions, as shown here:

```
@Test
@DisplayName("returns address by given existing ID")
public void getAddressByOrderIdWhenExists() throws Exception {
  given(service.getAddressesById(id))
      .willReturn(Optional.of(entity));
  // when
  ResultActions result = mockMvc.perform(
```

```
    get("/api/v1/addresses/a1b9b31d-e73c-4112-af7c-
        b68530f38222")
        .contentType(MediaType.APPLICATION_JSON)
        .accept(MediaType.APPLICATION_JSON));
// then
result.andExpect(status().isOk());
verifyJson(result);
}
```

https://github.com/PacktPublishing/Modern-API-Development-with-Spring-6-and-Spring-Boot-3/tree/dev/Chapter08/src/test/java/com/packt/modern/api/controller/AddressControllerTest.java

You have captured the MockHttpResponse instance from the mockMvc.perform() call in the previous test example – that is, testGetShipmentByOrderId(). This time, you will directly use the returned value of the mockMvc.perform() call rather than calling an extra andReturn(). getResponse() on it.

The ResultAction class provides the andExpect() assertion method, which takes ResultMatcher as an argument. The StatusResultMatchers.status(). isOk() result matcher evaluates the HTTP status returned by the perform() call. The VerifyJson() method evaluates the JSON response object, as shown in the following code:

```
// AddressControllerTest.java
private void verifyJson(final ResultActions result)
    throws Exception {
  final String BASE_PATH = "http://localhost";
  result
      .andExpect(jsonPath("id",
          is(entity.getId().toString())))
      .andExpect(jsonPath("number", is
          (entity.getNumber())))
      .andExpect(jsonPath("residency",
          is(entity.getResidency())))
      .andExpect(jsonPath("street", is
          (entity.getStreet())))
      .andExpect(jsonPath("city", is(entity.getCity())))
      .andExpect(jsonPath("state", is(entity.getState())))
      .andExpect(jsonPath("country", is
          (entity.getCountry())))
      .andExpect(jsonPath("pincode", is
          (entity.getPincode())))
      .andExpect(jsonPath("links[0].rel", is("self")))
      .andExpect(jsonPath("links[0].href",
```

```
                is(BASE_PATH + "/" + entity.getId())))
        .andExpect(jsonPath("links[1].rel", is("self")))
        .andExpect(jsonPath("links[1].href",
            is(BASE_PATH + URI + "/" + entity.getId()))));
}
```

Here, the `MockMvcResultMatchers.jsonPath()` result matcher takes two arguments – a JSON path expression and a matcher. Therefore, first, you must pass the JSON field name and then the Hamcrest matcher known as `Is.is()`, which is a shortcut for `Is.is(equalsTo(entity.getCity()))`.

Writing the unit test for a service is much easier compared to writing one for the controller because you don't have to deal with `MockMvc`.

You will learn how to test private methods in the next subsection.

Testing private methods

Unit testing a private method is a challenge. The Spring test library provides the `ReflectionTestUtils` class, which provides a method called `invokeMethod`. This method allows to you invoke private methods. The `invokeMethod` method takes three arguments – the target class, the method's name, and the method's arguments (using variable arguments). Let's use it to test the `AddressServiceImpl.toEntity()` private method, as shown in the following code block:

```
@Test
@DisplayName("returns an AddressEntity when private method
    toEntity() is called with Address model")
public void convertModelToEntity() {
 // given
 AddressServiceImpl srvc = new AddressServiceImpl
    (repository);
 // when
 AddressEntity e = ReflectionTestUtils.invokeMethod(
    srvc, "toEntity", addAddressReq);
 // then
 then(e).as("Check address entity is returned & not null")
    .isNotNull();
 then(e.getNumber()).as("Check house/flat number is set")
    .isEqualTo(entity.getNumber());
 then(e.getResidency()).as("Check residency is set")
    .isEqualTo(entity.getResidency());
 then(e.getStreet()).as("Check street is set")
    .isEqualTo(entity.getStreet());
 then(e.getCity()).as("Check city is set")
```

```
        .isEqualTo(entity.getCity());
    then(e.getState()).as("Check state is set")
        .isEqualTo(entity.getState());
    then(e.getCountry()).as("Check country is set")
        .isEqualTo(entity.getCountry());
    then(e.getPincode()).as("Check pincode is set")
        .isEqualTo(entity.getPincode());
  }
```

https://github.com/PacktPublishing/Modern-API-Development-with-Spring-6-and-Spring-Boot-3/tree/dev/Chapter08/src/test/java/com/packt/modern/api/service/AddressServiceTest.java

In the preceding code, you can see that when you call `ReflectionTestUtils.invokeMethod()` with the given arguments, it returns the `AddressEntity` instance, which has been converted using the given argument's `AddAddressReq` model instance.

Here, you are using a third kind of assertion using AssertJ's `BDDAssertions` class. The `BDDAssertions` class provides methods that resonate with the BDD style. `BDDAssertions.then()` takes the actual value that you want to verify. The `as()` method describes the assertion and should be added before you perform the assertion. Finally, you perform verification using AssertJ's assertion methods, such as `isEqualTo()`.

You will learn how to test void methods in the next subsection.

Testing void methods

A method that returns a value can easily be stubbed, but how can we stub a method that returns nothing? Mockito provides the `doNothing()` method for this. It has a wrapper `willDoNothing()` method in the `BDDMockito` class that internally uses `doNothing()`.

This is very handy, especially when you want such methods to do nothing while you're spying, as shown here:

```
List linkedList = new LinkedList();
List spyLinkedList = spy(linkedList);
doNothing().when(spyLinkedList).clear();
```

Here, `linkedList` is a real object and not a mock. However, if you want to stub a specific method, then you can use `spy()`. Here, when the `clear()` method is called on `spyLinkedList`, it will do nothing.

Let's use `willDoNothing` to stub the void method and see how it helps test void methods:

```
// AddressServiceTest.java
@Test
```

```
@DisplayName("delete address by given existing id")
public void deleteAddressesByIdWhenExists() {
  given(repository.findById(UUID.fromString(nonExistId)))
      .willReturn(Optional.of(entity));
  willDoNothing().given(repository)
      .deleteById(UUID.fromString(nonExistId));
  // when
  service.deleteAddressesById(nonExistId);
  // then
  verify(repository, times(1))
      .findById(UUID.fromString(nonExistId));
  verify(repository, times(1))
      .deleteById(UUID.fromString(nonExistId));
}
```

In the preceding code, `AddressRepository.deleteById()` is being stubbed using Mockito's `willDoNothing()` method. Now, you can use the `verify()` method of Mockito, which takes two arguments – the mock object and its verification mode. Here, the `times()` verification mode is used, which determines how many times a method is invoked.

We'll learn how to unit-test exceptional scenarios in the next subsection.

Testing exceptions

Mockito provides `thenThrow()` for stubbing methods with exceptions. BDDMockito's `willThrow()` is a wrapper that uses it internally. You can pass the `Throwable` argument and test it like so:

```
// AddressServiceTest.java
@Test
@DisplayName("delete address by given non-existing id,
    should throw ResourceNotFoundException")
public void deleteAddressesByNonExistId() throws Exception {
  given(repository.findById(UUID.fromString(nonExistId)))
      .willReturn(Optional.empty())
      .willThrow(new ResourceNotFoundException(String.format(
  "No Address found with id %s.", nonExistId)));
  // when
  try { service.deleteAddressesById(nonExistId);
  } catch (Exception ex) {
  // then
    assertThat(ex)
      .isInstanceOf(ResourceNotFoundException.class);
    assertThat(ex.getMessage())
```

```
        .contains("No Address found with id " + nonExistId);
    }
    // then
    verify(repository, times(1))
        .findById(UUID.fromString(nonExistId));
    verify(repository, times(0))
        .deleteById(UUID.fromString(nonExistId));
}
```

Here, you basically catch the exception and perform assertions on it.

With that, you have explored the unit tests that you can perform for both controllers and services. You can make use of these examples and write unit tests for the rest of the classes.

Executing unit tests

You can run the following command to execute unit tests:

```
$ ./gradlew clean test
```

This will generate the unit test reports at `Chapter08/build/reports/tests/test/index.html`.

A generated test report will look like this:

Figure 8.1 – Unit test report

You can click on the links to drill down further. If the test fails, it also shows the cause of the error.

Let's move on to the next section to learn how to configure code coverage for unit tests.

Code coverage

Code coverage provides important metrics, including line and branch coverage. You are going to use the **JaCoCo** tool to perform and report your code coverage.

First, you need to add the `jacoco` Gradle plugin to the `build.gradle` file, as shown in the following code:

```
plugins {
    id 'org.springframework.boot' version '3.0.4'
    id 'io.spring.dependency-management' version '1.1.0'
    id 'java'
    id 'org.hidetake.swagger.generator' version '2.19.2'
    id 'jacoco'
}
```

https://github.com/PacktPublishing/Modern-API-Development-with-Spring-6-and-Spring-Boot-3/tree/dev/Chapter08/build.gradle

Next, configure the `jacoco` plugin by providing its version and reports directory:

```
// build.gradle
jacoco {
    toolVersion = "0.8.8"
    reportsDirectory = layout.buildDirectory.dir(
        "$buildDir/jacoco")
}
```

Next, create a new task called `jacocoTestReport` that depends on the `test` task because code coverage can only be evaluated after test execution. You don't want to calculate coverage for auto-generated code, so add the `exclude` block. Exclusion can be added by configuring `afterEvaluate`, as shown in the following code block:

```
// build.gradle
jacocoTestReport {
    dependsOn test
    afterEvaluate {
        classDirectories.setFrom(
            files(classDirectories.files.collect {
                fileTree(
                    dir: it,
```

```
                    exclude: [
                        'com/packt/modern/api/model/*',
                        'com/packt/modern/api/*Api.*',
                        'com/packt/modern/api/security
                            /UNUSED/*',
                    ])
            }))
        }
    }
```

Next, you need to configure `jacocoTestCoverageVerification`, which defines the violation rules. We have added instructions to cover the ratio rule in the following code block. This will set the expected ratio to a minimum of 90%. If the ratio is below 0.9, then it will fail the build. You can find out more about such rules at `https://docs.gradle.org/current/userguide/jacoco_plugin.html#sec:jacoco_report_violation_rules`:

```
// build.gradle
jacocoTestCoverageVerification {
    violationRules {
        rule {
            limit { minimum = 0.9 }
        }
    }
}
```

Next, add `finalizedBy(jacocoTestReport)` to the test task, which ensures that the `jacocoTestReport` task will execute after performing the tests:

```
test {
    useJUnitPlatform()
    finalizedBy(jacocoTestReport)
}
```

Let's run the following command to generate the code coverage report:

```
$ ./gradlew clean build
```

The previous command will not only run the test but also generate the code coverage report, along with the test reports. The code coverage report is available at `Chapter08/build/jacoco/test/html/index.html` and looks like this:

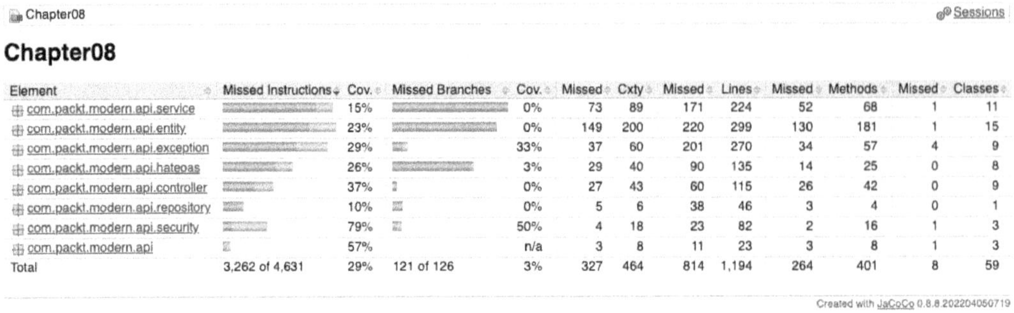

Chapter08

Sessions

Chapter08

Element	Missed Instructions	Cov.	Missed Branches	Cov.	Missed	Cxty	Missed	Lines	Missed	Methods	Missed	Classes
com.packt.modern.api.service		15%		0%	73	89	171	224	52	68	1	11
com.packt.modern.api.entity		23%		0%	149	200	220	299	130	181	1	15
com.packt.modern.api.exception		29%		33%	37	60	201	270	34	57	4	9
com.packt.modern.api.hateoas		26%		3%	29	40	90	135	14	25	0	8
com.packt.modern.api.controller		37%		0%	27	43	60	115	26	42	0	9
com.packt.modern.api.repository		10%		0%	5	6	38	46	3	4	0	1
com.packt.modern.api.security		79%		50%	4	18	23	82	2	16	1	3
com.packt.modern.api		57%		n/a	3	8	11	23	3	8	1	3
Total	3,262 of 4,631	29%	121 of 126	3%	327	464	814	1,194	264	401	8	59

Created with JaCoCo 0.8.8.202204050719

Figure 8.2 – Code coverage report

Here, you can see that our instruction coverage is only at 29%, while our branch coverage is only at 3%. You can add more tests and increase these percentages.

You will learn about integration testing in the next section.

Integration testing

Once you have the automated integration tests in place, you can ensure that any changes you make won't produce bugs, provided you cover all the testing scenarios. You don't have to add any additional plugins or libraries to support integration testing in this chapter. The Spring test library provides all the libraries required to write and perform integration testing.

Let's add the configuration for integration testing in the next subsection.

Configuring the Integration testing

First, you need a separate location for your integration tests. This can be configured in `build.gradle`, as shown in the following code block:

```
sourceSets {
    integrationTest {
        java {
            compileClasspath += main.output + test.output
            runtimeClasspath += main.output + test.output
            srcDir file('src/integration/java')
        }
        resources.srcDir file('src/integration/resources')
    }
}
```

Here, you add the integration tests and their resources to source sets. Gradle then picks the tests when a relevant Gradle command (`integrationTest`, `build`) gets executed.

Next, you can configure the integration test's implementation and runtime so that it's extended from the test's implementation and runtime, as shown in the following code block:

```
configurations {
    integrationTestImplementation.extendsFrom
        testImplementation
    integrationTestRuntime.extendsFrom testRuntime
}
```

Finally, create a task called `integrationTest` that will not only use the JUnit Platform but also use our `classpath` and test `classpath` from `sourceSets.integrationTest`.

Finally, configure the check task so that it depends on the `integrationTest` task and run `integrationTest` after the test task. You can remove the last line in the following code block if you want to run `integrationTest` separately:

```
tasks.register('integrationTest', Test) {
    useJUnitPlatform()
    description = 'Runs the integration tests.'
    group = 'verification'
    testClassesDirs = sourceSets.integrationTest
        .output.classesDirs
    classpath = sourceSets.integrationTest.runtimeClasspath
}

check.dependsOn integrationTest
integrationTest.mustRunAfter test
```

Now, we can start writing the integration tests. Before writing integration tests, first, let's write the supporting Java classes in the next subsection. First, let's create the `TestUtils` class. This will contain a method that returns an instance of `ObjectMapper`. It will contain a method to check whether the JWT has expired.

Writing supporting classes for integration tests

The `ObjectMapper` instance was retrieved from the `AppConfig` class and added an extra configuration so that we can accept a single value as an array. For example, a JSON string field value might be `{ [{...}, {...}] }`. If you take a closer look at it, you will see that it is an array wrapped as a single value. When you convert this value into an object, `ObjectMapper` treats it as an array. The complete code for this class is as follows:

```
public class TestUtils {
    private static ObjectMapper objectMapper;
```

```java
   public static boolean isTokenExpired(String jwt)
       throws JsonProcessingException {
     var encodedPayload = jwt.split("\\.")[1];
     var payload = new String(Base64.getDecoder()
         .decode(encodedPayload));
     JsonNode parent = new ObjectMapper().readTree(payload);
     String expiration = parent.path("exp").asText();
     Instant expTime = Instant.ofEpochMilli(
         Long.valueOf(expiration) * 1000);
     return Instant.now().compareTo(expTime) < 0;
   }
   public static ObjectMapper objectMapper() {
     if (Objects.isNull(objectMapper)) {
       objectMapper = new AppConfig().objectMapper();
       objectMapper.configure(DeserializationFeature
           .ACCEPT_SINGLE_VALUE_AS_ARRAY, true);
     }
     return objectMapper;
   }
 }
```

https://github.com/PacktPublishing/Modern-API-Development-with-Spring-6-and-Spring-Boot-3/tree/dev/Chapter08/src/integration/java/com/packt/modern/api/TestUtils.java

Next, you need a client that lets you log in so that you can retrieve the JWT. RestTemplate is an HTTP client in Spring that provides support for making HTTP calls. The AuthClient class makes use of TestRestTemplate, which is a replica of RestTemplate from a testing perspective.

Let's write this AuthClient class, as follows:

```java
public class AuthClient {
  private final TestRestTemplate restTemplate;
  private final ObjectMapper objectMapper;
  public AuthClient(TestRestTemplate restTemplate,
      ObjectMapper objectMapper) {
    this.restTemplate = restTemplate;
    this.objectMapper = objectMapper;
  }
  public SignedInUser login(String username,
      String password) {
    SignInReq signInReq = new SignInReq()
                      .username(username).password(password);
```

```
   return restTemplate
     .execute("/api/v1/auth/token",HttpMethod.POST,
       req -> {
         objectMapper.writeValue(req.getBody(),
             signInReq);
         req.getHeaders().add(HttpHeaders.CONTENT_TYPE,
             MediaType.APPLICATION_JSON_VALUE);
         req.getHeaders().add(HttpHeaders.ACCEPT,
             MediaType.APPLICATION_JSON_VALUE);
       },
       res -> objectMapper.readValue(res.getBody(),
         SignedInUser.class)
     );
  }
}
```

https://github.com/PacktPublishing/Modern-API-Development-with-Spring-6-and-Spring-Boot-3/tree/dev/Chapter08/src/integration/java/com/packt/modern/api/AuthClient.java

The Spring test library provides MockMvc, WebTestClient, and TestRestTemplate for performing integration testing. You have already used MockMvc in unit testing. The same approach can be used for integration testing as well. However, instead of using mocks, you can use the actual objects by adding the @SpringBootTest annotation to the test class. @SpringBootTest, along with SpringExtension, provides all the necessary Spring context, such as the actual application.

@TestPropertySource provides the location of the test properties file.

WebTestClient is used to test the reactive applications. However, to test REST services, you must use TestRestTemplate, which is a replica of RestTemplate.

The integration test you are going to write is a fully fleshed-out test that doesn't contain any mocks. It will use Flyway scripts, like the actual application, which we added to src/integration/resources/db/migration. The integration test will also have its own application.properties located in src/integration/resources.

Therefore, the integration test will be as good as if you are hitting the REST endpoints from REST clients such as cURL or Postman. These Flyway scripts create the tables and data required in the H2 memory database. This data will then be used by the RESTful web service. You can also use other databases, such as Postgres or MySQL, using their test containers.

Let's create a new integration test called AddressControllerIT in src/ integration/java in an appropriate package and add the following code:

```
@ExtendWith(SpringExtension.class)
@SpringBootTest( webEnvironment = WebEnvironment.RANDOM_PORT,
```

```
        properties = "spring.flyway.clean-disabled=false")
@TestPropertySource(locations =
    "classpath:application-it.properties")
@TestMethodOrder(OrderAnnotation.class)
@TestInstance(TestInstance.Lifecycle.PER_CLASS)
public class AddressControllerIT {
  private static ObjectMapper objectMapper;
  private static AuthClient authClient;
  private static SignedInUser signedInUser;
  private static Address address;
  private static String idOfAddressToBeRemoved;
  @Autowired
  private AddressRepository repository;
  @Autowired
  private TestRestTemplate restTemplate;
  @BeforeAll
  public static void init(@Autowired Flyway flyway) {
    objectMapper = TestUtils.objectMapper();
    address = new Address().id(
     "a731fda1-aaad-42ea-bdbc-a27eeebe2cc0").
       number("9I-999")
     .residency("Fraser Suites Le Claridge")
     .street("Champs-Elysees").city("Paris").state(
       "Île-de-France").country("France").pincode("75008");
    flyway.clean();
    flyway.migrate();
  }
  @BeforeEach
  public void setup(TestInfo info)
      throws JsonProcessingException {
    if (Objects.isNull(signedInUser) ||
        Strings.isNullOrEmpty(signedInUser.getAccessToken())
        || isTokenExpired(signedInUser.getAccessToken())) {
      authClient = new AuthClient
        (restTemplate, objectMapper);
      if (if (info.getTags().contains("NonAdminUser")) {
        signedInUser = authClient.login("scott", "tiger");
      } else {
        signedInUser = authClient.login("scott2", "tiger");
      }
    }
  }
```

```
https://github.com/PacktPublishing/Modern-API-Development-with-
Spring-6-and-Spring-Boot-3/blob/main/Chapter08/src/integration/java/
com/packt/modern/api/controller/AddressControllerIT.java
```

Here, `SpringExtension` is now being used to run the unit test on the JUnit Platform. The `SpringBootTest` annotation provides all the dependencies and context for the test class. A random port is being used to run the test server. You are also using @TestMethodOrder, along with the @Order annotation, to run the test in a particular order. You are going to execute the test in a particular order so that the POST HTTP method on the addresses resource is only called before the DELETE HTTP method on the addresses resource. This is because you are passing the newly created address ID in the DELETE call. Normally, tests run in a random order. If the DELETE call is made before the POST call, then the build will fail, without testing the proper scenarios.

@TestInstance sets the life cycle of the test instance to per class (TestInstance.Lifecycle. PER_CLASS) because we want to clean and migrate the database before integration test execution.

The static init() method is annotated with @BeforeAll and will be run before all the tests. You are setting up objectMapper and the address model in this method. You are also making use of the Flyway instance for cleaning the database schema and recreating the schema using the migrate command.

The method's setup will be run before each test is executed because it is marked with the @BeforeEach annotation. Here, you are making sure that the login call will only be made if signedInUser is null or the token has expired. The TestInfo instance helps us to assign different users – scott2 (admin) and scott (non-admin) – for different tests.

Let's add an integration test that will verify the GET /api/v1/addresses REST endpoint, as shown in the following code:

```
@Test
@DisplayName("returns all addresses")
@Order(6)
public void getAllAddress() throws IOException {
  // given
  MultiValueMap<String, String> headers =
      new LinkedMultiValueMap<>();
  headers.add(HttpHeaders.CONTENT_TYPE,
                          MediaType.APPLICATION_JSON_VALUE);
  headers.add(HttpHeaders.ACCEPT,
                          MediaType.APPLICATION_JSON_VALUE);
  headers.add("Authorization", "Bearer " +
      signedInUser.getAccessToken());
  // when
  ResponseEntity<JsonNode> addressResponseEntity =
    restTemplate.exchange("/api/v1/addresses",
```

```
    HttpMethod.GET,
        new HttpEntity<>(headers), JsonNode.class);
// then
assertThat(addressResponseEntity.getStatusCode())
  .isEqualTo(HttpStatus.OK);
JsonNode n = addressResponseEntity.getBody();
List<Address> addressFromResponse = objectMapper
  .convertValue(n,new TypeReference
    <ArrayList<Address>>(){});
assertThat(addressFromResponse).hasSizeGreaterThan(0);
assertThat(addressFromResponse.get(0))
  .hasFieldOrProperty("links");
assertThat(addressFromResponse.get(0))
  .isInstanceOf(Address.class);
}
```

https://github.com/PacktPublishing/Modern-API-Development-with-Spring-6-and-Spring-Boot-3/blob/main/Chapter08/src/integration/java/com/packt/modern/api/controller/AddressControllerIT.java

First, you must set the headers in the given section. Here, you are using the signedInUser instance to set the bearer token. Next, you must call the exchange method of TestRestTemplate, which takes four arguments – the URI, the HTTP method, HttpEntity (which contains the headers and payload if required), and the type of the returned value. You can also use optional fifth argument if the template is being used to set urlVariables, which expands the template.

Then, you must use the assertions to perform the verification process. Here, you can see that it replicates the actual calls.

Run the tests using the following command:

```
$ gradlew clean integrationTest
# or
$ gradlew clean build
```

After this, you can find the test report in Chapter08/build/ reports/tests/integrationTest. The test report should look like this:

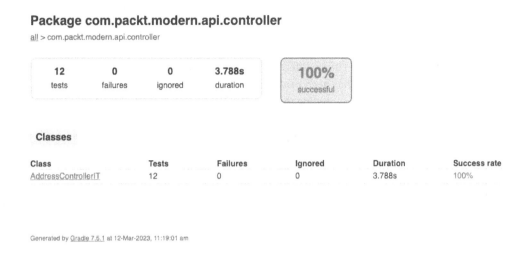

Package com.packt.modern.api.controller

all > com.packt.modern.api.controller

| 12 | 0 | 0 | 3.788s | 100% |
| tests | failures | ignored | duration | successful |

Classes

Class	Tests	Failures	Ignored	Duration	Success rate
AddressControllerIT	12	0	0	3.788s	100%

Generated by Gradle 7.5.1 at 12-Mar-2023, 11:19:01 am

Figure 8.3 – Integration test report

You can find all the test address resources in `AddressControllerIT.java`, which contains tests for success, errors, authentication, and authorization. It has tests for all types of operations, including `create`, `read`, and `delete` operations.

You have now learned how to write integration tests. You can make use of this skill to write integration tests for other REST resources.

Summary

In this chapter, you explored both manual and automated testing. You learned how to write unit and integration tests using JUnit, the Spring test libraries, AssertJ, and Hamcrest. You also learned how to use the Gherkin `Given > When > Then` language to make tests more readable. You then learned how to separate unit and integration tests.

Finally, you learned about various test automation skills by automating unit and integration tests. This will help you to automate your tests and catch bugs and gaps before you deliver the code to quality analysts or customers.

In the next chapter, you will learn how to containerize an application and deploy it in Kubernetes.

Questions

1. What is the difference between unit testing and integration testing?
2. What is the advantage of having separate unit and integration tests?
3. What is the difference between mocking and spying on an object?

Answers

1. Unit testing is done to test the smallest code unit, such as a method, whereas integration testing is performed where either different layers or multiple modules are involved. In this chapter, integration testing has been performed for the entire application, which involves all the layers of the application, including the database, whereas unit testing has been performed class-wise for each of the methods. In the context of this chapter, unit testing is white-box testing, whereas API integration testing is a kind of black-box testing because you verify the API's functional requirement.

2. Having separate unit and integration tests (and including their source location) allows you to manage tests easily. You can also have a configurable build setup that will perform unit testing during development or on demand because unit tests are faster. You can run only unit tests by using the `gradlew clean build -x integrationTest` command, whereas on merge request builds, you can execute the integration tests to verify the merge request. The default build (`gradlew clean build`) will execute both unit and integration tests.

3. When you use `Mockito.mock()` or `@Mock`, it creates a complete fake object of the given class, and then you can stub its method based on the test requirement, whereas `Mockito.spy()` or `@Spy` creates the real object, on which you can stub the required methods. If stubbing is not done on the `spy` object, then its real methods will be called during the test.

Further reading

- JUnit: `https://junit.org/`
- AssertJ: `https://assertj.github.io/doc/`
- Hamcrest: `http://hamcrest.org/`
- Mockito: `https://site.mockito.org/`
- *Test Automation Engineering Handbook*: `https://www.packtpub.com/product/test-automation-engineering-handbook/9781804615492`

9
Deployment of Web Services

In this chapter, you will learn about the fundamentals of containerization, Docker, and Kubernetes. You will then use these concepts to containerize a sample e-commerce app using Docker. This container will then be deployed as a Kubernetes cluster. You will use Minikube for Kubernetes, which makes learning and Kubernetes-based development easier.

After completing this chapter, you will be able to perform containerization and container deployment in a Kubernetes cluster.

In this chapter, you'll explore the following topics:

- Exploring the fundamentals of containerization
- Building a Docker image
- Deploying an application in Kubernetes

Technical requirements

You will need the following to develop and execute the code in this chapter:

- Docker (the container engine)
- Kubernetes (Minikube)
- An internet connection to clone the code (`https://github.com/PacktPublishing/Modern-API-Development-with-Spring-6-and-Spring-Boot-3/tree/dev/Chapter09`) and download the dependencies and Gradle
- Insomnia/cURL or any API client for API testing

Let's begin!

What is containerization?

One problem that's encountered frequently by teams while developing large, complex systems is that the code that works on one machine doesn't work on another. The main reason behind these kinds of scenarios is a mismatch of dependencies (such as different versions of Java, a certain web server, or OS), configurations, or files.

Also, setting up a new environment to deploy new products sometimes takes a day or more. This is unacceptable in today's environment and slows down your development turnaround. These kinds of issues can be solved by containerizing the application.

In containerization, an application is bundled, configured, and wrapped with all the required dependencies and files. This bundle can then be run on any machine that supports the containerization process. This bundling ensures that the application displays the exact same behavior in all environments. As a result, bugs related to misconfigurations or dependencies can be resolved, and the deployment time can be reduced to a few minutes or less.

This bundle, which sits on top of a physical machine and its OS, is called a container. This container shares the kernel, as well as the libraries and binaries of its host OS, in read-only mode. Therefore, the container is lightweight. In this chapter, you will use Docker and Kubernetes for containerization and container deployment.

A related concept is virtualization – the process of creating a virtual environment using the existing hardware system by splitting it into different parts. Each part acts as a separate, distinct, individual system. These systems are called **virtual machines** (**VMs**). Each VM runs on its own unique OS with its own binaries, libraries, and apps. VMs are heavyweight and can be many **gigabytes** (**GB**) in size. A hardware system can have VMs with different OSes, such as Unix, Linux, and Windows. The following diagram depicts the difference between VMs and containers:

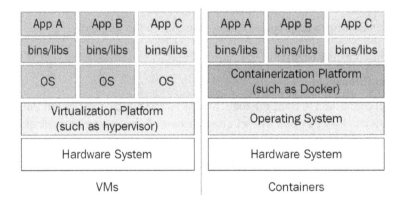

Figure 9.1 – VMs versus containers

Sometimes, people think that virtualization and containerization are the same thing, but they are not. VMs are created on top of the host system, which shares its hardware with the VMs, whereas containers are executed as isolated processes on top of the hardware and its OS. Containers are lightweight and are only a few MB, sometimes GB, whereas VMs are heavyweight and many GB in size. Containers run faster than VMs, and they are also more portable.

We'll explore containers in more detail by building a Docker image in the next section.

Building a Docker image

At this point, you know the benefit of containerization and why it is becoming popular – you create an application, product, or service, bundle it using containerization, and give it to the QA team, customer, or DevOps team to run without any issues.

In this section, you'll learn how to use Docker as a containerization platform. Let's learn about it before creating a Docker image of a sample e-commerce app.

What is Docker?

Launched in 2013, Docker is a leading container platform and an open source project. Ten thousand developers tried it after its interactive tutorial was launched in August 2013. It was downloaded 2.75 million times by the time of its 1.0 release in June 2013. Many large corporations have signed a partnership agreement with Docker Inc., including Microsoft, Red Hat, HP, and OpenStack, as well as service providers such as AWS, IBM, and Google.

Docker makes use of Linux kernel features to ensure resource isolation and the packaging of an application, along with its dependencies, such as `cgroups` and `namespaces`. Everything in a Docker container executes natively on the host and uses the host kernel directly. Each container has its own user namespace – a **process identifier** (**PID**) for process isolation, a **network** (**NET**) to manage network interfaces, **inter-process communication** (**IPC**) to manage access to IPC resources, a **mount point** (**MNT**) to manage filesystem mount points, and **Unix Time Sharing** (**UTS**) namespaces to isolate kernel and version identifiers. This packaging of dependencies enables an application to run as expected across different Linux OSes and distributions by supporting a level of portability.

Furthermore, this portability allows developers to develop an application in any language and then easily deploy it from any computer, such as a laptop, to different environments, such as test, stage, or production. Docker runs natively on Linux. However, you can also run Docker on Windows and macOS.

Containers are comprised of just an application and its dependencies, including the basic OS. This makes the application lightweight and efficient in terms of resource utilization. Developers and system administrators are interested in a container's portability and efficient resource utilization.

We'll explore Docker's architecture in the next subsection.

Understanding Docker's architecture

As specified in its documentation, Docker uses a client-server architecture. The Docker client (**Docker**) is basically a **command-line interface** (**CLI**) that is used by an end user; clients communicate back and forth with the Docker server (read as a Docker daemon). The Docker daemon does the heavy lifting, in that it builds, runs, and distributes your Docker containers. The Docker client and the daemon can run on the same system or different machines.

The Docker client and daemon communicate via sockets or through a RESTful API. Docker registers are public or private Docker image repositories that you can upload or download images from – for example, Docker Hub (`hub.docker.com`) is a public Docker registry.

The primary components of Docker are as follows:

- **Docker image**: A Docker image is a read-only template. For example, an image can contain an Ubuntu OS with an Apache web server and your web application installed on it. Docker images are the building components of Docker, and images are used to create Docker containers. Docker provides a simple way to build new images or update existing images. You can also use images created by others and/or extend them.

- **Docker container**: A Docker container is created from a Docker image. Docker works so that the container can only see its own processes, and it has its own filesystem layered on a host filesystem and a networking stack, which pipes to the host-networking stack. Docker containers can be run, started, stopped, moved, or deleted. Docker also provides commands such as `docker stats` and `docker events` for container usage statics, such as CPU and memory usage, and for activities that are performed by the Docker daemons, respectively. These commands help you monitor Docker in a deployed environment.

Docker container life cycle

You also need to be aware of Docker's container life cycle, which is as follows:

1. **Creates a container**: Docker creates a container from the Docker image using the `docker create` command.

2. **Runs the container**: Docker runs the container that was created in *step 1* using the `docker run` command.

3. **Pauses the container (optional)**: Docker pauses the process running inside the container using the `docker pause` command.

4. **Un-pauses the container (optional)**: Docker un-pauses the processes running inside the container using the `docker unpause` command.

5. **Starts the container**: Docker starts the container using the `docker start` command.

6. **Stops the container**: Docker stops the container and processes running inside the container using the `docker stop` command.

7. **Restarts the container**: Docker restarts the container and processes running inside it, using the `docker restart` command.

8. **Kills the container**: Docker kills the running container using the `docker kill` command.

9. **Destroys the container**: Finally, Docker removes the stopped containers using the `docker rm` command. Therefore, this should only be performed for the containers in a stopped state.

At this point, you might be eager to use the Docker container life cycle, but first, you'll need to install Docker by going to `https://docs.docker.com/get-docker/`.

Once you've installed Docker, go to `https://docs.docker.com/get-started/#start-the-tutorial` to execute the first Docker command. You can refer to `https://docs.docker.com/engine/reference/commandline/docker/` to learn more about Docker commands.

For more information, you can look at the overview of Docker that is provided by Docker (`https://docs.docker.com/get-started/overview/`).

Let's make the necessary code changes so that we can create a Docker image for a sample e-commerce app.

Coding to build an image by adding the Actuator dependency

I advise you to refer back to *Chapter 8*, *Testing APIs*, as a basis for this chapter's code. You don't need any additional libraries to create a Docker image. However, you do need to add the Spring Boot Actuator dependency, which provides production-ready features for the sample e-commerce app we'll create.

The dependency's features help you to monitor and manage applications using HTTP REST APIs and **Java Management Extensions** (JMX). These endpoints can be found in their respective documentation (`https://docs.spring.io/spring-boot/docs/current/reference/html/actuator.html#actuator.endpoints`). In this chapter, however, we will only use the `/actuator/health` endpoint, which tells us about the application's health status. For the purpose of this exercise, It is sufficient to find out the health of the services/applications running inside the Docker container.

You can add Actuator by performing the following steps:

1. Add the Actuator dependency to `build.gradle` (`https://github.com/PacktPublishing/Modern-API-Development-with-Spring-6-and-Spring-Boot-3/tree/dev/Chapter09/build.gradle`):

   ```
   runtimeOnly 'org.springframework.boot:spring-boot-starter-actuator'
   ```

2. Next, you need to remove all security from the /actuator endpoints. Let's add a constant to Constants.java (https://github.com/PacktPublishing/Modern-API-Development-with-Spring-6-and-Spring-Boot-3/tree/dev/Chapter09/src/main/java/com/packt/modern/api/security/Constants.java) for the Actuator URL, as shown here:

```
public static final String ACTUATOR_URL_PREFIX = "/actuator/**";
```

3. Now, you can update the security configuration in SecurityConfig.java (https://github.com/PacktPublishing/Modern-API-Development-with-Spring-6-and-Spring-Boot-3/tree/dev/Chapter09/src/main/java/com/packt/modern/api/security/SecurityConfig.java), as shown here:

```
// rest of the code
req.requestMatchers(toH2Console()).permitAll()
    .requestMatchers(new AntPathRequestMatcher(
        ACTUATOR_URL_PREFIX)).permitAll()
    .requestMatchers(new AntPathRequestMatcher
        (TOKEN_URL, HttpMethod.POST.name())).permitAll()
// rest of the code
```

With that, you have added a matcher with Actuator endpoints. This allows all Actuator endpoints to be accessed with and without authentication and authorization.

Now, you can configure the Spring Boot plugin's task, called bootBuildImage, to customize the name of the Docker image. We will do this in the next subsection.

Configuring the Spring Boot plugin task

The Spring Boot Gradle plugin already provides a command (bootBuildImage) to build Docker images. It becomes available when the Java plugin is applied in the plugins section. The bootBuildImage task is only available for building a .jar file and is not available for building a .war file.

You can customize an image's name by adding the following code block to the build.gradle file:

```
bootBuildImage {
    imageName = "192.168.1.2:5000/${project.name}:${
        project.version}"
}
```

Here, change the IP address and port of the local Docker registry. The configuration of the Docker registry is explained in the next section. A Docker image will be built based on your project's name and version. The project version is already defined in the `build.gradle` file's top section. The project name, on the other hand, is picked from the `settings.gradle` file (`https://github.com/PacktPublishing/Modern-API-Development-with-Spring-6-and-Spring-Boot-3/tree/dev/Chapter09/settings.gradle`). Let's rename it, as shown in the following code snippet:

```
rootProject.name = 'packt-modern-api-development-chapter09'
```

In *Chapter 8*, *Testing APIs*, the value of `rootProject.name` contains a capital letter, so the Docker image build failed. This is because the plugin has a validation check for capital letters. Therefore, Docker image names should only be in lowercase.

For more information and customization options, please refer to the plugin documentation (`https://docs.spring.io/spring-boot/docs/current/gradle-plugin/reference/htmlsingle/#build-image`).

Now that you have configured the code, you can use this to build an image after configuring the Docker registry. You will do this in the next subsection.

Configuring the Docker registry

If you have Docker Desktop installed, by default, when you build an image (`gradlew bootBuildImage`), it will be called `docker.io/library/packt-modern-api-development-chapter09:0.0.1-SNAPSHOT`. Here, the name refers to `docker.io/library/packt-modern-api-development-chapter09` and the version refers to `0.0.1-SNAPSHOT`. You may be wondering why the name is prefixed with `docker.io/library`. This is because if you don't specify the Docker registry, it takes the `docker.io` registry by default. You need a Docker registry where you can pull and push images from. It is like an artifact repository, where you push and pull artifacts such as Spring libraries.

Once the image has been built, you can push it to Docker Hub by applying your Docker Hub login credentials. Then, you can fetch the image from Docker Hub for deployment in your Kubernetes environment. For development purposes, this is not an ideal scenario. The best option is to configure the local Docker registry and then use it for Kubernetes deployment.

> **Using Git Bash on Windows**
>
> You can use Git Bash on Windows to run these commands; it emulates Linux commands.

Let's execute the following commands to check whether Docker is up and running:

```
$ docker version
Client:
 Cloud integration: v1.0.22
 Version:           20.10.11
 API version:       1.41
 Go version:        go1.16.10
 Git commit:        dea9396
 Built:             Thu Nov 18 00:36:09 2021
 OS/Arch:           darwin/amd64
 Context:           default
 Experimental:      true
Server: Docker Engine - Community
 Engine:
  Version:          20.10.11
  API version:      1.41 (minimum version 1.12)
  Go version:       go1.16.9
// Output truncated for brevity
```

Here, Docker is installed. Therefore, when you run docker version, it displays the output. A version output without any error confirms Docker is up and running.

Now, you can pull and start the Docker registry by using the following command:

```
$ docker run -d -p 5000:5000 -e REGISTRY_STORAGE_DELETE_ENABLED=true
--restart=always --name registry registry:2
Unable to find image 'registry:2' locally
2: Pulling from library/registry
ef5531b6e74e: Pull complete
a52704366974: Pull complete
dda5a8ba6f46: Pull complete
eb9a2e8a8f76: Pull complete
25bb6825962e: Pull complete
Digest: sha256:41f413c22d6156587e2a51f3e80c09808b8c70e82be149b82b5e01
96a88d49b4
Status: Downloaded newer image for registry:2
bca056bf9653abb14ee6c461612a999c7c61ab45ea8837ecfa1c4b1ec5e5f047
```

Here, when you run the Docker registry for the first time, it downloads the Docker registry image before running it. The execution of the Docker registry creates the container, called registry, on port 5000. If port 5000 is used by other services in your machine, then you can use a different port such as 5001. There are two port entries – one internal container port and another exposed external port. Both are set to 5000. The –restart=always flag tells Docker to start the registry container every time Docker is restarted. The REGISTRY_STORAGE_DELETE_ENABLED flag, as its name suggests, is used to remove any images from `registry`, as it is set to `true`. The default value of this flag is `false`.

Now, let's check the containers:

```
$ docker ps
CONTAINER ID    IMAGE        COMMAND
CREATED              STATUS            PORTS               NAMES
bca056bf9653    registry:2   "/entrypoint.sh /etc…"
11 minutes ago   Up 11 minutes   0.0.0.0:5000->5000/tcp   registry
```

This shows that the Docker container registry is up and running and was created using the `registry:2` image.

The host's name is necessary when you're using the containers. Therefore, you'll use the IP number instead of the local hostname for the registry. This is because the container will refer to its localhost, rather than the localhost of your system when you use `localhost` as the hostname. In a Kubernetes environment, you need to provide a registry host, so you will need to use the IP or a proper hostname in place of `localhost`.

Let's find out what IP we can use by running the following command:

```
# For Mac
$ echo $(osascript -e "Ipv4 address of (system info)")
192.168.1.2
# For Windows
$ ipconfig
Windows IP Configuration
Ethernet adapter Ethernet:
   Media State . . . . . . . . . . . : Media disconnected
   Connection-specific DNS Suffix . :
Ethernet adapter vEthernet (Default Switch):
   Connection-specific DNS Suffix . :
   Link-local Ipv6 Address . . . . . : ef80::2099:f848:8903:f996%81
   Ipv4 Address. . . . . . . . . . . : 192.168.1.2
   Subnet Mask . . . . . . . . . . . : 255.255.240.0
   Default Gateway . . . . . . . . . :
```

You can find your system's IP address in the row highlighted in the preceding output. You can use a similar command on Linux to find out the IP address of your system.

We haven't yet configured the **Transport Layer Security** (**TLS**) for our system host, so this registry is an insecure registry. Docker only supports secure registries by default. We must configure Docker so that it can use insecure registries. Refer to the Docker documentation to learn how to configure an insecure registry (https://docs.docker.com/registry/insecure/-deploy-a-plain-http-registry).

> **Adding an insecure registry to daemon.json**
>
> `daemon.json` can be found in `/etc/docker/daemon.json` on Linux.
>
> 1. For Docker Desktop on Mac/Windows, navigate to **Docker app** | **Settings** | **Docker Engine**.
>
> 2. Add the `insecure-registries` entry to the JSON:
>
> ```
> {
>     "features": {
>         "buildkit": true
>     },
>     "insecure-registries": [
>         "192.168.1.2:5000"
>     ],
>     ...
>
>     ...
>
> }
> ```
>
> 3. Restart Docker.

Note that to build and publish the image successfully, the Docker configuration must be performed with a local registry, as explained previously.

> **Note**
>
> Don't use an insecure registry in any environment other than a local or development one for security purposes.

Now, let's create a Docker image for a sample e-commerce app.

Executing a Gradle task to build an image

You need to make a change to the `bootBuildImage` task so that the image's name contains the local Docker registry's prefix. Spring Boot's `bootBuildImage` uses Paketo Buildpacks to build the Docker image. It supports **long-term support** (**LTS**) Java releases and only current non-LTS Java releases. This means, when non-LTS Java 20 is released, then it will remove support for Java 19. Similarly, when Java 21 gets released, it will remove the Java 20 support. However, it won't remove Java 17 support because Java 17 is an LTS release. We can make this change like so:

```
bootBuildImage {
    imageName = "192.168.1.2:5000/${project.name}:${
        project.version}"
    environment = ["BP_JVM_VERSION" : "17"]
}
```

Here, you have customized the name of the Docker image according to the local Docker registry. You should change the IP address and port as per your system and configuration. You have also used the environment property to set the Paketo Buildpacks variables. You have set the JVM version to 17. It is recommended to use Java 17 (or any future LTS release). You can find all the supported Paketo Buildpacks environment variables at `https://github.com/paketo-buildpacks/bellsoft-liberica#configuration`. At the time of writing, Paketo Buildpacks does not provide official support to build images for ARM. However, there are alternative builders available such as `https://github.com/dashaun/paketo-arm64`, which supports building on ARM.

Now, you can build the image by executing the following commands from your project's home directory:

```
$ ./gradlew clean build
    # build the jar file of app after running the tests
$ ./gradlew bootBuildImage
> Task :bootBuildImage
Building image '192.168.1.2:5000/packt-modern-api-development-
chapter09:0.0.1-SNAPSHOT'
 > Pulling builder image
 'docker.io/paketobuildpacks/builder:base'
 ...........................................
 > Pulled builder image
 'paketobuildpacks/builder@sha256:e2bf5f2355b0daddb61c6c7ed3e55e58ab581
900da63f892949ded8b772048ee'
 > Pulling run image 'docker.io/paketobuildpacks/run:base-cnb'
 ...........................................
 > Pulled run image
 'paketobuildpacks/run@sha256:4a2fbf87a81964ef1a95445f343938ed19406fff
da142586a35c9e20904a3315'
 > Executing lifecycle version v0.16.0
 > Using build cache volume 'pack-cache-2fdc28fe99dc.build'
// continue...
```

The Spring Boot Gradle plugin uses the *Paketo BellSoft Liberica Buildpack* (`docker. io/paketobuildpacks`) to build an application image. First, it pulls the image from Docker Hub and then runs its container, as shown here:

```
> Running creator
    [creator]    ===> ANALYZING
    [creator]    Previous image with name
                 "192.168.1.2:5000/packt-modern-api-development-
                 chapter09:0.0.1-SNAPSHOT" not found
    [creator]    ===> DETECTING
    // truncated output for brevity
    [creator]    ===> RESTORING
    [creator]    ===> BUILDING
```

```
// truncated output for brevity
[creator]
[creator]   Paketo Buildpack for BellSoft Liberica 9.11.0
[creator]   https://github.com/paketo-buildpacks/bellsoft-
            liberica
// truncated output for brevity
[creator]   Using Java version 17 from BP_JVM_VERSION
[creator] BellSoft Liberica JRE 17.0.6: Contributing to layer
[creator]   Downloading from https://github.com/bell-sw/Liberica/
releases/download/17.0.6+10/bellsoft-jre17.0.6+10-linux-amd64.tar.gz
[creator]   Verifying checksum
[creator]   Expanding to /layers/paketo-
            buildpacks_bellsoft-liberica/jre
// truncated output for brevity
```

Here, the Spring Boot plugin uses Bellsoft's JRE 17.0.6 with Linux as a base image to build images. It uses finely grained filesystem layers inside the container to do so:

```
[creator]   Launch Helper: Contributing to layer
[creator]   Creating /layers/paketo-buildpacks_bellsoft-
            liberica/helper/exec.d/active-processor-count
[creator]   Creating /layers/paketo-buildpacks_bellsoft-
            liberica/helper/exec.d/java-opts
// truncated output for brevity
[creator]   Paketo Buildpack for Syft 1.26.0
[creator]   https://github.com/paketo-buildpacks/syft
[creator]   Downloading from
            https://github.com/anchore/syft/releases/
            download/v0.75.0/syft_0.75.0_linux_amd64.tar.gz
// truncated output for brevity
[creator]   Paketo Buildpack for Executable JAR 6.6.2
[creator] https://github.com/paketo-buildpacks/executable-jar
[creator]        Class Path: Contributing to layer
// truncated output for brevity
```

The plugin continues to add the layers and then the labels. Finally, it creates the Docker image:

```
[creator]   Paketo Buildpack for Spring Boot 5.23.0
[creator]   https://github.com/paketo-buildpacks/spring-boot
// truncated output for brevity
[creator]       ===> EXPORTING
[creator]   Adding layer 'paketo-buildpacks/ca-
            certificates:helper'
// truncated output for brevity
```

```
[creator]    Adding layer 'paketo-buildpacks/executable-
             jar:classpath'
[creator]    Adding layer 'paketo-buildpacks/spring-
             boot:helper'
[creator]    Adding layer 'paketo-buildpacks/spring-
             boot:spring-cloud-bindings'
[creator]    Adding layer 'paketo-buildpacks/spring-boot:web-
             application-type'
[creator]    Adding 5/5 app layer(s)
[creator]    Adding layer 'buildpacksio/lifecycle:launcher'
// truncated output for brevity
[creator]    Adding label 'org.springframework.boot.version'
[creator]    Setting default process type 'web'
[creator]    *** Images (9cc6ef620b7c):
[creator]    192.168.1.2:5000/packt-modern-api-development-
             chapter09:0.0.1-SNAPSHOT
Successfully built image '192.168.1.2:5000/packt-modern-api-
development-chapter09:0.0.1-SNAPSHOT'
BUILD SUCCESSFUL in 1m 22s
```

You can learn more about Spring Boot, Docker, and Kubernetes and their configuration at https://github.com/dsyer/kubernetes-intro.

Now that the Docker image has been built, you can use this image to run the sample e-commerce app locally using the following command:

```
$ docker run -p 8080:8080 192.168.1.2:5000/packt-modern-api-
development-chapter09:0.0.1-SNAPSHOT
```

This command will run the application on port 8080 inside the container. Because it has been exposed on port 8080, you can access the sample e-commerce app on 8080 outside the container too, once the app is up and running. You can test the application by running the following command in a separate terminal tab/window once the application container is up and running:

```
$ curl localhost:8080/actuator/health
{"status":"UP"}
$ curl localhost:8080/actuator
{
  "_links": {
    "self": {
      "href": "http://localhost:8080/actuator",
      "templated": false },
    "health-path": {
      "href": "http://localhost:8080/actuator/ health/{*path}",
      "templated": true },
```

```
    "health": {
      "href": "http://localhost:8080/actuator/health",
      "templated": false }
    }
  }
```

The `curl localhost:8080/actuator` command returns the available Actuator endpoints, such as `health` and `health-path`.

You can also list the containers and their statuses by using the following command:

```
$ docker ps
CONTAINER ID    IMAGE
                  COMMAND                      CREATED          STATUS
          PORTS                    NAMES
62255c54ab52    192.168.1.2:5000/packt-modern-api-development-chapter
09:0.0.1-SNAPSHOT    "/cnb/process/web"         7 minutes ago    Up 7
minutes    0.0.0.0:8080->8080/tcp    elated_ramanujan
bca056bf9653    registry:2
                     "/entrypoint.sh /etc..."   58 minutes ago   Up 58
minutes    0.0.0.0:5000->5000/tcp    registry
```

Next, let's find out the available Docker images by running the following command:

```
$ docker images
REPOSITORY
TAG              IMAGE ID        CREATED         SIZE
paketobuildpacks/run
base-cnb         68c538f4e078    5 hours ago     87MB
registry                                                         2
                 0d153fadf70b    5 weeks ago     24.2MB
paketobuildpacks/builder
base             38446f68a5f8    43 years ago    1.26GB
192.168.1.2:5000/packt-modern-api-development-chapter09
0.0.1-SNAPSHOT   9cc6ef620b7c    43 years ago    311MB
```

Now, you can tag and push the application image using the following commands:

```
$ docker tag 192.168.1.2:5000/packt-modern-api-development-
chapter09:0.0.1-SNAPSHOT 192.168.1.2:5000/packt-modern-api-
development-chapter09:0.0.1-SNAPSHOT
$ docker push 192.168.1.2:5000/packt-modern-api-development-
chapter09:0.0.1-SNAPSHOT
...

b7e0fa7bfe7f: Pushed
0.0.1-SNAPSHOT: digest: sha256:bde567c41e57b15886bd7108beb26b5de7b44c6
6cdd3500c70bd59b8d5c58ded size: 5327
```

Similarly, you can also query the local Docker registry container. First, let's run the following command to find all the published images in the registry (the default value is 100):

```
$ curl -X GET http://192.168.1.2:5000/v2/_catalog
{"repositories":["packt-modern-api-development-chapter09"]}
```

Similarly, you can find out what all the available tags are for any specific image by using the following command:

```
$ curl -X GET http://192.168.1.2:5000/v2/packt-modern-api-development-
chapter09/tags/list
{"name":"packt-modern-api-development-chapter09","tags":["0.0.1-
SNAPSHOT"]}
```

For these commands, you can also use `localhost` instead of the IP, if you run a local registry container.

We'll deploy this image to Kubernetes in the next section.

Deploying an application in Kubernetes

Docker containers are run in isolation. You need a platform that can execute multiple Docker containers and manage or scale them. Docker Compose does this for us. However, this is where Kubernetes helps. It not only manages the container but also helps you scale the deployed containers dynamically.

You will use Minikube to run Kubernetes locally. You can use it on Linux, macOS, and Windows. It runs a single-node Kubernetes cluster, which is used for learning or development purposes. You can install it by referring to the respective guide (`https://minikube.sigs.k8s.io/docs/start/`).

Once Minikube is installed, you need to update Minikube's local insecure registry because, by default, Minikube's registry uses Docker Hub. Adding an image to Docker Hub and then fetching it for local usage is cumbersome for development. You can add a local insecure registry to your Minikube environment by adding your host IP and local Docker registry port to Minikube's config at *HostOptions | EngineOptions | InsecureRegistry* in `~/.minikube/machines/minikube/config.json` (note that this file is only generated after Minikube has been started once; therefore, start Minikube before modifying `config.json`):

```
$ vi ~/.minikube/machines/minikube/config.json
 41      ...
 42      "DriverName": "qemu2",
 43      "HostOptions": {
 44          "Driver": "",
 45          "Memory": 0,
 46          "Disk": 0,
 47          "EngineOptions": {
 48              "ArbitraryFlags": null,
```

```
49              "Dns": null,
50              "GraphDir": "",
51              "Env": null,
52              "Ipv6": false,
53              "InsecureRegistry": [
54                  "10.96.0.0/12",
55                  "192.168.1.2:5000"
56              ],
57         ...
```

Once the insecure registry has been updated, you can start Minikube using the following command:

```
$ minikube start --insecure-registry="192.168.80.1:5000"
😊   minikube v1.29.0 on Darwin 13.1
✦   Using the qemu2 driver based on existing profile
🔥   Starting control plane node minikube in cluster minikube
🔄   Restarting existing qemu2 VM for "minikube" ...
🐳   Preparing Kubernetes v1.26.1 on Docker 20.10.23 ...
🔗   Configuring bridge CNI (Container Networking Interface) ...
    ▪ Using image gcr.io/k8s-minikube/storage-provisioner:v5
🔎   Verifying Kubernetes components...
🌟   Enabled addons: default-storageclass, storage-provisioner
🏃   Done! kubectl is now configured to use "minikube" cluster and
"default" namespace by default
```

Here, we have used the --insecure-registry flag while starting Minikube. This is important, as it makes the insecure registry work. The Kubernetes cluster uses the default namespace by default.

A **namespace** is a Kubernetes special object that allows you to divide the Kubernetes cluster resources among users or projects. However, you can't have nested namespaces. Kubernetes resources can only belong to single namespaces.

You can check whether Kubernetes works or not by executing the following command once Minikube is up and running:

```
$ kubectl get po -A
NAMESPACE     NAME                           READY     STATUS
RESTARTS          AGE
kube-system   coredns-787d4945fb-5hzc2       1/1       Running   3 (17m
ago)          30m
kube-system   etcd-minikube                  1/1       Running   5 (17m
ago)          32m
kube-system   kube-apiserver-minikube        1/1       Running   4 (17m
ago)          32m
```

```
kube-system   kube-controller-manager-minikube 1/1      Running   5
(3m58s ago)    32m
kube-system   kube-proxy-z4n66                 1/1      Running   4 (17m
ago)       31m
kube-system   kube-scheduler-minikube          1/1      Running   4 (17m
ago)       32m
kube-system   storage-provisioner              1/1      Running
2 (3m25s ago)   18m
```

The kubectl command is a command-line tool that's used to control a Kubernetes cluster, like the docker command for Docker. It is a Kubernetes client that uses Kubernetes REST APIs to perform various Kubernetes operations, such as deploying applications, viewing logs, and inspecting and managing cluster resources.

The get po and get pod parameters allow you to retrieve pods from your Kubernetes cluster. The -A flag instructs kubectl to retrieve objects from across namespaces. Here, you can see that all the pods are from the kube-system namespace.

These pods are created by Kubernetes and are part of its internal system.

Minikube bundles the Kubernetes dashboard as a UI for additional insight into your cluster's state. You can start it by using the following command:

```
$ minikube dashboard
    Enabling dashboard ...
      ▪ Using image docker.io/kubernetesui/dashboard:v2.7.0
      ▪ Using image docker.io/kubernetesui/metrics-
        scraper:v1.0.8
    Some dashboard features require the metrics-server addon. To
enable all features please run:
      minikube addons enable metrics-server
    Verifying dashboard health ...
    Launching proxy ...
    Verifying proxy health ...
    Opening http://127.0.0.1:56858/api/v1/namespaces/kubernetes-
dashboard/services/http:kubernetes-dashboard:/proxy/ in your default
browser...
```

Running the dashboard allows you to manage the Kubernetes cluster from the UI and looks as follows:

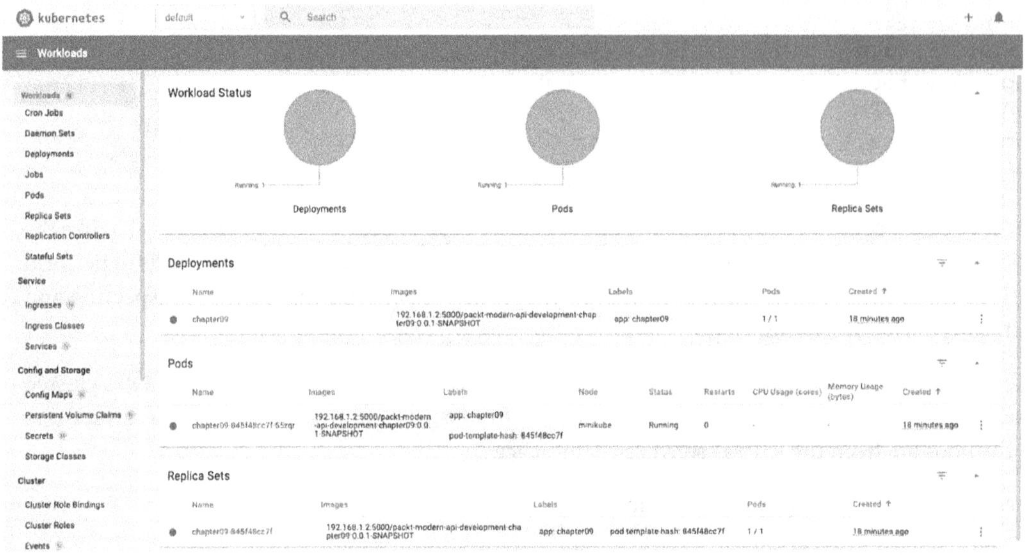

Figure 9.2 – The Kubernetes dashboard

Kubernetes uses YAML configuration to create objects. For example, you need a deployment and service object to deploy and access the sample e-commerce application. The deployment will create a pod in the Kubernetes cluster that will run the application container, and the service will allow it to access it. You can create these YAML files either manually or generate them using kubectl. You should typically use kubectl, which generates the files for you. If you need to, you can modify the content of the file.

Let's create a new directory (k8s) in the project's home directory so that we can store the Kubernetes deployment configuration. We can generate the deployment Kubernetes configuration file by using the following commands from the newly created k8s directory:

```
$ kubectl create deployment chapter09
--image=192.168.1.2:5000/packt-modern-api-developmentchapter09:0.0.1-
SNAPSHOT --dry-run=client -o=yaml > deployment.yaml
$ echo --- >> deployment.yaml
$ kubectl create service clusterip chapter09 --tcp=8080:8080
--dry-run=client -o=yaml >> deployment.yaml
```

Here, the first command generates the deployment configuration in the deployment.yaml file using the create deployment command. A Kubernetes deployment defines the scale at which you want to run your application. You can see that the replica is defined as 1. Therefore, Kubernetes will run a single replica of this deployment. Here, you pass the name (chapter09) of the deployment, the image name of the application to deploy, the `--dry-run=client` flag to preview the object that will be sent to the cluster, and the `-o=yaml` flag to generate the YAML output.

The second command appends `---` to the end of the `deployment.yaml` file.

Finally, the third command creates the service configuration in `deployment.yaml`, with a value of `8080` for both internal and external ports.

Here, you have used the same file for both deployment and service objects. However, you can create two separate files for these – `deployment.yaml` and `service.yaml`. In this case, you need to apply these objects separately in your Kubernetes cluster.

Let's have a look at the content of the `deployment.yaml` file, which was generated by the previous code block:

```yaml
apiVersion: apps/v1
kind: Deployment
metadata:
  creationTimestamp: null
  labels:
    app: chapter09
  name: chapter09
spec:
  replicas: 1
  selector:
    matchLabels:
      app: chapter09
  strategy: {}
  template:
    metadata:
      creationTimestamp: null
      labels:
        app: chapter09
    spec:
      containers:
      - image: 192.168.1.2:5000/
        packt-modern-api-developmentchapter09:0.0.1-SNAPSHOT
        name: packt-modern-api-developmentchapter09
        resources: {}
status: {}
---
apiVersion: v1
kind: Service
metadata:
```

```
  creationTimestamp: null
  labels:
    app: chapter09
  name: chapter09
spec:
  ports:
  - name: 8080-8080
    port: 8080
    protocol: TCP
    targetPort: 8080
  selector:
    app: chapter09
  type: ClusterIP
status:
  loadBalancer: {}
```

https://github.com/PacktPublishing/Modern-API-Development-with-Spring-6-and-Spring-Boot-3/tree/dev/Chapter09/k8s/deployment.yaml

Now, you can deploy the sample e-commerce application using the file created previously in the deployment.yaml file, with the following command running from the project root directory:

```
$ kubectl apply -f k8s/deployment.yaml
deployment.apps/chapter09 created
service/chapter09 created
```

This will deploy a sample e-commerce app (https://github.com/PacktPublishing/Modern-API-Development-with-Spring-6-and-Spring-Boot-3/tree/dev/Chapter09) in Kubernetes on successful creation.

Alternatively, you can perform the following steps to publish the Docker image to Minikube. Start a new terminal and execute the following commands (the same terminal window should be used here, since the eval commands are only valid in an active terminal):

1. Execute eval $(minikube docker-env) to align the Minikube environment with your Docker configuration.

2. Execute gradle bootBuildImage to generate an image based on the Minikube environment.

3. Execute the following commands:

    ```
    $ docker tag 192.168.1.2:5000/packt-modern-api-development-
    chapter09:0.0.1-SNAPSHOT 192.168.80.1:5000/ packt-modern-api-
    development-chapter09:0.0.1-SNAPSHOT
    $ docker push 192.168.1.2:5000/library/packt-modern-api-
    development-chapter09:0.0.1-SNAPSHOT
    ```

4. Execute `minikube stop` and `minikube start` to ensure that the new configuration is applied.

5. You can start the Minikube logs by using the following commands:

```
$ minikube -p minikube docker-env
$ eval $(minikube -p minikube docker-env)
```

6. Afterward, deployment can be done with the `kubectl apply -f deploymentTest.yaml` command.

This will initiate the application deployment of `chapter09`. You can then either use the Kubernetes dashboard or the `kubectl get all` command to check the status of your pod and service. **Pods** are Kubernetes' smallest and most deployable objects. They contain one or more containers and represent a single instance of a running process in a Kubernetes cluster. A pod's IP address and other configuration details may change because Kubernetes keep track of these, and it may replace them if a pod goes down. Therefore, a **Kubernetes service** adds an abstraction layer over the pods it exposes the IP addresses of and manages mapping to internal pods.

Let's run the following command to find out the status of the pod and service:

```
$ kubectl get all
NAME    READY STATUS    RESTARTS    AGE
pod/chapter09-845f48cc7f-55zqr 1/1 Running 0    9m17s
NAME    TYPE CLUSTER-IP    EXTERNAL-IP    PORT(S)    AGE
service/chapter09    ClusterIP    10.100.135.86    <none>    8080/
TCP    9m18s
service/kubernetes    ClusterIP    10.96.0.1 <none>    443/
TCP    65m
NAME    READY UP-TO-DATE    AVAILABLE    AGE
deployment.apps/chapter09 1/1    1    1    9m18s
NAME    DESIRED CURRENT READY AGE
replicaset.apps/chapter09-845f48cc7f 1    1    1    9m17s
```

This returns all the Kubernetes resources in the default namespace. Here, you can see that it returns a running pod, a service, a deployment resource, and a ReplicaSet for `chapter09`. You need to run this command multiple times until you find a successful or erroneous response (such as "`image is not pullable`").

You can't access the application running inside Kubernetes directly, as you can see from the reponse of the following command:

```
$ curl localhost:8080/actuator/health
curl: (7) Failed to connect to localhost port 8080 after 0 ms:
Connection refused
```

You must either use some kind of proxy or SSH tunneling to access the application running inside the Kubernetes cluster. Let's quickly create an SSH tunnel using the following command:

```
$ kubectl port-forward service/chapter09 8080:8080
Forwarding from 127.0.0.1:8080 -> 8080
Forwarding from [::1]:8080 -> 8080
```

The application now runs on port 8080 inside the Kubernete's cluster. It is also mapped to the local machine's port8080. Due to this port mapping, you can access the application outside the Kubernetes cluster.

Let's try again accessing the application after opening a new terminal window:

```
$ curl localhost:8080/actuator/health
{"status":"UP","groups":["liveness","readiness"]}
```

With that, the application has been successfully deployed on our Kubernetes cluster. Now, you can use the Postman collection and run all the available REST endpoints.

Summary

In this chapter, you learned about containerization and how it is different from virtualization. You also learned about the Docker containerization platform and how to use the Spring Boot plugin to generate a Docker image for a sample e-commerce app.

Then, you learned about the Docker registry and how to configure a local insecure registry so that you can use it to push and pull images locally. The same commands can be used to push and pull images from a remote Docker registry.

You also learned about Kubernetes and its cluster operations by using Minikube. You configured it so that you can pull Docker images from insecure local Docker registries.

Now, you have the necessary skills to build a Docker image of a Spring Boot application and deploy it to a Kubernetes cluster.

In the next chapter, you'll learn about the fundamentals of the gRPC APIs.

Questions

1. What is the difference between virtualization and containerization?

2. What is Kubernetes used for?

3. What is `kubectl`?

Answers

1. Virtualization is used to create VMs on top of the host system, which shares its hardware with the VMs, whereas containerization creates containers that are executed as an isolated process on top of the hardware and its OS. Containers are lightweight and need only a few MB (occasionally, GB). VMs are heavyweight and need many GB. Containers run faster and are more portable than VMs.

2. Kubernetes is a container orchestration system and is used to manage application containers. It keeps track of running containers. It shuts down containers when they are not used and restarts orphaned containers. A Kubernetes cluster is also used for scale. It can provision resources such as CPU, memory, and storage automatically when required.

3. `kubectl` is a Kubernetes **command-line interface** (**CLI**) utility that is used to run commands against a Kubernetes cluster. You can manage Kubernetes resources using `kubectl`. You used `kubectl`'s `apply` and `create` commands in this chapter.

Further reading

- *Mastering Kubernetes – Fourth Edition*: `https://www.packtpub.com/product/kubernetes-basics-and-beyond-fourth-edition/9781804611395`

- Docker documentation: `https://docs.docker.com/get-started/overview/`

- Minikube documentation: `https://minikube.sigs.k8s.io/docs/start/`

Part 3 – gRPC, Logging, and Monitoring

In this part, you will learn about gRPC-based API development. On the completion of this part, you will be able to differentiate between REST and reactive APIs with gRPC-based APIs. You will be able to build a server and client using the Protobuf schema. Finally, you will be able to facilitate distributed logging and tracing, collecting the logs as an Elasticsearch index that will be used for debugging and analysis on the Kibana app.

This part contains the following chapters:

10
Getting Started with gRPC

gRPC is an open source framework for general-purpose **Remote Procedure Calls** (**RPCs**) across a network. RPCs allow a remote procedure (hosted on a different machine) to call as if it were calling a local procedure in connected systems without coding the remote interaction details. RPC has a constant meaning in the *gRPC* abbreviation. It seems logical that the *g* in gRPC would refer to *Google* because it was initially developed there. But the meaning of the *g* has changed with every release. For its first release, version 1.0, the *g* in gRPC stood for gRPC itself. That is, in version 1.0, it stood for **gRPC Remote Procedure Call**. In this chapter, you are going to use gRPC version 1.54, where the *g* stands for **gracious**. Therefore, you can refer to gRPC as **gracious Remote Procedure Call** (for version 1.54). You can find out all the meanings of the *g* for different versions at `https://github.com/grpc/grpc/blob/master/doc/g_stands_for.md`.

In this chapter, you'll learn the fundamentals of gRPC such as its architecture, service definitions, life cycle, server, and client. This chapter will provide you with a foundation that you can use to implement gRPC-based APIs. These fundamentals will help you to implement inter-service communication in a sample e-commerce app.

You will use gRPC-based APIs to develop a basic payment gateway for processing payments in an e-commerce app in the next chapter.

> **Note**
> gRPC is pronounced *Jee-Arr-Pee-See*.

You will explore the following topics in this chapter:

- Introduction and gRPC architecture
- Understanding service definitions
- Exploring the gRPC life cycle
- Understanding the gRPC server and gRPC stub
- Handling errors

After completing this chapter, you will understand the gRPC basics, which will help you to implement a gRPC-based web service in the next chapter.

Technical requirements

This chapter contains only the theory of gRPC. However, you would generally need any gRPC API client such as Insomnia for the development and testing of gRPC-based web services.

You are going to learn the fundamentals of gRPC in this chapter, so this chapter doesn't have its own code repository. However, for actual code, you can refer to the *Chapter 11* code at `https://github.com/PacktPublishing/Modern-API-Development-with-Spring-6-and-Spring-Boot-3/tree/dev/Chapter11`.

How does gRPC work?

gRPC is an open source framework for general-purpose RPCs across a network. gRPC supports full-duplex streaming and is also mostly aligned with HTTP/2 semantics. It supports different media formats, such as **Protocol Buffers (Protobuf)**, JSON, XML, and Thrift. Protobuf is the default media format. The use of Protobuf aces the others because of higher performance.

gRPC brings the best of **REST (Representational State Transfer)** and RPC to the table and is well suited for distributed network communication through APIs. It offers some prolific features, as follows:

- It is designed for a highly scalable distributed system and offers *low latency*.
- It offers load balancing and failover.
- It can be integrated easily at the application layer for interaction with flow control because of its layered design.
- It supports cascade call cancellation.
- It offers wide communication — mobile app to server, web app to server, and any gRPC client app to the gRPC server app on different machines.

You're already aware of REST and its implementation. Let's find out the differences between REST and gRPC in the next subsection, which gives you a different perspective and allows you to choose between REST or gRPC based on your requirements and use case.

REST versus gRPC

gRPC is based on client-server architecture, whereas this is not true for REST.

Both gRPC and REST leverage the HTTP protocol. gRPC supports HTTP/2 specifications and full-duplex streaming communication in contrast to REST, which serves well for various scenarios such as voice or video calls.

You can pass payloads using query parameters, path parameters, and the request body in REST. This means that the request payload/data can be passed using different sources that lead to the parsing of the payload/data from different sources, which adds latency and complexity. On the other hand, gRPC performs better than REST as it uses the static paths and single source of the request payload.

As you know, the REST response error depends on HTTP status codes, whereas gRPC has formalized the set of errors to make it well aligned with APIs.

The REST API is more flexible in its implementation because it is purely dependent on HTTP. This gives you flexibility, but you need standards and conventions for strict verification and validation. But do you know why you need these strict verifications and validations? It is because you can implement an API in different ways. For example, you can delete a resource using any HTTP method instead of just using the HTTP DELETE method, and this simply sounds horrific.

On top of everything mentioned, gRPC is also built for supporting and handling call cancellations, load balancing, and failovers.

REST is mature and widely adopted, but gRPC brings its advantages. Therefore, you can choose between them based on their pros and cons. (Mind you, we haven't yet discussed GraphQL, which brings its own offerings. You will learn about GraphQL in *Chapter 13, Getting Started with GraphQL, and Chapter 14, GraphQL API Development and Testing.*)

Let's find out whether we can use gRPC for web communication like REST in the next subsection.

Can I call the gRPC server from web browsers and mobile apps?

Of course, you can. The gRPC framework is designed for communication in distributed systems and is mostly aligned with HTTP/2 semantics. You can call a gRPC API from a mobile app, just like calling any local object. That's the beauty of gRPC! It supports inter-service communication across the intranet and internet and calls from the mobile app and web browser to the gRPC server. Therefore, you can utilize it for all kinds of communications.

gRPC for web (that is, gRPC-web) was quite new in 2018, but now (in 2023), it is getting more recognition and is especially being used for **Internet of Things (IoT)** applications. Ideally, you should adopt it first for your internal inter-service communications and then for web/mobile server communication.

Let's find out more about its architecture in the next subsection.

Getting to know the gRPC architecture

gRPC is a general-purpose RPC-based framework. It works very well in the RPC style, which involves the following steps:

1. First, you define the service interface, which includes method signatures, with their parameters and return types.

2. Then, you implement the defined service interface as a part of the gRPC server. You are now ready to serve the remote calls.

3. Next, you need the stub for clients, which you can generate using the service interface. The client application calls the stub, which is a local call. In turn, the stub communicates with the gRPC server, and the returned value is passed to the gRPC client. This is shown in the following diagram:

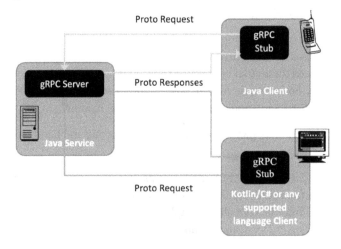

Figure 10.1 – gRPC client-server architecture

For client applications, it is just a local call to the stub to get the response. You can have a server on either the same machine or a different machine. This makes it easier to write distributed services. It is an ideal tool for writing microservices. gRPC is language-independent. You can write servers and clients in different languages. This provides a lot of flexibility for development.

gRPC is a layered architecture that has the following layers to make remote calling possible:

- **Stub**: You know the client calls the server through stubs. A stub is the topmost layer. Stubs are generated from the **interface definition language** (**IDL**) file, which contains service interfaces, methods, and messages. The IDL file will have a .proto extension if the interface is defined using Protobuf.

- **Channel**: The stub uses **application binary interfaces** (**ABIs**) to communicate with the server. The channel is the middle layer that provides these ABIs. In general, the channel provides the connection to the server on a specific host and port. That's the reason the channel has a status such as `connected` or `idle`.

- **Transport**: This is the lowest layer and uses HTTP/2 as its protocol. Therefore, gRPC provides full-duplex communication and multiplex parallel calls over the same network connection.

You can develop a gRPC-based service by following these steps:

1. Define the service interface using the `.proto` file (Protobuf).

2. Write the implementation of the service interface defined in *step 1*.

3. Create a gRPC server and register the service with it.

4. Generate the service stub and use it with the gRPC client.

You'll implement the actual gRPC service in the next chapter, *Chapter 11, gRPC API Development and Testing*.

---

gRPC stub

A stub is an object that exposes service interfaces. The gRPC client calls the stub method, hooks the call to the server, and gets the response back.

You need to understand Protobuf to define the service interfaces. Let's explore it in the next subsection.

How gRPC uses Protobuf

Protobuf was created in 2001 and was publicly made available in 2008. It was also used by Google's microservice-based system, Stubby.

gRPC also works well with JSON and other media types. However, you'll define the service interfaces using Protobuf because it is known for its performance. It allows formal contracts, better bandwidth optimization, and code generation. Protobuf is also the default format for gRPC. gRPC makes use of Protobuf not only for data serialization but also for code generation. Protobuf serializes data and, unlike JSON, YAML is not human-readable. Let's see how it is built.

Protobuf messages contain a series of key-value pairs. The key specifies the `message` field and its type. Let's examine the following `Employee` message:

```
message Employee {
   int64 id = 1;
   string firstName = 2;
}
```

Let's represent this message using Protobuf (with an id value of 299 and a firstName value of Scott), as shown in the following diagram:

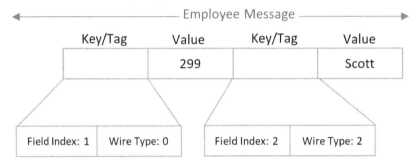

Figure 10.2 – Employee message representation using Protobuf

The Id and firstName fields are tagged with numbers, sequenced 1 and 2, respectively, which is required for serialization. The wire type is another aspect that provides information to find the length of the value.

The following table contains the wire types and their respective meanings:

Wire Type	Meaning	Used For
0	Var int (variable-length integer)	int32, int64, uint32, uint64, sint32, sint64, bool, enum
1	64-bit	fixed64, sfixed64, double
2	Length-delimited	string, bytes, embedded messages, packed repeated fields
3	Start group	groups (deprecated)
4	End group	groups (deprecated)
5	32-bit	fixed32, sfixed32, float

A Protobuf file is created with the .proto extension. You define service interfaces in the form of method signatures and messages (objects), which are referred to in method signatures. These messages can be method parameters or returned types. You can compile a defined service interface with the protoc compiler, which generates the classes for interfaces and given messages. Similarly, you can also generate the stubs for the gRPC client.

Let's have a look at the following sample .proto file:

Sample service interface of Employee

```
syntax = "proto3";
package com.packtpub;
option java_package = "com.packt.modern.api.proto";
option java_multiple_files = true;
message Employee {
  int64 id = 1;
  string firstName = 2;
  string lastName = 3;
  int64 deptId = 4;
  double salary = 5;
  message Address {
    string houseNo = 1;
    string street1 = 2;
    string street2 = 3;
    string city = 4;
    string state = 5;
    string country = 6;
    string pincode = 7;
  }
}
message EmployeeCreateResponse {
  int64 id = 1;
}
service EmployeeService {
  rpc Create(Employee) returns (EmployeeCreateResponse);
}
```

Let's understand this code line by line:

1. The first line represents the Protobuf version denoted by the syntax keyword. The value of
    syntax (proto3) tells the compiler that version 3 of Protobuf is used. The default version
    is proto2. Protobuf version 3 offers more features and simplified syntax and supports more
    languages. gRPC recommends using Protobuf version 3.

2. Next, you define the proto package name using the package keyword followed by the
    package name. It prevents name clashes among message types.

3. Next, you use the option keyword to define the Java package name using the java_
    package parameter.

4. Then, you use the `option` keyword again to generate a separate file for each root-level message type using the `java_multiple_files` parameter.

5. Then, you define the messages, which are nothing but objects, using the `messages` keyword. The message and its fields are defined using the strong types, which define the objects with exact specifications. You can define nested messages just like nested classes in Java. The last point contains the table of Protobuf types that you can use for defining the types of `message` fields.

6. You can use `Employee.Address` to define the `address` field in other messages.

7. The tagging of fields marked with a sequence number is required because it is used for serialization and parsing binary messages.

    Please note that you cannot change the message structure once it is serialized.

8. Service definitions are defined using the `service` keyword. A service definition contains the methods. You can define methods using the `rpc` keyword. Please refer to the `EmployeeService` service definition for reference. You'll explore more about service definitions in the next subsection.

9. Protobuf has predefined types (scalar types). A `message` field can have one of the Protobuf scalar types. When we compile the `.proto` file, it converts the `message` field into its respective language type. The following table defines the mapping between Protobuf types and Java types:

Protobuf types	Java types	Remarks
`Double`	`Double`	Like Java type `double`.
`Float`	`Float`	Like Java type `float`.
`int32`	`Int`	Use `sint32` if the field contains negative values because it uses variable-length encoding, which is inefficient for encoding negative numbers.
`int64`	`Long`	Use `sint64` if the field contains negative values because it uses variable-length encoding, which is inefficient for encoding negative numbers.
`uint32`	`Int`	It uses variable-length encoding. Use `fixed32` if values are greater than 2^{28}.
`uint64`	`Long`	It uses variable-length encoding. Use `fixed64` if values are greater than 2^{56}.
`sint32`	`Int`	More efficient for encoding negative numbers because it contains a signed `int` value. It uses variable-length encoding.

Protobuf types	Java types	Remarks
sint64	Long	More efficient for encoding negative numbers because it contains a signed int value. It uses variable-length encoding.
fixed32	int	Always 4 bytes.
fixed64	long	Always 8 bytes.
sfixed32	int	Always 4 bytes. More efficient for encoding values greater than 2^{28}.
sfixed64	long	Always 8 bytes. More efficient for encoding values greater than 2^{56}.
Bool	boolean	true or false.
String	String	Contains UTF-8 encoded string or 7-bit ASCII text, which should not be longer than 2^{32}.
Bytes	ByteString	Contains an arbitrary sequence of bytes, which should not be longer than 2^{32}.

Protobuf also allows you to define the enumeration types (using the enum keyword) and maps (using the map<keytype, valuetype> keyword). Please refer to the following code for examples of enumeration and map types:

```
… omitted
message Employee {
  … omitted
  enum Grade {
    I_GRADE = 1;
    II_GRADE = 2;
    III_GRADE = 3;
    IV_GRADE = 4;
  }
  map<string, int32> nominees = 1;
  … omitted
}
```

The preceding sample code creates the Employee message, which has a Grade enumeration field with values such as I_GRADE. The nominees field is a map that has a key with a string type and a value with an int32 type.

Let's explore service definitions further in the next section.

Understanding service definitions

You define a service by specifying its methods with the respective parameters and return types. These methods are exposed by the server, which can be called remotely. You defined the `EmployeeService` definition in the previous subsection, as shown in the next code block:

```
service EmployeeService {
    rpc Create(Employee) returns (EmployeeCreateResponse);
}
```

Here, `Create` is a method exposed by the `EmployeeService` service definition. Messages used in the `Create` service should also be defined as a part of the service definition. The `Create` service method is a unary service method because the client sends a single request object and receives a single response object in return from the server.

Let's dig further into the types of service methods offered by gRPC:

- **Unary**: We have already discussed the unary service method in the previous example. This would have a one-way response for a single request.

- **Server streaming**: In these types of service methods, the client sends a single object to the server and receives the stream response in return. This stream contains the sequence of messages. The stream is kept open until the client receives all the messages. The message sequence order is guaranteed by gRPC. In the following example, the client will keep receiving the live score messages until the match is over:

  ```
  rpc LiveMatchScore(MatchId) returns (stream MatchScore);
  ```

- **Client streaming**: In these types of service methods, the client sends a sequence of messages to the server and receives a response object in return. The stream is kept open until the client sends all the messages. The message sequence order is guaranteed by gRPC. Once all the messages are sent by the client, it waits for the server's response. In the following example, a client sends the data messages to the server until all the data records are sent, and it then waits for the report:

  ```
  rpc AnalyzeData(stream DataInput) returns (Report);
  ```

- **Bidirectional streaming**: This is the simultaneous execution of client and server streaming. It means both the server and client send a sequence of messages using a read-write stream. Here, the order of the sequence is preserved. However, these two streams operate independently. Therefore, each can read and write in whatever order they like. The server can read and reply to the messages one by one or at once or can have any combination. In the following example, processed records can be sent immediately one by one or can be sent later in different batches:

  ```
  rpc BatchProcessing(stream InputRecords)
      returns (stream Response);
  ```

Now that you have learned about the gRPC service definitions, let's explore the RPC life cycle in the next section.

Exploring the RPC life cycle

In the previous section, you learned about four types of service definitions. Each type of service definition has its own life cycle. Let's find out more about the life cycle of each service definition in this section:

- **The life cycle of a unary RPC**: A unary RPC is the simplest form of service method. Both the client and the server send a single object. Let's find out how it works. A unary RPC is initiated by the client. The client calls a stub method, which notifies the server that the RPC has been invoked. A `stub` also provides the server client's metadata, the method name, and the specified deadline, if applicable, with notification.

    Metadata is data about the RPC in the form of key-value pairs, such as timeout and authentication details.

    Next, in response, the server sends back its initial metadata. Whether the server sends initial metadata immediately or after receiving the client's request message depends on the application. But the server must send it before any response.

    The server works on the request and prepares the response after receiving the client's request message. The server sends back the response with the status (code and optional message) and optional trailing metadata for successful calls.

    The client receives a response and completes the call (for a status of OK, such as HTTP status 200).

- *The life cycle of a server-streaming RPC*: The life cycle of a server-streaming RPC is almost the same as a unary RPC. It follows the same steps. The only difference is the way the response is sent because of the stream response. The server sends messages as streams until all the messages are sent. In the end, the server sends back the response with the status (code and optional message) and optional trailing metadata and completes the server-side processing. The client completes the life cycle once it has all the server's messages.

- *The life cycle of a client-streaming RPC*: The life cycle of a client-streaming RPC is almost the same as a unary RPC. It follows the same steps. The only difference is the way the request is sent because of the stream request. The client sends messages as streams until all the messages are sent to the server. The server sends back the single message response with the status (code and optional message) and optional trailing metadata for successful calls. The server sends the response after receiving all the client's messages in idle scenarios. The client completes the life cycle once it receives the server message.

- *The life cycle of a bidirectional streaming RPC*: The first two steps in the life cycle of a bidirectional streaming RPC are the same as a unary RPC. Streaming processing from both sides is application specific. Both the server and client can read and write messages in any order because the two streams are independent of each other.

  The server can process streams of request messages sent by the client in any order. For example, the server and client can play ping-pong: the client sends the request message and the server processes it. Again, the client sends the request message, and the server processes it, and the process, as you know, goes on. Or the server waits until it receives all the client's messages before it writes its messages.

  The client completes the life cycle once it receives all the server messages.

Events that impact the life cycle

The following events may impact the life cycle of the RPC:

- **Deadlines/timeouts**: gRPC supports deadlines/timeouts. Therefore, a client would wait for the defined deadline/timeout to get the response from the server. If the wait exceeds the defined deadline/timeout, then it throws the DEADLINE_EXCEEDED error. Similarly, the server can query to find out whether a particular RPC has timed out, or how much time is left to complete the RPC.

  Timeout configuration is language-specific. Some language APIs support timeouts (durations of time), and some support deadlines (a fixed point in time). APIs may have a default value of deadline/timeout, and some may not.

- **RPC termination**: There are a few scenarios where the RPC gets terminated because both the client and server make independent and local determinations of the success of the call, and their conclusions may not match. For example, the server may finish its part by sending all its messages, but it may fail from the client's side because responses have arrived after the timeout. Another scenario would be when a server decides to complete the RPC before the client sends all messages.

- **Canceling an RPC**: gRPC has a provision to cancel the RPC at any time by either the server or client. This terminates the RPC immediately. However, changes made before the cancellation are not rolled back.

Let's explore the gRPC server and stub a bit more in the next section.

Understanding the gRPC server and gRPC stub

If you closely observe *Figure 10.1*, you'll find that the gRPC server and gRPC stub are core parts of the implementation because gRPC is based on the client-server architecture. Once you define the service, you can generate both service interfaces and the stub using the Protobuf compiler, protoc, with the gRPC Java plugin. You'll find a practical example in *Chapter 11*.

The following types of files are generated by the compiler:

- **Models**: It generates all the messages (that is, models) defined in the service definition file, which contains the Protobuf code to serialize, deserialize, and fetch the types of request and response messages.

- **gRPC Java files**: It contains the service base interface and stubs. The base interface is implemented and then used as a part of the gRPC server. Stubs are used by the clients for communication with the server.

First, you need to implement the interface, as shown in the following code for `EmployeeService`:

```
public class EmployeeService extends EmployeeServiceImplBase {
  // some code
  @Override
  public void create(Employee request,
      io.grpc.stub.StreamObserver<Response> responseObserver) {
    // implementation
  }
}
```

Once you implement the interface, you can run the gRPC server to serve the requests from gRPC clients:

```
public class GrpcServer {
  public static void main(String[] arg) {
    try {
      Server server = ServerBuilder.forPort(8080)
          .addService(new EmployeeService()).build();
      System.out.println("Starting gRPC Server Service...");
      server.start();
      System.out.println("Server has started at port: 8080");
      System.out.println("Following services are
          available: ");
      server.getServices().stream().forEach( s ->
        System.out.println("Service Name: " +
        s.getServiceDescriptor().getName())
      );
      server.awaitTermination();
    } catch (Exception e) {
      // error handling
    }
  }
}
```

For clients, first, you need to create the channel using `ChannelBuilder`, and then you can use the created channels to create stubs, as shown in the following code:

```
public EmployeeServiceClient(ManagedChannelBuilder<?> channelBuilder)
{
    channel = channelBuilder.build();
    blockingStub = EmployeeServiceGrpc.
        newBlockingStub(channel);
    asyncStub = EmployeeServiceGrpc.newStub(channel);
}
```

Here, both blocking and asynchronous stubs have been created using the channel built using the `ManageChannelBuilder` class.

Let's explore error handling in the next section.

Handling errors and error status codes

Unlike REST, which makes use of the HTTP status codes, gRPC uses a `Status` model, which contains its error codes and optional error message (string).

If you remember, you have used the special class called `Error` to contain the error details because HTTP error codes contain limited information. Similarly, the gRPC error `Status` model is limited to code and an optional message (string). You don't have sufficient error details for the client to use to handle the error or retry. You can make use of the richer error model as described at `https://cloud.google.com/apis/design/errors#error_model`, which allows you to pass detailed error information back to the client. You can also find the error models in the next code block for quick reference:

```
package google.rpc;
message Status {
    // actual error code is defined by `google.rpc.Code`.
    int32 code = 1;
    // A developer-facing human-readable error message
    string message = 2;
    // Additional error information that the client
        code can use
    // to handle the error, such as retry info or a
        help link.
    repeated google.protobuf.Any details = 3;
}
```

The `details` field contains extra information, and you can use it to pass relevant information such as `RetryInfo`, `DebugInfo`, `QuotaFailure`, `ErrorInfo`, `PreconditionFailure`, `BadRequest`, `RequestInfo`, `ResourceInfo`, `Help`, and `LocalizedMethod`. All these message types are available at `https://github.com/googleapis/googleapis/blob/master/google/rpc/error_details.proto`.

These richer error models are described using Protobuf. If you would like to use richer error models, you must make sure that support libraries are aligned with the practical use of APIs, as described for Protobuf.

Like REST, errors can be raised by the RPC for various reasons, such as network failure or data validation. Let's have a look at the following REST error codes and their respective gRPC counterparts:

HTTP Status Code	gRPC Status Code	Notes
400	INVALID_ARGUMENT	For invalid arguments.
400	FAILED_PRECONDITION	The action could not be performed due to a failed pre-condition.
400	OUT_OF_RANGE	If an invalid range is specified by the client.
401	UNAUTHENTICATED	If the client's request is not authenticated, such as having a missing or expired token.
403	PERMISSION_DENIED	The client does not have the sufficient permission.
404	NOT_FOUND	The requested resource was not found.
409	ABORTED	A conflict for read-write operations or any concurrency conflict.
409	ALREADY_EXISTS	If the request is for creating a new resource that already exists.
429	RESOURCE_EXHAUSTED	If the request reaches the API rate limiting.
499	CANCELLED	If the request is canceled by the client.
500	DATA_LOSS	For unrecoverable data loss or corruption.
500	UNKNOWN	For an unknown error at the server.
500	INTERNAL	For an internal server error.
501	NOT_IMPLEMENTED	API not implemented by the server.
502	N/A	Error due to unreachable network or network misconfiguration.

HTTP Status Code	gRPC Status Code	Notes
503	UNAVAILABLE	The server is down or unavailable due to any reason. The client can perform a retry on such errors.
504	DEADLINE_EXCEEDED	Either request doesn't finish within the deadline.

gRPC error codes are more readable as you don't need mapping to understand the number codes.

Summary

In this chapter, you explored Protobuf, IDL, and the serialization utility. You also explored gRPC fundamentals such as service definitions, messages, server interfaces, and methods. You compared gRPC with REST. I hope this has given you enough perspective to understand gRPC.

You also learned about the gRPC life cycles, servers, and clients with stubs. You covered Protobuf, gRPC architecture, and gRPC fundamentals, which will allow you to develop gRPC-based APIs and services.

You will make use of the fundamentals you learned in this chapter in the next chapter to implement the gRPC server and client.

Questions

1. What is RPC?
2. How is gRPC different in comparison to REST and which one should be used?
3. Which type of service method is useful when you want to view the latest tweets or do similar types of work?

Answers

1. RPC stands for Remote Procedure Call. A client can call an exposed procedure on a remote server, which is just like calling a local procedure, but it gets executed on a remote server. An RPC is best suited for inter-service communication in connected systems.
2. gRPC is based on the client-server architecture, whereas this is not true for REST. gRPC also supports full-duplex streaming communication in contrast to REST. gRPC performs better than REST as it uses static paths and a single source of the request payload.

A REST response error depends on HTTP status codes, whereas gRPC has formalized the set of errors to make it well aligned with APIs. gRPC has also been built to support and handle call cancellations, load balancing, and failovers. For more information, please refer to the *REST versus gRPC* subsection.

3. You should use the server-streaming RPC method because you want to receive the latest messages from the server, such as tweets.

Further reading

You can find out more at the following links:

- gRPC documentation: `https://grpc.io/`

- *Practical gRPC*: `https://www.packtpub.com/in/web-development/practical-grpc`

gRPC API Development and Testing

You will learn how to implement gRPC-based APIs in this chapter. You will learn how to write the gRPC server and client along with writing APIs based on gRPC. In the later part of this chapter, you will be introduced to microservices and see how they can help you to design modern, scalable architecture.

You will also go through the implementation of two services – the gRPC server and the gRPC client. gRPC-based APIs are more popular and preferred over REST APIs for service-to-service communication in a microservice-based system. Hence, gRPC development skills are important in the API space.

After completing this chapter, you will be well versed in the gRPC server and client development, gRPC-based API testing automation, and microservice concepts.

You will explore the following topics in this chapter:

- Writing an API
- Developing the gRPC server
- Handling errors
- Developing the gRPC client
- Learning microservice concepts

Technical requirements

This chapter contains a great deal of theory on gRPC. However, you will also undertake the development and testing of gRPC-based web services, for which you will need the following:

- Any Java IDE, such as NetBeans, IntelliJ, or Eclipse
- **Java Development Kit (JDK)** 17

- An internet connection to clone the code and download the dependencies and Gradle
- Postman/cURL (for API testing)

Please visit the following link to check the code: `https://github.com/PacktPublishing/ Modern-API-Development-with-Spring-6-and-Spring-Boot-3/tree/dev/ Chapter11`

So, let's begin!

Writing an API

In this section, we will write the API using **Protocol Buffer** (**Protobuf**) for a payment service. If you recall, this is the piece that you haven't yet implemented in the sample e-commerce app.

Before writing the API, let's set up the Gradle project.

Setting up the project

The code for this chapter will contain three projects under the `Chapter11` directory – the API, server, and client:

- **API**: This is a library project that contains the `.proto` file and its generated Java classes packaged in a JAR file. This project will generate the `payment-gateway-api-0.0.1.jar` library artifact, which you will publish in a local repository. This library will then be used in both the server and client projects.
- **Server**: This project represents the gRPC server, which will implement the gRPC services and serve the gRPC requests.
- **Client**: This project contains the gRPC client, which will call the gRPC server. To kick off the inter-service communication between the gRPC server and client applications, you are going to implement a REST call, which will call the gRPC server internally to serve the HTTP request.

Let's first create the server and client projects.

Creating the gRPC server and client projects

Either you can clone the *Chapter 11* code from the Git repository (`https://github.com/ PacktPublishing/Modern-API-Development-with-Spring-6-and-Spring-Boot-3/tree/dev/Chapter11`) or you can start by creating the new Spring project from scratch using **Spring Initializr** (`https://start.spring.io/`) for the server and client with the following options (later, you will create a gRPC `api` library project separately):

- **Project**: `Gradle - Groovy`
- **Language**: `Java`

- **Spring Boot**: 3.0.8.

  The preferred version is *3.0+*. Please choose the version that is available. You can modify it manually in the build.gradle file later too.

- **Project metadata**:

  - **Group**: com.packt.modern.api

  - **Artifact**: chapter11

  - **Name**: Chapter11

  - **Description**: Chapter 11 code of book Modern API Development with Spring and Spring Boot Ed 2

- **Package name**: com.packt.modern.api.

- **Packaging**: Jar.

- **Java**: 17.

  You can opt for any new version, such as *20*. It can be modified in the build.gradle file later too, as shown in the following code block:

  ```
  sourceCompatibility = JavaVersion.VERSION_20
  ```

- **ADD DEPENDENCIES**: Spring Web.

Then, you can click on **GENERATE** and download the project.

The downloaded project can be used to create both the server and the client. Then, create separate server and client directories under the Chapter11 directory. After creating the directories, copy the extracted content from the downloaded zipped project into them.

You can configure the server and client projects later. Let's first create the gRPC API library project as this library is going to be used in both the server and client projects.

Creating the gRPC API library project

Create a new directory, called api, in the Chapter11 directory. Then, use Gradle to create a new Gradle project using the following command executed from the Chapter11 directory. It will ask for a few options. The following block is executed after setting the JAVA_HOME environment variable to Java 17 and adding Java 17 to the path. You may find the order of questions a bit different in some systems. You should select the options highlighted in the following terminal interface output:

```
$ mkdir api
$ cd api
(you can also use gradlew from other chapter's code)
$ ../server/gradlew init
Select type of project to generate:
```

```
   1: basic
   2: application
   3: library
   4: Gradle plugin
Enter selection (default: basic) [1..4] 3
Select implementation language:
   1: C++
   2: Groovy
   3: Java
   4: Kotlin
   5: Scala
   6: Swift
Enter selection (default: Java) [1..6] 3
Select build script DSL:
   1: Groovy
   2: Kotlin
Enter selection (default: Groovy) [1..2] 1
Generate build using new APIs and behavior (some features may change
in the next minor release)? (default: no) [yes, no] no
Select test framework:
   1: JUnit 4
   2: TestNG
   3: Spock
   4: JUnit Jupiter
Enter selection (default: JUnit Jupiter) [1..4] 4
Project name (default: api): api
Source package (default: api): com.packt.modern.api
> Task :init
BUILD SUCCESSFUL in 1m 41s
2 actionable tasks: 2 executed
```

The project is bootstrapped by Gradle. Next, you will configure the api project.

Configuring the gRPC API library project

Here, you will configure the plugins section in api/libs/build.gradle with the Protobuf and Maven Publish plugins. These plugins and setting their configuration are key steps. Let's do this as follows:

1. Modify api/settings.gradle in the project's root directory:

   ```
   rootProject.name = 'payment-gateway-api'
   ```

2. Next, modify the api/lib/build.gradle file. Add the Protobuf and Maven Publish Gradle plugins. Also, replace the java-library plugin with java, as shown next:

   ```
   plugins {
       id 'java'
   ```

```
        id 'maven-publish'
        id "com.google.protobuf" version "0.9.2"
    }
```

The Maven Publish plugin will be used to publish the generated `Jar` artifact to the local Maven repository.

3. Add the group name, version, and source compatibility in `api/libs/build.gradle`, as shown in the following code block. The group and version will be used by the Maven Publish plugin to name the published artifact:

```
group = 'com.packt.modern.api'
version = '0.0.1'
sourceCompatibility = JavaVersion.VERSION_17
```

4. Next, add the following dependencies, which are required for Protobuf and gRPC (check the highlighted part). You can remove the existing dependencies added while generating the project using the `gradlew init` command and keep the dependencies mentioned next:

```
def grpcVersion = '1.54.0'
dependencies {
  implementation "io.grpc:grpc-protobuf:${grpcVersion}"
  implementation "io.grpc:grpc-stub:${grpcVersion}"
  implementation "io.grpc:grpc-netty:${grpcVersion}"
  implementation 'javax.annotation:javax.annotation-
     api:1.3.2'
  testImplementation 'org.junit.jupiter:junit-jupiter-
     api:5.9.2'
  testRuntimeOnly 'org.junit.jupiter:junit-
    jupiter-engine'
}
```

5. Next, let's configure the Protobuf Gradle plugin using the `protoc` command-line compiler. The plugin searches for the `protoc` executable in the system path by default. However, you can add a Protobuf compiler artifact to the plugin, which will make the build file self-sufficient as far as the gRPC compile task is concerned. Let's configure it as shown in the following code block by adding a `protobuf` section to the `api/libs/build.gradle` file:

```
protobuf {
  protoc {
    artifact = "com.google.protobuf:protoc:3.22.2"
  }
  plugins {
    grpc {
      artifact = "io.grpc:protoc-gen-grpc-java:1.54.0"
    }
```

```
  }
  generateProtoTasks {
    all()*.plugins {
      grpc { }
    }
  }
}
```

In the preceding code, you have configured the artifact used by the Protobuf compiler (`protoc`) and its Java plugin (`protoc-gen-grpc-java`), which will generate the Java code based on `.proto` files.

When you run the `gradlew build` command for the first time, Gradle will download the `protoc` and `protoc-gen-grpc-java` executables based on the OS.

6. The Protobuf Gradle plugin works with the configuration shared hitherto in this subsection. It works when you run the `build` command from the command line. However, the IDE may give a compilation error if you don't add the following block to the `api/libs/build.gradle` file to add the generated source files to `sourceSets`:

```
sourceSets {
  main {
    proto {
      // In addition to the default "src/main/proto"
      srcDir "src/main/grpc"
    }
  }
}
task sourcesJar(type: Jar, dependsOn: classes) {
    archiveClassifier = "sources"
    from sourceSets.main.allSource
}
```

7. Finally, you will add the following block to configure the Maven Publish plugin:

```
publishing {
  publications {
    mavenJava(MavenPublication) {
      artifactId = 'payment-gateway-api'
      from components.java
    }
  }
}
```

Here, you have configured the `api` project. You can find more information about the Protobuf Gradle plugin at `https://github.com/google/protobuf-gradle-plugin`.

Now that the `api` project setup is done, we are ready to write the service definitions using Protobuf in the next subsection. You haven't yet implemented the payment functionality for our sample e-commerce app. This is because it needs to be integrated with a payment gateway service such as Stripe or PayPal. Therefore, you are going to write the sample payment gateway service definition using gRPC in the next section.

Writing the payment gateway functionalities

Before you write the payment gateway service definition, let's first understand the basic functionality of the payment gateway system in easy terms.

The payment gateway provides a way to capture and transfer a payment from a customer to online sellers and then returns Accepted/Declined as a response to the customer. It performs various other actions here, such as verification, security, encryption, and communication with all participants.

The following are the actors who participate in this transaction:

- **Payment gateway**: A web interface that allows the processing of online payments and coordinates with all other actors. This is very similar to physical **point-of-sale** (**POS**) terminals.

- **Merchant**: Merchants are online sellers or service providers, such as Amazon, Uber, and Airbnb.

- **Customer**: This is you, the customer, who performs the buy/pay transaction for products or services and uses credit/debit cards, digital wallets, or online banking.

- **Issuing bank**: The party that provides the functionality to perform online money transfers, such as Visa, Mastercard, AmEx, PayPal, Stripe, or traditional banks.

- **Acquirer or acquiring bank**: The institution that holds the merchant account. It passes the transaction to the issuing bank to receive payment.

You are going to create two gRPC services – `ChargeService` and `SourceService` – as part of the payment gateway service. Don't get confused with the web service, which is an executable/deployable artifact. `ChargeService` and `SourceService` are part of the service component of Protobuf's **interface definition language** (**IDL**) file, the same as we learned about in the `EmployeeService` example in the last chapter (the *How gRPC uses Protobuf* section of *Chapter 10*, *Getting Started with gRPC*). Both services are inspired by Stripe public REST APIs.

Let's understand the transaction flow before we jump into creating the service components of a gRPC-based payment gateway service.

Online payment workflow steps

The following steps are performed when an online transaction takes place:

1. First, the customer should have a payment source (read method) created before initiating the payment. If not, then the customer creates a source, such as their card details.

2. Payment is initiated by creating a charge against the payment source (read method).

3. The payment gateway performs all the necessary validation and verification steps and then allows the charge to be captured. These steps trigger the fund transfer from the issuing bank to the merchant account.

You can observe that there are two objects (resources) involved in this workflow (aka source and charge). Therefore, you are going to write two services that function around these two objects. There are various other functionalities performed by the payment gateway, such as disputes, refunds, and payouts. However, you are going to implement only two services, charge and source, in this chapter.

Writing the payment gateway service definitions

Writing a Protobuf-based IDL is very similar to the way you defined the OpenAPI Specification for REST APIs. In REST, you define the models and API endpoints, whereas in gRPC, you define the messages and RPC procedures wrapped in the service. Let's write our payment gateway service IDL using the following steps:

1. First, let's create a new file, `PaymentGatewayService.proto`, in the `api/lib/src/main/proto` directory under the root directory of the `api` project.

2. After creating a new file, you can add the metadata, as shown in the following code block:

```
syntax = "proto3";                                      //1
package com.packtpub.v1;                                //2
option java_package = "com.packt.modern.api.grpc.v1";   //3
option java_multiple_files = true;                      //4
```

Let's understand the preceding code in detail:

- Line *1* tells the compiler to use version 3 of Protobuf by using the syntax specifier. If you don't specify this, then the compiler will use version 2 of Protobuf.

- Line *2* uses the optional package specifier to attach the namespace to message types. This prevents name clashes among message types. We must postfix it with a package version that allows us to create new versions of APIs with backward compatibility.

- Line *3* uses the `java_package` option specifier. This specifies the Java package to be used in the generated Java files. If you don't use this option specifier and declare the `package` specifier, then the value of `package` will be used as a Java package in the generated Java files instead.

- Line *4* declares the `java_multiple_files` option specifier, which is a Boolean option. It is set to `false` by default. If it is set to `true`, then it generates separate Java files for each top-level message type, enumeration (`enum`), and service.

3. Next, let's add the `ChargeService` service, which contains the operations required for charge functionality denoted by `rpc` (as shown in the following code block). Charge objects

get created for charging the card, bank account, or digital wallet. Let's add the charge service to the Protobuf (.proto) file:

```
service ChargeService {
  rpc Create(CreateChargeReq)
      returns(CreateChargeReq.Response);
  rpc Retrieve(ChargeId)
      returns (ChargeId.Response);
  rpc Update(UpdateChargeReq)
      returns(UpdateChargeReq.Response);
  rpc Capture(CaptureChargeReq)
      returns(CaptureChargeReq.Response);
  rpc RetrieveAll(CustomerId)
      returns (CustomerId.Response);
}
```

https://github.com/PacktPublishing/Modern-API-Development-with-Spring-6-and-Spring-Boot-3/tree/dev/Chapter11/api/lib/src/main/proto/PaymentGatewayService.proto

Each of these procedures in `ChargeService` will perform the following operations:

- **Create**: This procedure creates a new `Charge` object.

- **Retrieve**: This procedure retrieves the `Charge` object based on the given charge ID that was previously created.

- **Update**: This procedure updates the `Charge` object identified by the given charge ID by setting the values of the parameters passed. Any parameters not provided will be left unchanged.

- **Capture**: This procedure captures the payment of an existing, uncaptured charge. This is the payment workflow step, where you first create a charge with the `capture` option set to `false`. Uncaptured payments expire precisely seven days after they are created. If they are not captured by that point in time, they will be marked as refunded and capture will no longer be allowed.

- **RetrieveAll**: This procedure returns the list of charges that belong to the given customer ID.

Empty request or response type

You can use `google.protobuf.Empty` for void/empty request and response types. This can be used in .proto files. You just must place the following `import` statement before any message/service is defined:

`import "google/protobuf/timestamp.proto";`.

Then, you can use it as shown next:

`rpc delete(SourceId) returns (google.protobuf. Empty);`.

4. The amount is charged to a source, which could be a card, bank account, or digital wallet. A variety of payment methods can be used by the customer using a `Source` object. Therefore, you need a service that will allow you to perform operations on the `source` resource. Let's add the `Source` service and its operations to the Protobuf (`.proto`) file:

```
service SourceService {
  rpc Create(CreateSourceReq)
      returns (CreateSourceReq.Response);
  rpc Retrieve(SourceId)
      returns (SourceId.Response);
  rpc Update(UpdateSourceReq)
      returns (UpdateSourceReq.Response);
  rpc Attach(AttachOrDetachReq)
      returns (AttachOrDetachReq.Response);
  rpc Detach(AttachOrDetachReq)
      returns (AttachOrDetachReq.Response);
}
```

Each of these procedures in `SourceService` will perform the following operations:

-   **Create**: This procedure creates a new `Source` object.

-   **Retrieve**: This procedure allows you to retrieve the `Source` object based on the given source ID.

-   **Update**: This procedure allows you to update certain fields of the `Source` object passed using the `UpdateSourceReq` object. Any field that is not part of `UpdateSourceReq` will remain unchanged.

-   **Attach**: This procedure attaches the `Source` object to the customer. The `AttachOrDetachReq` parameter contains the IDs of both the source and the customer. However, the `Source` object should be in the `CHARGEABLE` or `PENDING` state to perform the attached operation.

-   **Detach**: This procedure will detach the `Source` object from the customer. It will also change the state of the `Source` object to `consumed` and it can no longer be used to create the charge. The `AttachOrDetachReq` parameter contains the IDs of both the source and the customer.

The recommended approach for defining the request and response types

It is recommended to always use the wrapper request and response types. This allows you to add another field to the request or response types.

5. Now that the service definitions are done, you can define the given parameters and the returned types of these procedures. Let's first define the parameters and returned types of `ChargeService`. First, you will define the `Charge` message type, as shown in the following code block:

```
message Charge {
  string id = 1;
  uint32 amount = 2;
  uint32 amountCaptured = 3;
  uint32 amountRefunded = 4;
  string balanceTransactionId = 5;
  BillingDetails billingDetails = 6;
  string calculatedStatementDescriptor = 7;
  bool captured = 8;
  uint64 created = 9;
  string currency = 10;
  string customerId = 11;
  string description = 12;
  bool disputed = 13;
  uint32 failureCode = 14;
  string failureMessage = 15;
  string invoiceId = 16;
  string orderId = 17;
  bool paid = 18;
  string paymentMethodId = 19;
  PaymentMethodDetails paymentMethodDetails = 20;
  string receiptEmail = 21;
  string receiptNumber = 22;
  bool refunded = 23;
  repeated Refund refunds = 24;
  string statementDescriptor = 25;
  enum Status {
    SUCCEEDED = 0;
    PENDING = 1;
    FAILED = 2;
  }
  Status status = 26;
  string sourceId = 27;
}
```

Here, the `Charge` message contains the following fields:

- `id`: The unique identifier of the `Charge` object.

- `amount`: The amount is a positive number or zero, referring to the amount of the payment.

- `amountCaptured`: This is the captured amount (a positive number or zero). It can be less than the value of the `amount` field if a partial capture is made.

- `amountRefunded`: The amount refunded (a positive number or zero). It can be less than the value of the `amount` field if a partial refund is issued.

- `balanceTransactionId`: The ID of the balance transaction, which describes the impact of this charge on your account balance (not including refunds or disputes).

- `billingDetails`: The object of the `BillingDetails` message type, which contains billing information associated with the payment method at the time of the transaction.

- `calculatedStatementDescriptor`: The statement description that is passed to card networks and is displayed on your customers' credit card and bank statements.

- `captured`: A Boolean field that represents whether a charge has since been captured. (It is possible to create a charge without capturing the charge details. Therefore, this field is added, which determines whether a charge will be captured or not.)

- `created`: The timestamp (measured in seconds since the Unix epoch) at which the object was created.

- `currency`: The three-letter ISO currency code.

- `customerId`: The ID of the customer owning the charge.

- `description`: A description of the charge displayed to the user.

- `disputed`: A Boolean field that represents whether the charge has been disputed.

- `failureCode`: The error code of the failure.

- `failureMessage`: A description of the failure. The reason may be stated if this option is available.

- `invoiceId`: The ID of the invoice this charge is for.

- `orderId`: The ID of the order this charge is for.

- `paid`: The Boolean value represents whether the charge succeeded or was successfully authorized for subsequent capture.

- `paymentMethodId`: The ID of the payment method.

- `paymentMethodDetails`: The object that contains the details of the payment method.

- `receiptEmail`: The email where receipt of the charge will be sent.

- `receiptNumber`: This represents the transaction number in the charge receipt that was sent by email. It should remain null until a charge receipt is sent.

- `refunded`: A Boolean field that represents whether the charge was refunded.

- refunds: This contains the list of refunds that have been issued. The repeated keyword is used to create a list of Refund objects.

- statementDescriptor: The description of a charge for a card.

- status: An object of the Status enumeration type (SUCCEEDED, PENDING, or FAILED) that represents the status of the charge.

- sourceId: ID of the Source object.

The UInt32 and string scalar types were discussed in the *How gRPC uses Protobuf* subsection under the *How does gRPC work?* section in the previous chapter (*Chapter 10, Getting Started with gRPC*). You can refer to it for further information.

> **Predefined well-known types**
>
> Apart from scalar types, Protobuf also provides predefined types such as Empty (which we saw earlier in *step 3*), Timestamp, and Duration. You can find the complete list at https://developers.google.com/protocol-buffers/docs/reference/google.protobuf.

6. Now, you can define the remaining message types of the other parameters (CreateChargeReq, ChargeId, UpdateChargeReq, CaptureChargeReq, and CustomerId) and return the ChargeList type of ChargeService, as shown in the following code block:

```
message CreateChargeReq {
  uint32 amount = 1;
  string currency = 2;
  string customerId = 3;
  string description = 4;
  string receiptEmail = 5;
  Source source Id = 6;
  string statementDescriptor = 7;
  message Response { Charge charge = 1; }
}
message UpdateChargeReq {
  string chargeId = 1;
  string customerId = 2;
  string description = 3;
  string receiptEmail = 4;
  message Response { Charge charge = 1; }
}
message CaptureChargeReq {
  string chargeId = 1;
  uint32 amount = 2;
```

```
      string receiptEmail = 3;
      string statementDescriptor = 4;
      message Response { Charge charge = 1; }
   }
   message ChargeId {
      string id = 1;
      message Response { Charge charge = 1; }
   }
   message CustomerId {
      string id = 1;
      message Response { repeated Charge charge = 1; }
   }
```

Here, the CreateChargeReq type contains the required attribute's charge amount (amount) and currency. It also contains several optional attributes – customerId, receiptEmail, source, and statementDescriptor.

UpdateChargeReq contains all the optional attributes – customerId, description, and receiptEmail.

CaptureChargeReq contains all the optional attributes – amount, receiptEmail, and statementDescriptor.

Less well-known Google common types

Money and Date (not Timestamp) are less commonly known types that can be used. However, you must copy the definitions instead of importing them (unlike what you do for Empty and Timestamp). You can copy the definitions from the following links: Money from https://github.com/googleapis/googleapis/blob/master/google/type/money.proto and Date from https://github.com/googleapis/googleapis/blob/master/google/type/date.proto. Other common types that you can use are also available in the repository.

---

7. Now, you can define the parameters and return the SourceService types. First, let's define the Source message type, as shown in the following code.

    The source uses a Flow value, which could be either REDIRECT, RECEIVER, CODEVERIFICATION, or NONE. Similarly, the Usage value could be REUSABLE or SINGLEUSE. Therefore, let's first create the Flow and Usage enumerations using enum:

```
enum Flow {
   REDIRECT = 0;
   RECEIVER = 1;
   CODEVERIFICATION = 2;
   NONE = 3;
}
```

```
enum Usage {
  REUSABLE = 0;
  SINGLEUSE = 1;
}
```

Now, you can use this `Flow` enum in the `Source` message:

```
message Source {
  string id = 1;
  uint32 amount = 2;
  string clientSecret = 3;
  uint64 created = 4;
  string currency = 5;
  Flow flow = 6;
  Owner owner = 7;
  Receiver receiver = 8;
  string statementDescriptor = 9;
  enum Status {
    CANCELLED = 0;
    CHARGEABLE = 1;
    CONSUMNED = 2;
    FAILED = 3;
    PENDING = 4;
  }
  Status status = 10;
  string type = 11;
  Usage usage = 12;
}
```

8. Now, you can define the remaining message types for the other parameters of `SourceService` – `CreateSourceReq`, `UpdateSourceReq`, `AttachOrDetachReq`, and `SourceId` – as shown in the following code block:

```
message CreateSourceReq {
  string type = 1;
  uint32 amount = 2;
  string currency = 3;
  Owner owner = 4;
  string statementDescriptor = 5;
  Flow flow = 6;
  Receiver receiver = 7;
  Usage usage = 8;
  message Response { Source source = 1; }
}
message UpdateSourceReq {
```

```
    string sourceId = 1;
    uint32 amount = 2;
    Owner owner = 3;
    message Response { Source source = 1; }
  }
  message SourceId {
    string id = 1;
    message Response { Source source = 1; }
  }
  message AttachOrDetachReq {
    string sourceId = 1;
    string customerId = 2;
    message Response { Source source = 1; }
  }
```

The other message types used in these messages can be referred to in the payment gateway definition file, located at `https://github.com/PacktPublishing/Modern-API-Development-with-Spring-6-and-Spring-Boot-3/tree/dev/Chapter11/api/lib/src/main/proto/PaymentGatewayService.proto`.

> **Multiple .proto files**
>
> You can also create a separate definition file for each service, such as `ChargeService.proto` and `SourceService.proto`, for modularity. You can then import these files into another Protobuf file using `import "SourceService.proto";`.
>
> You can find more information about importing at `https://protobuf.dev/programming-guides/proto3/#importing-definitions`.

You are now done with the payment gateway service definitions in the Protobuf file. Now, you can use this file to generate the gRPC server interface and stubs for the gRPC client.

Next, you will publish the Java classes generated from the Protobuf file packaged in the `Jar` file.

Publishing the payment gateway service gRPC server, stubs, and models

You can use the following command, which should be executed from the `api` project's root directory:

```
# Make sure to enable UTF-8 for file encoding because
# we are using UTF characters in Java files.
$ export JAVA_TOOL_OPTIONS="-Dfile.encoding=UTF8"

$ gradlew clean publishToMavenLocal
```

With the preceding command, you are setting the file encoding to UTF-8 first because we are using the UTF characters in Java files. Then, you are performing clean, build, and publish operations. The second command will first remove the existing files. Then, it will generate the Java files (the `generateProto` Gradle task) from the Protobuf file, build it (the `build` Gradle task), and publish the artifact to your local Maven repository (the `publishToMavenLocal` Gradle task).

The `generateProto` Gradle task will generate the two types of Java classes in two directories, as shown next:

- **Models**: This Protobuf Gradle plugin generates messages (aka models) and classes in separate Java files in the `/api/lib/build/generated/source/proto/main/java` directory, such as `Card.java` or `Address.java`. This directory will also contain the Java files of request and response objects used in operation contracts, such as `CreateChargeReq`, `CreateSourceReq`, `Charge.java`, and `Source.java`.

- **gRPC classes**: This Protobuf Gradle plugin generates the service definitions of both services (`ChargeServiceGrpc.java` and `SourceServiceGrpc .java`) in the `/api/lib/build/generated/source/proto/main /grpc` directory. Each of these gRPC Java files contains a base class, stub classes, and methods for each operation defined in the service descriptor for the `Charge` and `Source` services.

  The following key static classes are defined in `ChargeServiceGrpc`:

  - `ChargeServiceImplBase` (abstract base class)

  - Stubs: `ChargeServiceStub`, `ChargeServiceBlockingStub`, and `ChargeServiceFutureStub`

  Similarly, the following key static classes are defined in `SourceServiceGrpc`:

  - `SourceServiceImplBase` (abstract base class)

  - Stubs: `SourceServiceStub`, `SourceServiceBlockingStub`, and `SourceServiceFutureStub`

The abstract base classes described earlier contain the operations defined in the service block in the Protobuf file. You can use these base classes to implement the business logic for operations offered by these services, just like you implemented the REST endpoints from the Swagger-generated API Java interfaces.

These abstract classes should be implemented to provide the business logic implementations to the services offered by the gRPC server. Let's develop the gRPC server next.

Developing the gRPC server

You need to configure the `server` project before implementing these abstract classes. Let's configure the server project first.

The `server` project directory structure will look like the following. The project root directory contains the `build.gradle` and `settings.gradle` files:

```
├── server
│   ├── build.gradle
│   ├── gradle
│   │   └── wrapper
│   ├── gradlew
│   ├── gradlew.bat
│   ├── settings.gradle
│   └── src
│       ├── main
│       │   ├── java
│       │   │   └── com
│       │   │       └── packt
│       │   │           └── modern
│       │   │               └── api
│       │   └── resources
│       └── test
│           └── java
```

The `resources` directory will contain the `application.properties` file.

> **Using the Sprint Boot gRPC starters**
>
> There are two Spring Boot starter projects that you can use. However, we'll stick to the libraries provided by gRPC for a simplified solution and to aid understanding of the gRPC concepts. These libraries are available at the following links: `https://github.com/LogNet/grpc-spring-boot-starter` and `https://github.com/yidongnan/grpc-spring-boot-starter`.

Let's perform the following steps to configure the project:

1. First, you need to modify the project name in the `Chapter11/server/ settings. gradle` file to represent the server, as shown here:

    ```
    rootProject.name = 'chapter11-server'
    ```

2. Next, you can add the dependencies required for `server` projects to the `Chapter11/server/build.gradle` file:

    ```
    def grpcVersion = '1.54.1'
    dependencies {
      implementation 'com.packt.modern.api:payment-gateway-
          api:0.0.1'
    ```

```
implementation "io.grpc:grpc-protobuf:${grpcVersion}"
implementation "io.grpc:grpc-stub:${grpcVersion}"
implementation "io.grpc:grpc-netty:${grpcVersion}"
implementation 'com.google.api.grpc:googleapis-
    common-protos:0.0.3'

implementation 'org.springframework.boot:spring-boot-
    starter-web'
testImplementation 'org.springframework.boot:spring-
    boot-starter-test'
testImplementation "io.grpc:grpc-testing
    :${grpcVersion}"
}
```

3. The `payment-gateway-api` dependency is published in the local Maven repository. Therefore, you need to add the local Maven repository to the `repositories` section in `Chapter11/server/build.gradle`, as shown in the following code block:

```
repositories {
  mavenCentral()
  mavenLocal()
}
```

You are done with the Gradle configuration! Now, you can write the gRPC server. However, before writing the server, you need to implement the base abstract classes generated by Protobuf. Once the source and charge services (using base classes) are implemented, you can write the gRPC server code.

Implementation of the gRPC server

You are going to use the same layered architecture that you used in the REST implementation – persistence store > repository layer > service layer > API endpoint.

First, you need a persistence store where you can save the data, aka the first layer. You are going to use in-memory persistence (`ConcurrentHashMap`) for storing and retrieving the data. If you want, you can use the external database the way it is used in REST web services. This is done to keep the focus on gRPC server implementation.

First, create the in-memory persistence store for both the charge and source data stores. Create a new file, `server/src/main/java/com/packt/modern/api/server/ repository/ DbStore.java`, and add code, as shown in the following code block:

```
@Component
public class DbStore {
  private static final Map<String, Source> sourceEntities =
```

```
      new ConcurrentHashMap<>();
  private static final Map<String, Charge> chargeEntities =
      new ConcurrentHashMap<>();
  public DbStore() {
    Source source = Source.newBuilder().setId(
        RandomHolder.randomKey())
        .setType("card").setAmount(100)
        .setOwner(createOwner()).
        setReceiver(createReceiver())
        .setCurrency("USD").setStatementDescriptor("Statement")
        .setFlow(Flow.RECEIVER).setUsage(Usage.REUSABLE)
        .setCreated(Instant.now().getEpochSecond()).build();
    sourceEntities.put(source.getId(), source);

    Charge charge = Charge.newBuilder().setId(
        RandomHolder.randomKey()).setAmount(1000)
        .setCurrency("USD").setCustomerId("ab1ab2ab3ab4ab5")
        .setDescription("ChargeDescription")
        .setReceiptEmail("receipt@email.com")
        .setStatementDescriptor("Statement Descriptor")
        .setSourceId(source.getId())
        .setCreated(Instant.now().getEpochSecond()).build();
    chargeEntities.put(charge.getId(), charge);
  }
// continue …
```

```
https://github.com/PacktPublishing/Modern-API-Development-with-
Spring-6-and-Spring-Boot-3/tree/dev/Chapter11/server/src/main/
java/com/packt/modern/api/ server/repository/DbStore.java
```

Here, you create two ConcurrentHashMap objects for storing the Charge and Store objects, respectively. You create two seed objects of each of these in the constructor using the builder and store them in their respective hash maps.

According to the operations defined in the service contract, you create the methods in the database store to perform the operations. These operations are implemented with basic business logic to keep the flow and logic concise and to the point.

Let's now add the createSource() method to implement the create() contract of SourceService defined in the Protobuf file, as shown in the following code block:

```
public CreateSourceReq.Response
        createSource(CreateSourceReq req) {
    Source source = Source.newBuilder().setId(
        RandomHolder.randomKey()).setType(req.getType())
```

```
          .setAmount(req.getAmount()).setOwner(createOwner())
          .setReceiver(createReceiver())
          .setCurrency(req.getCurrency())
          .setStatementDescriptor(req.getStatementDescriptor())
          .setFlow(req.getFlow()).setUsage(req.getUsage())
          .setCreated(Instant.now().getEpochSecond()).build();
     sourceEntities.put(source.getId(), source);
     return CreateSourceReq.Response.newBuilder()
             .setSource(source).build();
}
```

This method creates a `source` object from the values received from the request object (`CreateSourceReq`). This newly created `Source` object is then saved in a hash map called `sourceEntities` and returned to the caller. You can enhance this method by adding validation that would validate the request object (`req`). Owner and receiver objects (highlighted in the code) should be retrieved from the request object. To keep the program simple, we have hardcoded these values here.

Similarly, you can implement other contract methods for `source` and `charge` along with their persistence. You can find the full source code of this class at `https://github.com/PacktPublishing/Modern-API-Development-with-Spring-6-and-Spring-Boot-3/tree/dev/Chapter11/server/src/main/java/com/packt/modern/api/server/repository/DbStore.java`.

Now, you have the in-memory persistence store – `DbStore`. Next, let's use this store in repository classes.

Writing repository classes

Now, you can implement the next layer – the repository layer. The in-memory persistence store (`DbStore`) can be consumed in the `ChargeRepositoryImpl` repository class to interact, as shown here:

```
@Repository
public class ChargeRepositoryImpl implements
    ChargeRepository {
  private DbStore dbStore;
  public ChargeRepositoryImpl(DbStore dbStore) {
    this.dbStore = dbStore;
  }
  @Override
  public CreateChargeReq.Response create(
      CreateChargeReq req) {
    return dbStore.createCharge(req);
  }
  // code truncated for brevity
```

```
https://github.com/PacktPublishing/Modern-API-Development-with-
Spring-6-and-Spring-Boot-3/tree/dev/Chapter11/server/src/main/java/
com/packt/modern/api/server/repository/ChargeRepositoryImpl.java
```

ChargeRepositoryImpl implements the ChargeRepository interface and makes
use of DbStore to perform the operations. The code of this repository interface is available
at https://github.com/PacktPublishing/Modern-API-Development-with-
Spring-6-and-Spring-Boot-3/tree/dev/Chapter11/server/src/main/
java/com/packt/modern/api/server/repository/ChargeRepository.java.

Similarly, you can create the SourceRepositoryImpl class, which implements
SourceRespository, as shown here:

```
@Repository
public class SourceRepositoryImpl
    implements SourceRepository {
  private DbStore dbStore;
  public SourceRepositoryImpl(DbStore dbStore) {
    this.dbStore = dbStore;
  }
  @Override
  public UpdateSourceReq.Response update
      (UpdateSourceReq req) {
    return dbStore.updateSource(req);
  }
}
// Other methods removed for brevity
```

```
https://github.com/PacktPublishing/Modern-API-Development-with-
Spring-6-and-Spring-Boot-3/tree/dev/Chapter11/server/src/main/java/
com/packt/modern/api/server/repository/SourceRepositoryImpl.java
```

Like ChangeRepositoryImpl, SourceRepositoryImpl too makes use of a persistence
store to persist the data. You can find the code for the SourceRepository interface at https://
github.com/PacktPublishing/Modern-API-Development-with-Spring-6-and-
Spring-Boot-3/tree/dev/Chapter11/server/src/main/java/com/packt/
modern/api/ server/repository/SourceRepository.java.

Methods of the Source and Charge repository classes are consumed by the service classes. Service
base classes are generated by gRPC (part of the api project). Service classes implement these abstract-
generated base classes (service base classes).

Let's write the service layer next.

Implementing service classes

Now you have the underlying implementation ready in the form of repository and database store classes, which can be used to implement the gRPC service's base classes.

Let's implement the `Source` service first, as shown next:

1. Create a new file, `SourceService.java`, in the `server/src/main/com/packt/modern/api/server/service` directory.

2. Add the implementations to operations defined in the `SourceService` abstract base class, as shown next:

```java
@Service
public class SourceService
        extends SourceServiceImplBase {
  private final SourceRepository repository;
  public SourceService(SourceRepository repository) {
    this.repository = repository;
  }

@Override
public void create(CreateSourceReq req,
  StreamObserver<CreateSourceReq.Response> resObserver) {
    CreateSourceReq.Response resp = repository.create
        (req);
    resObserver.onNext(resp);
    resObserver.onCompleted();
  }
}
// Other methods removed for brevity
```

https://github.com/PacktPublishing/Modern-API-Development-with-Spring-6-and-Spring-Boot-3/tree/dev/Chapter11/server/src/main/java/com/packt/modern/api/server/service/SourceService.java

Here, the `SourceServiceImplBase` abstract class is autogenerated by the Protobuf plugin, which contains the contract methods of the `Source` service. A unique part of the method signature generated is the second argument, `StreamObserver`. `StreamObserver` receives notifications from observable streams. It is being used here for service implementation. Similarly, it is also used in the client stubs. The gRPC library provides the `StreamObserver` argument for outgoing messages. However, you also must implement it for incoming messages.

`StreamObserver` arguments are not thread-safe, so you must take care of multithreading and should use synchronized calls.

3.   There are three primary methods of `StreamObserver`:

-   `onNext()`: This method receives the value from the stream. It can be called multiple times. However, it should not be called after `onCompleted()` or `onError()`. Multiple `onNext()` calls are required for streams when multiple datasets are sent to clients.

-   `onCompleted()`: This marks the completion of the stream and no further method calls are allowed after that. It can only be called once.

-   `onError()`: This method receives the termination error from the stream. Like `onCompleted()`, it can only be called once and no further method calls are allowed.

4.   Similarly, you can implement the other methods of an abstract class.

Next, you can implement the `Charge` service in the same way you have implemented the `Source` service. Let's do it:

1.   Create a new file, `ChargeService.java`, in the `server/src/main/com/packt/modern/api/server/service` directory.

2.   Add the implementations to operations defined in the `ChargeService` abstract base class, as shown here:

```
@Service
public class ChargeService
        extends ChargeServiceImplBase {
  private final ChargeRepository repository;
  public ChargeService(ChargeRepository repository) {
    this.repository = repository;
  }
  @Override
  public void create(CreateChargeReq req,
  StreamObserver<CreateChargeReq.Response>
    resObserver) {
    CreateSourceReq.Response resp =
                              repository.create(req);
    resObserver.onNext(resp);
    resObserver.onCompleted();
  }
// Other methods truncated for brevity
```

https://github.com/PacktPublishing/Modern-API-Development-with-Spring-6-and-Spring-Boot-3/tree/dev/Chapter11/server/src/main/java/com/packt/modern/api/server/service/ChargeService.java

This is along the same lines as the way the `SourceService` create method was implemented.

3. Similarly, you can implement the other methods of the abstract class. Please refer to the link to the source code after the preceding code block for the complete code implementation.

Now, you have the service layer implementation ready. Let's implement the API layer (gRPC server) next.

Implementation of the gRPC server class

The Spring Boot application runs on its own server. However, we want to run the gRPC server, which internally uses the Netty web server. Therefore, we first need to modify the Spring Boot configuration to stop running its web server. You can do that by modifying the `server/src/main/resources/application.properties` file, as shown in the following code block:

```
spring.main.web-application-type=none
grpc.port=8080
```

https://github.com/PacktPublishing/Modern-API-Development-with-Spring-6-and-Spring-Boot-3/tree/dev/Chapter11/server/src/main/resources/application.properties

Next, let's create the gRPC server. It will have three methods – `start()`, `stop()`, and `block()` – for starting up the server, stopping the server, and serving requests until a termination request is received, respectively.

Create a new file, `GrpcServer.java`, in the `server/src/main/com/packt/ modern/api/server` directory and the code, as shown in the following code block:

```
@Component
public class GrpcServer {
  @Value("${grpc.port:8080}");
  private int port;
  private Server server;
  private final ChargeService chargeService;
  private final SourceService sourceService;
  private final ExceptionInterceptor exceptionInterceptor;
  public GrpcServer(...) { // code removed for brevity }
  public void start() throws IOException,
      InterruptedException {
    server = ServerBuilder.forPort(port)
        .addService(sourceService).
            addService(chargeService)
        .intercept(exceptionInterceptor).build().start();
        server.getServices().stream().forEach(s ->
            Systen.out.println("Service Name: {}",
            s.getServiceDescriptor().getName())));
```

```
        Runtime.getRuntime().addShutdownHook(new Thread(() -> {
            GrpcServer.this.stop();
        }));
    }
    private void stop() {
        if (server != null) { server.shutdown(); }
    }
    public void block() throws InterruptedException {
        if (server != null) {
            // received the request until application is
               terminated
            server.awaitTermination();
        }
    }
}
```

https://github.com/PacktPublishing/Modern-API-Development-with-Spring-6-and-Spring-Boot-3/tree/dev/Chapter11/server/src/main/java/com/packt/modern/api/server/GrpcServer.java

The server library of gRPC provides the server builder for building the server. You can see that both services are added to the server. The builder also allows you to add interceptors that can intercept the incoming request and response. We are going to use the interceptor in the *Coding for handling errors* section.

The GrpcServer start() method has also added a shutdown hook that calls the stop() method, which internally calls the server.shutdown() method.

The server code is ready. Now, you need an interface to start the server. You are going to use the CommandLineRunner function interface to run the server.

Create a new file, GrpcServerRunner.java, in the same directory where you created the GrpcServer.java file and add the following code:

```
@Profile("!test")
@Component
public class GrpcServerRunner implements CommandLineRunner {
  private GrpcServer grpcServer;
  public GrpcServerRunner(GrpcServer grpcServer) {
    this.grpcServer = grpcServer;
  }
  @Override
  public void run(String... args) throws Exception {
    grpcServer.start();
```

```
    grpcServer.block();
  }
}
```

https://github.com/PacktPublishing/Modern-API-Development-with-Spring-6-and-Spring-Boot-3/tree/dev/Chapter11/server/src/main/java/com/packt/modern/api/server/GrpcServerRunner.java

Here, you override the `CommandLineRunner` `run()` method and call the `start` and `block` methods. Therefore, when you execute the `jar` file, `GrpcServerRunner` will be executed using its `run()` method and will start the gRPC server.

Another thing to remember is that you have marked the `GrpcServerRunner` class with the `@Profile` annotation with the `"!test"` value, which means that when the test profile is active, this class won't be loaded, and hence not executed.

You are now done with both service and server implementation, so let's test the gRPC server in the next subsection.

Testing the gRPC server

First of all, you need to set the active profile to `test` in your `test` classes because doing so will disable `GrpcServerRunner`. Let's do this and test it, as shown in the following code block:

```
@ActiveProfiles("test")
@SpringBootTest
@TestMethodOrder(OrderAnnotation.class)
class ServerAppTests {
  @Autowired
  private ApplicationContext context;
  @Test
  @Order(1)
  void beanGrpcServerRunnerTest() {
    assertNotNull(context.getBean(GrpcServer.class));
    assertThrows(NoSuchBeanDefinitionException.class,
        () -> context.getBean(GrpcServerRunner.class),
        "GrpcServerRunner should not be loaded during
          test");
  }
  // continue …
```

https://github.com/PacktPublishing/Modern-API-Development-with-Spring-6-and-Spring-Boot-3/tree/dev/Chapter11/server/src/test/java/com/packt/modern/api/ServerAppTests.java

The beanGrpcServerRunnerTest () method tests the loading of the GrpcServer class and GrpcServerRunner and the test should pass if the profile is set correctly.

Now, let's move on to test the gRPC services.

The gRPC test library provides a special class, GrpcCleanupRule, that manages the shutdown of registered servers and channels gracefully. You need to annotate it with the JUnit @Rule to make it effective. The gRPC test library also provides the InProcessServerBuilder builder class, which allows you to build the server, and the InProcessChannelBuilder builder class, which allows you to build the channel. These three classes are all you need to build and manage the server and channel.

Therefore, you first need to declare the required instances and then set up the method so that the execution environment is available before you fire the requests to the gRPC Source service.

Let's add the required class instances and test the setup () method in the following code:

```
@Rule
public final GrpcCleanupRule grpcCleanup = new
    GrpcCleanupRule();
private static
  SourceServiceGrpc.SourceServiceBlockingStub blockingStub;
@Autowired
private static String newlyCreatedSourceId = null;
@BeforeAll
public static void setup(@Autowired SourceService
      srcSrvc, @Autowired ChargeService chrgSrvc,
      @Autowired ExceptionInterceptor exceptionInterceptor)
      throws IOException {
  String sName = InProcessServerBuilder.generateName(); // 1
  grpcCleanup.register(InProcessServerBuilder
      .forName(sName).directExecutor().addService(srcSrvc)
      .intercept(exceptionInterceptor)
      .build().start());                                 // 2
  blockingStub = SourceServiceGrpc.newBlockingStub(
      grpcCleanup.register(InProcessChannelBuilder
      .forName(sName).directExecutor().build()));        // 3
}
```

Here, the setup method creates the server and channel with the Source service. Let's understand each of the lines mentioned in the setup () method:

- *Line 1* generates the unique name of the server.

- *Line 2* registers the newly created server and adds the Source service and server interceptor to it. We'll discuss ExceptionInterceptor in the *Coding for handling errors* section. Then, it starts the server for serving requests.

- *Line 3* creates the blocking stub, which will be used as a client for making the calls to the server. Here again, GrpcCleanUpRule is used to create the client channel.

Once the setup is executed, it provides us with the environment to carry out the tests. Let's test our first request, as shown in the following code block:

```
@Test
@Order(2)
@DisplayName("Creates source object using create RPC call")
public void SourceService_Create() {
  CreateSourceReq.Response response = blockingStub.create(
      CreateSourceReq.newBuilder().setAmount(100)
      .setCurrency("USD").build());
  assertNotNull(response);
  assertNotNull(response.getSource());
  newlyCreatedSourceId = response.getSource().getId();
  assertEquals(100, response.getSource().getAmount());
  assertEquals("USD", response.getSource().getCurrency());
}
```

All the complex aspects of the setup() method are complete. These tests now look pretty simple. You just use the blocking stub to make a call. You create the request object and use the stub to call the server. Finally, validate the server responses.

Similarly, you can test the validation error, as shown in the following code block:

```
@Test
@Order(3)
@DisplayName("Throws exception when invalid source id is
      passed to retrieve RPC call")
public void SourceService_RetrieveForInvalidId() {
  Throwable throwable = assertThrows(
      StatusRuntimeException.class, () -> blockingStub
      .retrieve(SourceId.newBuilder().setId("").build()));
  assertEquals("INVALID_ARGUMENT: Invalid Source ID passed"
      , throwable.getMessage());
}
```

You can also test for the valid response for source retrieval, as shown in the following code block:

```
@Test
@Order(4)
```

```
@DisplayName("Retrieves source obj created by
    createRPC call")
public void SourceService_Retrieve() {
  SourceId.Response response = blockingStub.retrieve
    (SourceId
      .newBuilder().setId(newlyCreatedSourceId).build());
  assertNotNull(response);
  assertNotNull(response.getSource());
  assertEquals(100, response.getSource().getAmount());
  assertEquals("USD", response.getSource().getCurrency());
}
```

This is the way in which you can write the test for the gRPC server and test the exposed RPC calls. You can use the same approach to write the rest of the test cases. After writing the test, you may have an idea of how the client is going to send the request to the server.

We have not yet discussed the exception interceptor that we have used in both the server code and test. Let's discuss this in the next section.

Coding for handling errors

You may have already gone through the theory-based *Handling errors and error status codes* section in *Chapter 10, Getting Started with gRPC*, where google.rpc.Status and gRPC status codes were discussed. You may want to revisit that section before going through this section as here you are going to write the actual code.

io.grpc.ServerInterceptor is a thread-safe interface for intercepting incoming calls that can be used for cross-cutting calls, such as authentication and authorization, logging, and monitoring. Let's use it to write ExceptionInterceptor, as shown in the following code block:

```
@Component
public class ExceptionInterceptor implements ServerInterceptor {
 @Override
 public <RQT, RST> ServerCall.Listener<RQT> interceptCall(
    ServerCall<RQT, RST> serverCall, Metadata
        metadata,
    ServerCallHandler<RQT, RST> serverCallHandler)
    ServerCall.Listener<RQT> listener = serverCallHandler
          .startCall(serverCall, metadata);
    return new ExceptionHandlingServerCallListener <>(
        listener, serverCall, metadata);
 }
// continue …
```

```
https://github.com/PacktPublishing/Modern-API-Development-with-
Spring-6-and-Spring-Boot-3/tree/dev/Chapter11/server/src/main/java/
com/packt/modern/api/server/interceptor/ExceptionInterceptor.java
```

Here, RQT represents the request type, and RST represents the response type.

We are going to use it for exception intercepting. An interceptor will pass the call to the server listener (ExceptionHandlingServerCallListener). ExceptionHandlingServer CallListener is a private class in ExceptionInterceptor that extends the ForwardingServerCallListener. SimpleForwardingServerCallListener abstract class.

The private listener class has overridden events, onHalfClose() and onReady(), which will catch the exception and pass the call to the handleException() method. The handleException() method will use the ExceptionUtils method to trace the actual exception and respond with error details. ExceptionUtils returns StatusRuntimeException, which is used to close the server call with an error status.

Let's see how this flow looks in code in the next code block:

```java
private class ExceptionHandlingServerCallListener<RQT, RST>
    extends ForwardingServerCallListener
        .SimpleForwardingServerCallListener<RQT> {
  private final ServerCall<RQT, RST> serverCall;
  private final Metadata metadata;
  ExceptionHandlingServerCallListener
      (ServerCall.Listener<RQT>
       lsnr,ServerCall<RQT, RST> serverCall, Metadata mdata) {
    super(lstnr);
    this.serverCall = serverCall;
    this.metadata = mdata;
  }
  @Override
  public void onHalfClose() {
    try { super.onHalfClose();}
    catch (RuntimeException e) {
      handleException(e, serverCall, metadata);
      throw e;
    }
  }
  @Override
  public void onReady() {
    try { super.onReady();}
    catch (RuntimeException e) {
```

```
        handleException(e, serverCall, metadata);
        throw e;
      }
    }
    private void handleException(RuntimeException e,
        ServerCall<RQT, RST> serverCall, Metadata metadata) {
      StatusRuntimeException status = ExceptionUtils.traceException(e);
      serverCall.close(status.getStatus(), metadata);
    }
  }
```

https://github.com/PacktPublishing/Modern-API-Development-with-Spring-6-and-Spring-Boot-3/tree/dev/Chapter11/server/src/main/java/com/packt/modern/api/server/interceptor/ExceptionInterceptor.java

Let's write the ExceptionUtils class next to complete the exception-handling core components. Then, you can use these components in a service implementation to raise the exceptions.

The ExceptionUtils class will have two types of overloaded methods:

- observerError(): This method will use StreamObserver to raise the onError() event
- traceException(): This method will trace the error from Throwable and return the StatusRuntimeException instance

You can use the following code to write the ExceptionUtils class:

```
@Component
public class ExceptionUtils {
private static final Logger LOG = LoggerFactory.getLogger
    (ExceptionInterceptor.class);
  public static StatusRuntimeException
      traceException(Throwable e) {
    return traceException(e, null);
  }
  public static <T extends GeneratedMessageV3> void
    observeError(StreamObserver<T>
        responseObserver, Throwable e) {
    responseObserver.onError(traceException(e));
  }
  public static <T extends GeneratedMessageV3> void
    observeError(StreamObserver<T> responseObserver,
      Exception
      e, T defaultInstance) {
```

```
    responseObserver.onError(
        traceException(e, defaultInstance));
}
// Continue …
```

https://github.com/PacktPublishing/Modern-API-Development-with-
Spring-6-and-Spring-Boot-3/tree/dev/Chapter11/server/src/main/java/
com/packt/modern/api/server/exception/ExceptionUtils.java

Here, you can see that the observerError() method is also calling traceException()
internally for onError events. Let's write the last overloaded method, traceException(), next:

```
public static <T extends
        com.google.protobuf.GeneratedMessageV3>
          StatusRuntimeException traceException(
            Throwable e, T defaultInstance) {
  com.google.rpc.Status status;
  StatusRuntimeException statusRuntimeException;
  if (e instanceof StatusRuntimeException) {
    statusRuntimeException = (StatusRuntimeException) e;
  } else {
    Throwable cause = e;
    if (cause != null && cause.getCause() != null &&
        cause.getCause() != cause) {
      cause = cause.getCause();
    }
    if (cause instanceof SocketException) {
      String errorMessage = "Sample exception message";
      status = com.google.rpc.Status.newBuilder()
          .setCode(com.google.rpc.Code.UNAVAILABLE_VALUE)
          .setMessage(errorMessage + cause.getMessage())
          .addDetails(Any.pack(defaultInstance))
          .build();
    } else {
      status = com.google.rpc.Status.newBuilder()
          .setCode(com.google.rpc.Code.INTERNAL_VALUE)
          .setMessage("Internal server error")
          .addDetails(Any.pack(defaultInstance))
          .build();
    }
    statusRuntimeException = StatusProto
        .toStatusRuntimeException(status);
  }
```

```
        return statusRuntimeException;
    }
```

Here, `SocketException` is shown by way of an example. You can add a check for another kind of exception here. You may notice that here we are using `com.google .rpc.Status` to build the status. Then, this instance of `Status` is passed to `toStatusRuntimeException()` of `StatusProto`, which converts the status to `StatusRuntimeException`.

Let's add the validation error in the `DbStore` class to make use of these exception-handling components, as shown in the following code block:

```
public SourceId.Response retrieveSource(String sourceId) {
    if (Strings.isBlank(sourceId)) {
        com.google.rpc.Status status =
            com.google.rpc.Status.newBuilder()
            .setCode(Code.INVALID_ARGUMENT.getNumber())
            .setMessage("Invalid Source ID is
                passed.").build();
        throw StatusProto.toStatusRuntimeException(status);
    }
    Source source = sourceEntities.get(sourceId);
    if (Objects.isNull(source)) {
        com.google.rpc.Status status =
            com.google.rpc.Status.newBuilder()
            .setCode(Code.INVALID_ARGUMENT.getNumber())
            .setMessage("Requested source is not available")
            .addDetails(Any.pack(
                SourceId.Response.getDefaultInstance())
                    ).build();
        throw StatusProto.toStatusRuntimeException(status);
    }
    return SourceId.Response.newBuilder()
            .setSource(source).build();
}
```

https://github.com/PacktPublishing/Modern-API-Development-with-Spring-6-and-Spring-Boot-3/tree/dev/Chapter11/server/src/main/java/com/packt/modern/api/server/repository/DbStore.java

You can similarly raise `StatusRuntimeException` in any part of the service implementation. You can also use the `addDetails()` method of `com.google.rpc.Status` to add more details to the error status, as shown in the `traceException(Throwable e, T defaultInstance)` code.

Finally, you can capture the error raised by the `retrieve()` method of `SourceService` in the `Service` implementation class, as shown next:

```
@Override
public void retrieve(SourceId sourceId, StreamObserver<SourceId.
Response> resObserver) {
  try {
    SourceId.Response resp =
                    repository.retrieve(sourceId.getId());
    resObserver.onNext(resp);
    resObserver.onCompleted();
  } catch (Exception e) {
    ExceptionUtils.observeError(resObserver, e,
                    SourceId.Response.getDefaultInstance());

  }
}
```

https://github.com/PacktPublishing/Modern-API-Development-with-Spring-6-and-Spring-Boot-3/tree/dev/Chapter11/server/src/main/java/com/packt/modern/api/server/service/SourceService.java

Exception handling is explained simply and constructively in this chapter. You can enhance it more as per your application requirements.

Now, let's write the gRPC client in the next section.

Developing the gRPC client

A client project's directory structure will look as follows. The project root directory contains the `build.gradle` and `settings.gradle` files, as shown in the following directory tree structure:

```
├── client
    ├── build.gradle
    ├── gradle
    │   └── wrapper
    ├── gradlew
    ├── gradlew.bat
    ├── settings.gradle
    └── src
        ├── main
        │   ├── java
        │   │   └── com
        │   │       └── packt
        │   │           └── modern
        │   │               └── api
        │   │
        │   │
```

```
|   └── resources
└── test
    └── java
```

The `resources` directory will contain the `application.properties` file.

Let's perform the following steps to configure the project:

1. First, you need to modify the project name in the `Chapter11/client/ settings. gradle` file to represent the server, as shown here:

    ```
    rootProject.name = 'chapter11-client'
    ```

2. Next, you can add the dependencies required for client projects to the `Chapter11/client/ build.gradle` file. The `grpc-stub` library provides the stubs-related APIs, and `protobuf- java-util` provides the utility methods for Protobuf and JSON conversions:

    ```
    def grpcVersion = '1.54.1'
    dependencies {
        implementation 'com.packt.modern.api:payment-
           gateway-api:0.0.1'
        implementation "io.grpc:grpc-stub:${grpcVersion}"
        implementation "com.google.protobuf:protobuf-java-
           util:3.22.2"
        implementation 'org.springframework.boot:
           spring-boot-starter-web'
        testImplementation 'org.springframework.boot:
           spring-boot-starter-test'
    }
    ```

    https://github.com/PacktPublishing/Modern-API-Development-with-
    Spring-6-and-Spring-Boot-3/tree/dev/Chapter11/client/build.gradle

3. The `payment-gateway-api` dependency is published in the local Maven repository. Therefore, you need to add the local Maven repository to the `repositories` section, as shown in the following code block:

    ```
    repositories {
      mavenCentral()
      mavenLocal()
    }
    ```

You are done with the Gradle configuration. Now, you can write the gRPC client.

Implementing the gRPC client

As you know, the Spring Boot application runs on its own server. Therefore, the client's application port should be different from the gRPC server port. Also, we need to provide the gRPC server host and port. These can be configured in `application.properties`:

```
server.port=8081
grpc.server.host=localhost
grpc.server.port=8080
```

https://github.com/PacktPublishing/Modern-API-Development-with-Spring-6-and-Spring-Boot-3/tree/dev/Chapter11/client/src/main/resources/application.properties

Next, let's create the gRPC client. This client will be used to configure the gRPC service stubs with the channel. The channel is responsible for providing the virtual connection to a conceptual endpoint in order to perform gRPC calls.

Create a new file, `GrpcClient.java`, in the `client/src/main/com/packt/modern/api/client` directory and add the code shown in the following code block:

```
@Component
public class GrpcClient {
  @Value("${grpc.server.host:localhost}")
  private String host;
  @Value("${grpc.server.port:8080}")
  private int port;
  private ManagedChannel channel;
  private SourceServiceBlockingStub sourceServiceStub;
  private ChargeServiceBlockingStub chargeServiceStub;
  public void start() {
    channel = ManagedChannelBuilder.forAddress(host, port)
        .usePlaintext().build();
    sourceServiceStub = SourceServiceGrpc
        .newBlockingStub(channel);
    chargeServiceStub = ChargeServiceGrpc
        .newBlockingStub(channel);
  }
  public void shutdown() throws InterruptedException {
    channel.shutdown().awaitTermination
      (1, TimeUnit.SECONDS);
  }
  public SourceServiceBlockingStub getSourceServiceStub() {
    return this.sourceServiceStub;
  }
```

```
    public ChargeServiceBlockingStub getChargeServiceStub() {
        return this.chargeServiceStub;
    }
}
```

https://github.com/PacktPublishing/Modern-API-Development-with-
Spring-6-and-Spring-Boot-3/tree/dev/Chapter11/client/src/main/java/
com/packt/modern/api/client/GrpcClient.java

Here, start() is the key that initialized the Source and Charge service stubs.
ManagedChannelBuilder is used to build ManagedChannel. ManagedChannel is a
channel that also provides life cycle management. This managed channel is passed to stubs.

You are using plain-text communication. However, it also provides encrypted communication.

We are now done with the client's code. Now, we need to call the start() method. You are going to
implement CommandLineRunner the way it was implemented for the GrpcServerRunner class.

It can be implemented as shown here:

```
@Profile("!test")
@Component
public class GrpcClientRunner implements CommandLineRunner {
    private static final Logger LOG = LoggerFactory.getLogger
        (GrpcClient.class);
    @Autowired
    GrpcClient client;
    @Override
    public void run(String... args) {
        client.start();
        Runtime.getRuntime().addShutdownHook(new Thread(() -> {
            try {
                client.shutdown();
            } catch (InterruptedException e) {
                System.out.println("error: {}", e.getMessage());
            }
        }));
    }
}
```

https://github.com/PacktPublishing/Modern-API-Development-with-
Spring-6-and-Spring-Boot-3/tree/dev/Chapter11/client/src/main/java/
com/packt/modern/api/client/GrpcClientRunner.java

This will initiate the stub instantiation following the start of the application. You can then call the stub methods.

Now, to call the stub methods, let's add a simple REST endpoint. This will demonstrate how to use the charge service stub to call its `retrieve` method.

You can create a new `ChargeController.java` file for the REST controller in the `src/main/java/com/packts/modern/api/controller` directory and add the code as shown here:

```
@RestController
public class ChargeController {
  private final GrpcClient client;
  public ChargeController(GrpcClient client) {
    this.client = client;
  }
  @GetMapping("/charges")
  public String getSources(@RequestParam(defaultValue =
      "ab1ab2ab3ab4ab5") String customerId)
        throws InvalidProtocolBufferException {
    var req = CustomerId.newBuilder()
        .setId(customerId).build();
    CustomerId.Response resp =
      client.getChargeServiceStub().retrieveAll(req);
    var printer = JsonFormat.printer()
        .includingDefaultValueFields();
      return printer.print(resp);
  }
}
```

`https://github.com/PacktPublishing/Modern-API-Development-with-Spring-6-and-Spring-Boot-3/tree/dev/Chapter11/client/src/main/java/com/packt/modern/api/controller/ChargeController.java`

Here, we have created a REST endpoint, `/charges`. This uses the `GrpcClient` instance to call the `retrieveAll()` RPC method of the `Charge` gRPC service using `ChargeServiceStub`.

Then, the response is converted into a JSON-formatted string using the `JsonFormat` class from the `protobuf-java-util` library and returned as a response. Generated JSON-formatted strings will also contain the fields with default values.

We are done with our development. Let's now test the complete flow in the next subsection.

Testing the gRPC service

Make sure that your gRPC server is up and running before testing the client. It is assumed that the gRPC `api` project has been built and that its latest artifacts have been published to the local Maven repository:

1. First, make sure that your `api` project library is published in the local Maven repository because it is required by both the `server` and `client` projects. Skip to *step 2* if you have already published the library. Java should be set to version 17. Execute the following commands from the `api` root project directory:

   ```
   # Make sure to enable UTF-8 for file encoding because
   # we are using UTF characters in Java files.
   $ export JAVA_TOOL_OPTIONS="-Dfile.encoding=UTF8"
   $ ./gradlew clean publishToMavenLocal
   Picked up JAVA_TOOL_OPTIONS: -Dfile.encoding=UTF8
   BUILD SUCCESSFUL in 6s
   10 actionable tasks: 10 executed
   ```

2. Then, start the server using the following command. Run it from the `server` project's root directory (Java should be set to version 17):

   ```
   // Server project root directory
   $ ./gradlew clean build
   $ java -jar build/libs/chapter11-server-0.0.1-SNAPSHOT.jar
   ... ... ...
   com.packt.modern.api.server.GrpcServer    : gRPC server started
   and listening on port: 8080.
   ```

3. Next, start the client using the following command in a new terminal window. Run it from the `client` project's root directory (Java should be set to version 17):

   ```
   // Client project root directory
   $ ./gradlew clean build
   $ java -jar build/libs/chapter11-client-0.0.1-SNAPSHOT.jar
   ... ... ...
   INFO 68732 Tomcat initialized with port(s): 8081 (http)
   INFO com.packt.modern.api.client.GrpcClient    : gRPC client
   connected to localhost:8080
   ```

4. Once the `server` and `client` services are up and running, open a new terminal window and execute the following command (the output is truncated):

   ```
   // calls the client service's charges API endpoint
   $ curl http://localhost:8081/charges
   {
   ```

```
            "charge": [{
              "id": "cle9e9oam6gajkkeivjof5pploq89ncp",
              "amount": 1000,
              "amountCaptured": 0,
              ...
              "created": "1679924425",
              "currency": "USD",
              "customerId": "ab1ab2ab3ab4ab5",
              "description": "Charge Description",
              ...
              "receiptEmail": "receipt@email.com",
              ...
              "status": "SUCCEEDED",
              "sourceId": "0ovjn416crgp9apr79bhpefme4dok3qf"
            }]
          }
```

A REST endpoint is used for demonstration purposes only. Similarly, you can use the gRPC client to call other services and their methods. gRPC is often used for inter-service communication, which is essential for microservice-based applications. However, it can also be used for web-based communication.

Let's learn a bit about microservices in the next section.

Understanding microservice concepts

Microservices are self-contained lightweight processes that communicate over a network. Microservices provide narrowly focused APIs to their consumers. These APIs can be implemented using REST, gRPC, or events.

Microservices are not new—they have been around for many years. For example, *Stubby*, a general-purpose infrastructure based on RPC, was used in Google data centers in the early 2000s to connect several services with and across data centers.

They have seen a recent rise in popularity and visibility. Before microservices became popular, monolithic architectures were mainly used for developing on-premises and cloud-based applications.

A monolithic architecture allows the development of different components, such as presentation, application logic, business logic, and **data access objects** (**DAOs**), and then you either bundle them together in an **enterprise archive** (**EAR**) or **web archive** (**WAR**) or store them in a single directory hierarchy (such as Rails or Node.js).

Many famous applications, such as Netflix, have been developed using a microservices architecture. Moreover, eBay, Amazon, and Groupon have evolved from monolithic architectures into microservices architectures. Nowadays, microservices-based application development is very common. The gRPC server that we have developed in this chapter could be called a microservice (obviously if you keep the scope of the server to either the Source service or Charge server).

Let's have a look at simple monolithic and microservices application designs in the next subsection.

In this section, we will take a look at the different system designs, which are designed using a monolithic design, an SOA monolithic design, and a microservices design. Let's discuss each of these in turn.

Traditional monolithic design

The following diagram depicts the traditional monolithic application design. This design was widely used before SOA became popular:

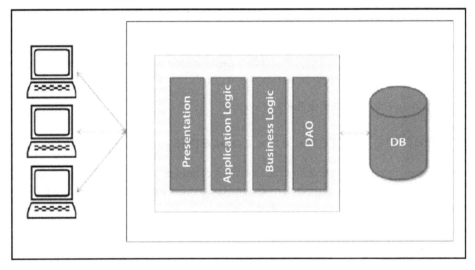

Figure 11.1 – Traditional monolithic application design

In a traditional monolithic design, everything is bundled in the same archive (all the presentation code is bundled in with the presentation archive, the application logic goes into the application logic archive, and so on), regardless of how it all interacts with the database files or other sources.

Monolithic design with services

After SOA, applications started being developed based on services, where each component provides services to other components or external entities. The following diagram depicts a monolithic application with different services; here, services are being used with a presentation component. All services, the presentation component, or any other components are bundled together:

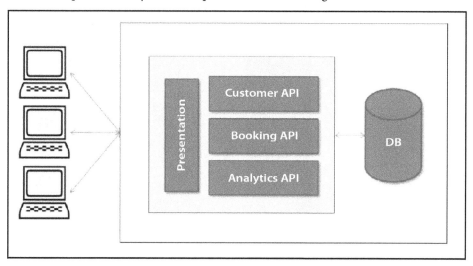

Figure 11.2 – Monolithic design with services

So, everything is bundled together in the form of EAR with a modules approach. A few SOA services may be deployed separately, but overall, it will be monolithic. However, the database is shared across the services.

Microservices design

The following diagram depicts the microservices design. Here, each component is autonomous. Each component can be developed, built, tested, and deployed independently. Here, even the application's UI component could also be a client and consume the microservices. For our example, the layer designed is used within the microservice.

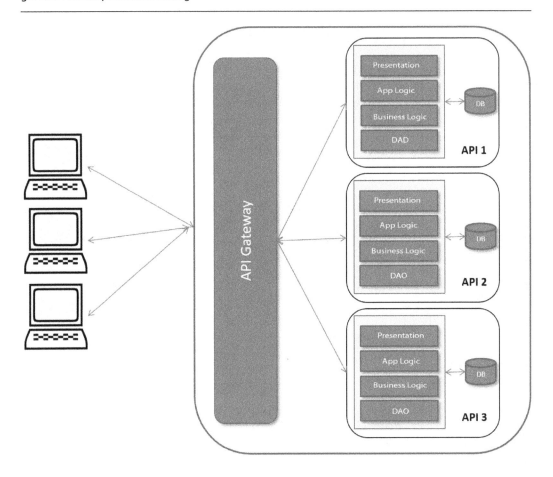

Figure 11.3 – Microservices design

The API gateway provides an interface where different clients can access the individual services and solve various problems, such as what to do when you want to send different responses to different clients for the same service. For example, a booking service could send different responses to a mobile client (minimal information) and a desktop client (detailed information), providing different details to each, before providing something different again to a third-party client.

A response may require the fetching of information from two or more services.

Each API service will be developed and deployed as a separate process and communication among these will happen based on exposed APIs.

For a sample e-commerce app, you can divide the application based on domains and bounded context and then develop a separate microservice for each of the domains. The following is a brief list of the provided microservices:

- Customers
- Orders
- Billing
- Shipping
- Invoicing
- Inventory
- Payment collection

You can develop each of these separately and use inter-process (inter-service) communication to stitch the solution together.

Summary

In this chapter, you explored Protobuf and gRPC-based service implementation. You developed the gRPC server and then consumed its services by developing a gRPC client. You learned about unit-testing the gRPC server and handling exceptions for gRPC-based services, and you also learned about the basic concepts of microservices.

You now have the skills to develop gRPC-based services (servers) and clients by defining the services using Protobuf.

In the next chapter, you will learn about distributed logging and tracing in web services.

Questions

1. Why should you use gRPC for binary large object transfers via HTTP/2?
2. You have implemented exception handling using `com.google.rpc.Status`. Can you do so without using this?
3. What is the difference between `com.google.rpc.Status` and `io.grpc.Status`?

Answers

1. Because, unlike HTTP libraries, gRPC libraries also provide the following features:

   - Interaction with flow control at the application layer

   - Cascading call cancellation

   - Load balancing and failover

2. Yes, you can. You can use the metadata shown in the following code block. However, making use of com.google.rpc.Status allows you to use the details (with a type of Any) object, which can capture more information:

```
Metadata.Key<SourceId.Response> key = ProtoUtils
    .keyForProto(SourceId.Response. getDefaultInstance);
Metadata metadata = new Metadata();
metadata.put(key, sourceIdResponse);
respObserver.onError(Status.INVALID_ARGUMENT
    .withDescription("Invalid Source ID")
    .asRuntimeException(metadata));
```

3. com.google.rpc.Status can include details of the Any type, which can be used to provide more error details. io.grpc.Status does not have a field that contains the error details. You must rely on another class's metadata to provide the error-related details, which may or may not contain only error specific information.

Further reading

- Protobuf version 3 documentation: https://developers.google.com/protocol-buffers/docs/proto3

- Protobuf's well-known types: https://developers.google.com/protocol-buffers/docs/reference/google.protobuf

- *Practical gRPC*: https://www.packtpub.com/in/web-development/practical-grpc

12

Adding Logging and Tracing to Services

In this chapter, you will learn about logging and tracing tools. We will use Spring Micrometer, Brave, the **Elasticsearch, Logstash, and Kibana (ELK)** stack, and Zipkin. ELK and Zipkin will be used to implement the distributed logging and tracing of the request/response of API calls. **Spring Micrometer** with **Actuator** will be used to inject tracing information into API calls. You will learn how to publish and analyze the logging and tracing of different requests and logs related to responses.

These aggregated logs will help you to troubleshoot web services. You will call one service (such as the gRPC client), which will then call another service (such as the gRPC server), and link them with a trace identifier. Then, using this trace identifier, you can search the centralized logs and debug the request flows. In this chapter, we will use this sample flow. However, the same tracing can be used when service calls require more internal calls. You will also use **Zipkin** to ascertain the performance of each API call.

Then, we will explore logging and monitoring tools, including the ELK stack and Zipkin, using Spring Micrometer. These tools (ELK and Zipkin) will then be used to implement the distributed logging and tracing of API requests and responses. Spring Micrometer will be used to inject the tracing information into API calls. You will learn how to publish and analyze the logging and tracing of different requests and logs related to responses.

You will explore the following topics in this chapter:

- Logging and tracing using the ELK stack
- Implementing logging and tracing in the gRPC code
- Distributed tracing with Zipkin and Micrometer

Technical requirements

You will need the following to develop and execute the code in this chapter:

- Any Java IDE, such as NetBeans, IntelliJ, or Eclipse

- **Java Development Kit (JDK) 17**

- An internet connection to clone the code and download the dependencies and Gradle

- Insomnia/cURL (for API testing)

- Docker and Docker Compose

You can find the code used in this chapter at `https://github.com/PacktPublishing/Modern-API-Development-with-Spring-6-and-Spring-Boot-3/tree/dev/Chapter12`.

So, let's begin!

Logging and tracing using the ELK stack

Today, products and services are divided into multiple small parts and executed as separate processes or deployed as separate services, rather than as a monolithic system. An API call may make several other internal API calls. Therefore, you need distributed and centralized logging to trace a request that spans multiple web services. This tracing can be done using the trace identifier (`traceId`), which can also be referred to as a correlation identifier (`correlationId`). This identifier is a collection of characters that forms a unique string, which is populated and assigned to an API call that requires multiple inter-service calls. Then, the same trace identifier is propagated to subsequent API calls for tracking purposes.

Errors and issues are imminent in the production system. You need to carry out debugging to ascertain the root cause. One of the key tools associated with debugging is logs. Logs can also give you warnings related to the system if the system is designed to do so. Logs also offer throughput, capacity, and monitoring of the health of the system. Therefore, you need an excellent logging platform and strategy that enables effective debugging.

There are different open source and enterprise tools available on the market for logging, including Splunk, Graylog, and the ELK stack. The ELK stack is the most popular of these, and you can use it for free if you are not going to provide ELK-based service as SaaS. We are going to use the ELK stack for logging throughout this chapter.

Let's understand the ELK stack in the next subsection.

Understanding the ELK stack

The ELK stack comprises three components – Elasticsearch, Logstash, and Kibana. All three products are part of Elasticsearch B.V. (https://www.elastic.co/). The ELK stack performs the aggregation, analysis, visualization, and monitoring of logs. The ELK stack provides a complete logging platform that allows you to analyze, visualize, and monitor all types of logs, including product and system logs.

You are going to use the following workflow to publish logs:

Figure 12.1 – Log flows in the ELK stack

Let's understand the diagram:

- Services/system logs are pushed to Logstash on the TCP port
- Logstash pushes the logs to Elasticsearch for indexing
- Kibana then uses the Elasticsearch index to query and visualize the logs

In an ideal production system, you should use one more layer. A broker layer such as Redis, Kafka, or RabbitMQ should be placed between the service logs and Logstash. This prevents data loss and can handle the sudden spike in input load.

Tips for ELK stack configuration

The ELK stack is fully customizable and comes with a default configuration. However, if you are using an Elasticsearch cluster (more than one instance of Elasticsearch is deployed), it is better to use an odd number of Elasticsearch nodes (instances) to avoid the split-brain problem.

It is recommended to use the appropriate data type for all the fields (input in JSON format for the logs). This will allow you to perform logical checks and comparisons while querying the log data. For example, the http_status < 400 check would work only if the http_status field type is a number and may fail if the http_status field type is a string.

If you are already familiar with the ELK stack, you can skip this introduction and move on to the next section. Here, you'll find a brief introduction to each of the tools in the ELK stack.

Elasticsearch

Elasticsearch is one of the most popular enterprise full-text search engines. It is based on Apache Lucene and developed using Java. Elasticsearch is also a high-performance, full-featured text search engine library. Recent changes in the licensing terms have made it restricted open source software, which prevents you from offering Elasticsearch or the ELK stack as SaaS. It is distributable and supports multi-tenancy. A single Elasticsearch server stores multiple indexes (each index represents a database), and a single query can search the data of multiple indexes. It is a distributed search engine and supports clustering.

It is readily scalable and can provide near-real-time searches with a latency of one second. Elasticsearch APIs are extensive and very elaborate. Elasticsearch provides JSON-based schema-less storage and represents data models in JSON. Elasticsearch APIs use JSON documents for HTTP requests and responses.

Logstash

Logstash is an open source data collection engine with real-time pipeline capabilities. It performs three major operations – it collects the data, filters the information, and outputs the processed information to data storage, in the same way as Elasticsearch does. It allows you to process any event data, such as logs from a variety of systems, because of its data pipeline capabilities.

Logstash runs as an agent that collects the data, parses it, filters it, and sends the output to a designated data store, such as Elasticsearch, or as simple standard output on a console.

On top of that, it has a rich set of plugins.

Kibana

Kibana is an open source web application that is used for visualizing and performing information analytics. It interacts with Elasticsearch and provides easy integration with it. You can perform searching and display and interact with the information stored in Elasticsearch indices.

It is a browser-based web application that lets you perform advanced data analysis and visualize your data in a variety of charts, tables, and maps. Moreover, it is a zero-configuration application. Therefore, it does not require any coding or additional infrastructure following installation.

Next, let's learn how to install the ELK stack.

Installing the ELK stack

You can use various methods to install the ELK stack, such as installing individual components as per the operating system, downloading the Docker images and running them individually, or executing the Docker images using Docker Compose, Docker Swarm, or Kubernetes. You are going to use Docker Compose in this chapter.

Let's understand the grammar of a Docker Compose file before we create the ELK stack Docker Compose file. A Docker Compose file is defined using YAML. The file contains four important top-level keys:

- `version`: This denotes the version of the Docker Compose file format. You can use the appropriate version based on the installed Docker Engine. You can check `https://docs.docker.com/compose/compose-file/` to ascertain the mapping between the Docker Compose file version and the Docker Engine version.

- `services`: This contains one or more service definitions. The service definition represents the service executed by the container and contains the container name (`container_name`), Docker image (`image`), environment variables (`environment`), external and internal ports (`port`), the command to be executed when running the container (`command`), the network to be used for communicating with other services (`networks`), mapping of the host filesystem with a running container (`volume`), and the container to be executed once the dependent service has started (`depends_on`).

- `networks`: This represents the (top-level) named network that needs to be created to establish a communication channel among the defined services. Then, this network is used by the service to communicate based on the `networks` key of the defined service. The top-level network key contains the driver field, which can be `bridge` for a single host and `overlay` when used in Docker Swarm. We will use `bridge`.

- `volumes`: A top-level `volumes` key is used to create the named volume that mounts the host path. Make sure to use it only if required by the multiple services; otherwise, you can use the `volumes` key inside the service definition, which would be service-specific.

Now, let's create the Docker Compose file, `docker-compose.yaml`, in the `Chapter12` directory to define the ELK stack. Then, you can add the following code to this file:

```yaml
version: "3.2"
services:
  elasticsearch:
    container_name: es-container
    image: docker.elastic.co/elasticsearch/
        elasticsearch:8.7.0
    environment:
      - xpack.security.enabled=false
      - "discovery.type=single-node"
    networks:
      - elk-net
    ports:
      - 19200:9200
```

https://github.com/PacktPublishing/Modern-API-Development-with-Spring-6-and-Spring-Boot-3/tree/dev/Chapter12/docker-compose.yaml

First, we defined the version of the Docker Compose file. Then, we created the `services` key section, which contains the `elasticsearch` service. The service contains the container name, Docker image, environment variables, and network (because you want ELK components to communicate with one another). Finally, ports are defined in `external:internal` format. You are going to use port `19200` from the browser to access it. However, other services will use port `9200` for communicating with Elasticsearch.

Similarly, you can define the `logstash` service next, as shown in the following code block:

```
logstash:
  container_name: ls-container
  image: docker.elastic.co/logstash/logstash:8.7.0
  environment:
    - xpack.security.enabled=false
  command: logstash -e 'input { tcp { port => 5001 codec
  => "json" }} output { elasticsearch { hosts =>
  "elasticsearch:9200" index => "modern-api" }}'
  networks:
    - elk-net
  depends_on:
    - elasticsearch
  ports:
    - 5002:5001
```

The Logstash configuration contains two extra service keys:

- First, a command key that contains the `logstash` command with a given configuration (using `-e`). The Logstash configuration normally contains three important parts:

  - `input`: Logstash input channels, such as `tcp` or `File`. We are going to use a TCP input channel. This means that gRPC server and client services will push the logs in JSON format (a JSON-coded plugin is used) to `logstash` on port `5001`.

  - `filter`: The `filter` key contains various filter expressions using different means, such as `grok`. You don't want to filter anything from logs, so you should opt out of using this key.

  - `output`: Where to send the input data after filtering out the information. Here, we are using Elasticsearch. Logstash pushes the received log information to Elasticsearch on port `9200` and uses the `modern-api` Elasticsearch index. This index is then used on Kibana for querying, analyzing, and visualizing the logs.

- A second key, `depends_on`, tells Docker Compose to start Elasticsearch before executing the `logstash` service.

Next, let's add the final service, `kibana`, as shown in the following code block:

```
kibana:
    container_name: kb-container
    image: docker.elastic.co/kibana/kibana:8.7.0
    environment:
      - ELASTICSEARCH_HOSTS=http://es-container:9200
    networks:
      - elk-net
    depends_on:
      - elasticsearch
    ports:
      - 5600:5601
networks:
  elk-net:
    driver: bridge
```

The service's `kibana` definition is in line with other defined services. It uses the `ELASTICSEARCH_ HOSTS` environment variable to connect to Elasticsearch.

At the end of the Docker Compose file, you define the `elk-net` network, which uses the `bridge` driver.

You are done with configuring the ELK stack Docker Compose file. Let's now start Docker Compose using the following command. If you run this for the very first time, local images of Elasticsearch, Logstash, and Kibana will also be fetched. You can use the `-f` flag if you've used other filenames apart from `docker-compose.yaml` or `docker-compose.yml` in the following command:

```
$ docker-compose up -d
Creating network "chapter12_elk-net" with driver "bridge"
Creating es-container ... done
Creating ls-container ... done
Creating kb-container ... done
```

Here, the `-d` option is used, which will start Docker Compose in the background. It starts the `es-container` Elasticsearch container first based on dependencies (the `depends_on` key).

> **Note**
>
> Elasticsearch uses 2 GB of heap size by default. Docker also uses 2 GB of memory by default in some systems, such as Mac.
>
> This may cause an error such as `error-137`. Therefore, you should increase the default Docker memory to at least 8 GB (more the better) and swap memory to at least 2 GB to avoid such issues.
>
> Please refer to `https://docs.docker.com/config/containers/resource_ constraints/#memory` for Docker memory configurations.

You can see that Docker Compose first creates the network, and then creates the service containers and starts them. Once all the containers are up, you can hit the URL `http://localhost:19200/` (which contains the external port defined for the `Elasticsearch` service) in the browser to check whether the Elasticsearch instance is up or not.

Once you hit the URL, you may find the following type of JSON response if the Elasticsearch service is up:

```
{
  "name" : "1bfa291e20b2",
  "cluster_name" : "docker-cluster",
  "cluster_uuid" : "Lua_MmozTS-grM0ZeJ5EBA",
  "version" : {
    "number" : "8.7.0",
    "build_flavor" : "default",
    "build_type" : "docker",
    "build_hash" :
        "09520b59b6bc1057340b55750186466ea715e30e",
    "build_date" : "2023-03-27T16:31:09.816451435Z",
    "build_snapshot" : false,
    "lucene_version" : "9.5.0",
    "minimum_wire_compatibility_version" : "7.17.0",
    "minimum_index_compatibility_version" : "7.0.0"
  },
  "tagline" : "You Know, for Search"
}
```

Next, let's check the Kibana dashboard by hitting the URL `http://localhost:5600` (which contains the external port defined for the `kibana` service) in the browser. This should load the home page of Kibana, as shown in the following screenshot:

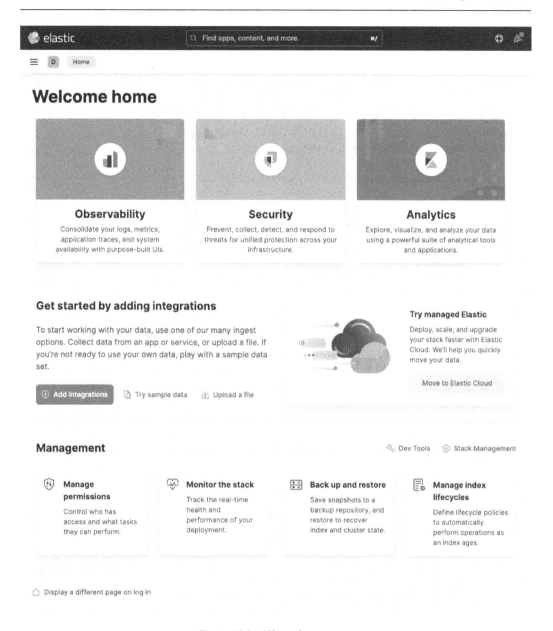

Figure 12.2 – Kibana home page

You may be wondering how the logs can be viewed since you have used the -d option. You can use the docker-compose logs [service name] command. If you don't provide the service name, then it will show the logs of all the services. You can use the --tail flag to filter the number of lines. The --tail="all" flag will show all the lines:

```
// Don't use flag -t as it is a switch that turns on the timestamp
$ docker-compose logs --tail="10" elasticsearch
$ docker-compose logs --tail="10" kibana
```

You can use the following command to stop Docker Compose:

```
$ docker-compose down
Stopping ls-container ... done
Stopping kb-container ... done
Stopping es-container ... done
Removing ls-container ... done
Removing kb-container ... done
Removing es-container ... done
Removing network chapter12_elk-net
```

The output may vary a bit. However, it should stop and remove all the running containers. The command stops the containers according to the dependencies provided in the docker-compose.yaml file based on the depends_on property and then removes them. Finally, it removes the network.

Next, let's make the code changes to integrate the application with the ELK stack.

Implementing logging and tracing in the gRPC code

Logging and tracing go hand in hand. Logging in the application code is already taken care of by default. You use Logback for logging. Logs are either configured to display on the console or pushed to the filesystem. However, you also need to push the logs to the ELK stack for indexing and analysis. For this purpose, you make certain changes to the Logback configuration file, logback-spring. xml, to push the logs to Logstash. On top of that, these logs should also contain tracking information.

Correlation/trace identifiers should be populated and propagated in distributed transactions for tracing purposes. A distributed transaction refers to the main API call that internally calls other services to serve the request. Before Spring Boot 3, Spring provided distributed tracing support through the **Spring Cloud Sleuth** library; now, tracing support is provided by Spring Micrometer. It generates the trace ID along with the span identifier. The trace ID gets propagated to all the participant services during the distributed transaction. The span ID also participates in a distributed transaction. However, the scope of the span identifier belongs to its service (the one it populates).

As communicated earlier, you are going to use Zipkin; therefore, you'll use Brave, a distributed tracing instrumentation library by Zipkin.

You can copy and enhance the code from *Chapter 11, gRPC API Development and Testing*, found at `https://github.com/PacktPublishing/Modern-API-Development-with-Spring-6-and-Spring-Boot-3/tree/dev/Chapter11`, to implement logging and tracing, or refer to this chapter's code at `https://github.com/PacktPublishing/Modern-API-Development-with-Spring-6-and-Spring-Boot-3/tree/dev/Chapter12` for changes.

First, you'll make the changes to the gRPC server code in the following subsection.

Changing the gRPC server code

To enable tracing and the publishing of logs to the ELK stack, you need to make the following code changes, as demonstrated in the following steps:

1. Add the following dependencies to the `build.gradle` file:

    ```
    implementation 'net.logstash.logback:
        logstash-logback-encoder:7.3'
    implementation 'io.micrometer:
        micrometer-tracing-bridge-brave'

    implementation 'org.springframework.boot:
        spring-boot-starter-actuator'
    ```

    `https://github.com/PacktPublishing/Modern-API-Development-with-Spring-6-and-Spring-Boot-3/tree/dev/Chapter12/server/build.gradle`

    Here, you are adding the following four dependencies:

    - **logstash-logback-encoder**: This library provides the Logback encoder, which publishes the logs to Logstash. This will be configured in the `spring-logback.xml` file.

    - **micrometer-tracing-bridge-brave**: This dependency takes care of managing the trace and span IDs. Micrometer Tracing Bridge Brave is an abstraction library for Zipkin Brave that publishes the collected tracing information to Zipkin using Brave.

    - **spring-boot-starter-actuator**: You used this library earlier in *Chapter 9, Deployment of Web Services*, to provide the health endpoint. It also provides metric endpoints. On top of that, it performs the metrics and tracing autoconfiguration.

2. Next, add/modify the `spring-logback.xml` file with the following content:

    ```xml
    <?xml version="1.0" encoding="UTF-8"?>
    <configuration>
      <springProperty scope="context"
    ```

```
          name="applicationName" source="spring.application
             .name"/>
   <springProperty scope="context"
      name="logstashDestination"
      source="logstash.destination" />
   <property name="LOG_PATTERN"
      value="%d{yyyy-MM-dd HH:mm:ss.SSS}
      %5p [${applicationName},%X{traceId:-},%X{spanId:-}]
      ${PID:-} --- [%15.15t] %-40.40logger{39} :
         %msg%n"/>
   <property name="LOG_FILE" value="${chapter12-grpc-
      server.service.logging.file:-chapter12-grpc-server-
      logs}"/>
   <property name="LOG_DIR" value="${chapter12-grpc-
      server.service.logging.path:-chapter12-grpc-server-
      logs}"/>
   <property name="SERVICE_ENV" value="${service.env:-
      dev}"/>
   <property name="LOG_BASE_PATH"
      value="${LOG_DIR}/${SERVICE_ENV}"/>
   <property name="MAX_FILE_SIZE"
    value="${chapter12.service.
      logging.rolling.maxFileSize:-
      100MB}"/>
   <!-- other configuration has been removed for brevity -->
```

https://github.com/PacktPublishing/Modern-API-Development-with-Spring-6-and-Spring-Boot-3/tree/dev/Chapter12/server/src/main/resources/logback-spring.xml

Here, you have defined the properties. The values of two of them are taken from the Spring configuration file (application.properties or application.yaml).

Let's now add the Logstash encoder, as shown in the following code block:

```
<appender name="STASH"
   class="net.logstash.logback.appender
      .LogstashTcpSocketAppender">
   <destination>${logstashDestination}</destination>
   <encoder
   class="net.logstash.logback.encoder.LogstashEncoder" />
</appender>
<!-- other configuration has been removed for brevity -->
```

Here, the STASH appender is defined and uses the TCP socket to push the logs to Logstash. It contains the destination element, which is used to assign Logstash's <HOST>:<TCP Port> value. Another element encoder contains the fully qualified class name, LogstashEncoder.

Finally, you will add the STASH appender to the root element, as shown next:

```
<!-- other configuration has been removed for brevity -->
<root level="INFO">
  <appender-ref ref="STDOUT"/>
  <appender-ref ref="STASH"/>
  <appender-ref ref="FILE"/>
</root>
```

The root level is set as INFO because you want to simply print the information logs.

Test configuration for Logstash

To disable the LOGSTASH (STASH) appender sending the logs to the Logstash instance, do the following:

1. You may copy logback-spring.xml into the test/resources directory.

2. Rename it test/resources/logback-test.xml.

3. Remove the LOGSTASH (STASH) appender and its entry from <ROOT>.

---

3. Next, let's add the Spring properties used in this logback-spring.xml file to application.properties, as shown in the following code block:

```
spring.application.name=grpc-server
spring.main.web-application-type=none
grpc.port=8080
logstash.destination=localhost:5002
management.tracing.sampling.probability=1.0
```

https://github.com/PacktPublishing/Modern-API-Development-with-Spring-6-and-Spring-Boot-3/tree/dev/Chapter12/server/src/main/resources/application.properties

Here, the Logstash destination host is set to localhost. If you are running on a remote machine, change the host accordingly. The Logstash TCP port is set to the same as the Logstash external port set in the Docker Composer file.

Default probability sampling for tracing is only 10% of the actual requests. Therefore, we are setting the management.tracing.sampling.probability property to 1.0. Now, it collects tracing information for 100% of the requests. Therefore, every request will be traced.

4. The required dependencies and configurations are now set. You can add the tracing server interceptor to the gRPC server. (Note: you don't need a tracing interceptor if you are using the RESTful web service as Spring's autoconfiguration mechanism takes care of this.)

First of all, let's define a new bean in a configuration file, as shown here:

```
@Configuration
public class Config {
  @Bean
  public ObservationGrpcServerInterceptor
        interceptor(ObservationRegistry registry) {
    return new
        ObservationGrpcServerInterceptor(registry);
  }
}
```

```
https://github.com/PacktPublishing/Modern-API-Development-with-
Spring-6-and-Spring-Boot-3/blob/dev/Chapter12/server/src/main/
java/com/packt/modern/api/server/Config.java
```

Here, you are creating an `ObservationGrpcServerInterceptor` bean, which is required for creating the tracing server interceptor. Before Spring Boot 3, the `RpcTracing` bean was provided by Spring Sleuth. Now, autoconfiguration of the `RpcTracing` bean is not available because Spring Boot 3 supports Spring Micrometer over Sleuth. You'll add this bean in the gRPC server as an interceptor.

5. Let's modify the gRPC server Java file (`GrpcServer.java`) to add the tracing server interceptor to the gRPC server, as shown in the following code block:

```
@Component
public class GrpcServer {
  // code truncated for brevity
  private final ObservationGrpcServerInterceptor
        oInterceptor;
  public GrpcServer(SourceService sourceService,
      ChargeService chargeService,
      ExceptionInterceptor
      exceptionInterceptor,
      ObservationGrpcServerInterceptor oInterceptor) {
    this.sourceService = sourceService;
    this.chargeService = chargeService;
    this.exceptionInterceptor = exceptionInterceptor;
    this.oInterceptor = oInterceptor;
  }
```

```
public void start()throws IOException,
    InterruptedException {
  server = ServerBuilder.forPort(port)
      .addService(sourceService)
      .addService(chargeService)
      .intercept(exceptionInterceptor)
      .intercept(oInterceptor)
      .build().start();
// code truncated for brevity
```

https://github.com/PacktPublishing/Modern-API-Development-with-Spring-6-and-Spring-Boot-3/tree/dev/Chapter12/server/src/main/java/com/packt/modern/api/server/GrpcServer.java

Here, you can see that the bean created in the configuration file in the previous step is injected using the constructor. Later, the `oInterceptor` bean has been used to create the gRPC server interceptor.

The changes required to enable log publishing to the ELK stack and tracing are made to the gRPC server. You can rebuild the gRPC server and run its JAR file to see the changes in effect. Check the following command-line output for the reference:

```
// Commands from Chapter12/server directory
$ ./gradlew clean build
// You may want to up docker-compose before running the server
// to avoid Logstash connect errors.
$ java -jar build/libs/chapter12-server-0.0.1-SNAPSHOT.jar
// Logs truncated for brevity
2023-04-23 21:30:42.120      INFO [grpc-server,,]      49296 ---
[           main] com.packt.modern.api.server.GrpcServer   : gRPC
server is starting on port: 8080.
```

You can see that the logs are following the pattern configured in `logback-spring.xml`. The log block printed after `INFO` contains the application/service name, as well as the trace and span IDs. The highlighted line is showing a *blank* trace ID and span ID because no external call is made that involves the distributed transaction. The trace and span IDs only get added to logs if the distributed transaction (service-to-service communication) is called.

Similarly, you can add the logging and tracing implementation in the gRPC client next.

Changing the gRPC client code

To enable tracing and the publishing of logs to the ELK stack, you need to make code changes in the gRPC client as well, which are very similar to the changes implemented in the gRPC server code. Refer to the following steps for more information:

1. Add the following dependencies to the `build.gradle` file:

    ```
    implementation 'net.logstash.logback:
        logstash-logback-encoder:7.3'
    implementation 'io.micrometer:micrometer-tracing-bridge-brave'

    implementation 'org.springframework.boot:spring-boot-starter-
    actuator'
    ```

    https://github.com/PacktPublishing/Modern-API-Development-with-Spring-6-and-Spring-Boot-3/tree/dev/Chapter12/client/build.gradle

    These are the same dependencies you added to the gRPC server.

2. Next, you can add the `logback-spring.xml` file in the same way you added it to the gRPC server code to configure the logging. Make sure to use `chapter12-grpc-client` in place of `chapter12-grpc-server` in the XML file.

3. Next, let's add the following Spring properties to `application.properties`. A few of these properties are referred to in the `logback-spring.xml` file as well:

    ```
    spring.application.name=grpc-client
    server.port=8081
    grpc.server.host=localhost
    grpc.server.port=8080
    logstash.destination=localhost:5002
    management.tracing.sampling.probability=1.0
    ```

    https://github.com/PacktPublishing/Modern-API-Development-with-Spring-6-and-Spring-Boot-3/tree/dev/Chapter12/client/src/main/resources/application.properties

4. The required dependencies and configurations are now set. Now, you can add tracing to the gRPC client. `ObservationGrpcClientInterceptor`, provided by the Micrometer library, provides the interceptor for the gRPC client.

> **Note**
>
> You don't need additional tracing changes if you are using the RESTful web service; Spring autoconfiguration takes care of this.

First of all, let's define a new bean, `ObservationGrpcClientInterceptor`, in a configuration file, as shown next:

```
@Configuration
public class Config {
  @Bean
  public ObservationGrpcClientInterceptor interceptor
      (ObservationRegistry registry) {
    return new ObservationGrpcClientInterceptor
      (registry);
  }
}
```

```
https://github.com/PacktPublishing/Modern-API-Development-
with-Spring-6-and-Spring-Boot-3/tree/dev/Chapter12/client/
src/main/java/com/packt/modern/api/Config.java
```

5. Now, you can modify the gRPC client Java file to add the `ObservationGrpcClientInterceptor` interceptor to the gRPC client:

```
@Component
public class GrpcClient {
  @Autowired
  private ObservationGrpcClientInterceptor
      observationGrpcClientInterceptor;
  // code truncated for brevity
  public void start() {
   channel = ManagedChannelBuilder.forAddress
       (host, port)
               .intercept(observationGrpcClientInterceptor)
     .usePlaintext().build();
   sourceServiceStub = SourceServiceGrpc
       .newBlockingStub(channel);
   chargeServiceStub = ChargeServiceGrpc
       .newBlockingStub(channel);
  }
  // code truncated for brevity
```

```
https://github.com/PacktPublishing/Modern-API-Development-
with-Spring-6-and-Spring-Boot-3/tree/dev/Chapter12/client/
src/main/java/com/packt/modern/api/client/GrpcClient.java
```

Here, you can see that the bean created in the configuration file in the previous step is autowired. Later, the `observationGrpcClientInterceptor` interceptor is added to the client.

The changes required to facilitate log publishing to the ELK stack and tracing are also done for the gRPC client. You can now rebuild the gRPC client and run its JAR file to see the changes in effect. Check the following command-line output for reference:

```
// Commands from Chapter12/client directory
$ ./gradlew clean build
// You may want to up docker-compose before running the server
// to avoid Logstash connect errors.
$ java -jar build/libs/chapter12-client-0.0.1-SNAPSHOT.jar
// log truncated for brevity
2023-04-23 23:02:35.297        INFO [grpc-client,,]       51746 ---
[           main] com.packt.modern.api.ClientApp           : Started
ClientApp in 3.955 seconds (process running for 4.611)
2023-04-23 23:02:35.674        INFO [grpc-client,,]       51746 ---
[           main] com.packt.modern.api.client.GrpcClient    : gRPC
client connected to localhost:8080
```

You can see that the logs follow the pattern configured in `logback-spring.xml`. The log block printed after `INFO` contains the application/service name, trace ID, and span ID. The trace and span IDs are blank because they only get added to logs if distributed transactions (service-to-service communication) are called.

The changes required to enable log aggregation and distributed tracing in both the gRPC server and client services are now complete.

Next, you'll test the changes and view the logs in Kibana.

Testing the logging and tracing changes

Before beginning testing, make sure that the ELK stack is up and running. Also, make sure that you first start the gRPC server and then the gRPC client service.

You can add the appropriate log statements to your services for verbose logs.

Let's run the following command in the new terminal window. This will call the `/charges` REST endpoint in the gRPC client service:

```
$ curl http://localhost:8081/charges
```

It should respond with the following JSON output:

```
{
  "charge": [{
    "id": "aibn4f45m49bojd3u0p16erbi5lnelui",
    "amount": 1000,
    "amountCaptured": 0,
    "amountRefunded": 0,
```

```
        "balanceTransactionId": "",
        "calculatedStatementDescriptor": "",
        "receiptEmail": "receipt@email.com",
output truncated for brevity
        "refunded": false,
        "refunds": [],
        "statementDescriptor": "Statement Descriptor",
        "status": "SUCCEEDED",
        "sourceId": "inccsjg6gogsvi4rlprdbvvfq2ft2e6c"
    }]
}
```

The previous `curl` command should generate logs like the following one in the gRPC client:

```
2023-04-23 23:10:37.882        INFO [grpc-client,64456d940c51e3e2baec07f
7448beee6,baec07f7448beee6]       51746 --- [nio-8081-exec-1] brave.Trac
er                             : {"traceId":"64456d940c51e3e2baec07f
7448beee6","parentId":"baec07f7448beee6","id":"0645e686d86968b6",
"kind":"CLIENT","name":"com.packtpub.v1.ChargeService/RetrieveAll"
,"timestamp":1682271636300866,"duration":1578184,"localEndpoint":
{"serviceName":"unknown","ipv4":"192.168.1.2"},"tags":{"rpc.service"
:"com.packtpub.v1.ChargeService","rpc.method":"RetrieveAll"}}
2023-04-23 23:10:37.886        INFO [grpc-client,64456d940c51e3e2baec07f
7448beee6,baec07f7448beee6]       51746 --- [nio-8081-exec-1] c.p.m.api.
controller.ChargeController    : Server response received in Json
Format: charge {
  id: "iivpc3i9el2dso9s2s2rqf9j3s2pomlm"
  amount: 1000
  created: 1682265641
  currency: "USD"
  customerId: "ab1ab2ab3ab4ab5"
  description: "Charge Description"
  receiptEmail: receipt@email.com
  statementDescriptor: "Statement Descriptor"
  sourceId: "6ufgh93stkjod1ih2vhkmamj9l1m0hvv"
}
```

Here, the blocks highlighted the application name (`grpc-client`), trace ID (`64456d940c51e3e2baec07f7448beee6`), and span ID (`baec07f7448beee6`).

The command should also generate the following logs in the gRPC server:

```
2023-04-23 23:10:37.821        INFO [grpc-server,64456d940c51
e3e2baec07f7448beee6,182159d509ce0714]       49296 ---
[ault-executor-0] brave.Tracer                             :
{"traceId":"64456d940c51e3e2baec07f7448beee6","parentId":"0645e686d869
68b6","id":"182159d509ce0714","kind":"SERVER","name":"com.packtpub.v1
.ChargeService/RetrieveAll","timestamp":1682271637683829,"duration"
```

```
:127229,"localEndpoint":{"serviceName":"unknown","ipv4":
"192.168.1.2"},"tags":{"rpc.service":"com.packtpub.
v1.ChargeService","rpc.method":"RetrieveAll"}}
```

Here, the blocks highlighted the application name (`grpc-server`), trace ID
(`64456d940c51e3e2baec07f7448beee6`), and span ID (`182159d509ce0714`). The
trace ID is the same as what is displayed in the gRPC client logs. The span IDs are different from
the gRPC client service because span IDs belong to their respective individual services. This is how
the trace/correlational ID helps you trace the call for requests across different services because it
will be propagated to all the services it involves.

Tracing this request was simple as the logs contain just a few lines and are scattered across only two
services. What if you have a few gigabytes of logs scattered across various services? Then, you can
make use of the ELK stack to search the log index using different query criteria. However, we are
going to use the trace ID for this purpose.

First, open the Kibana home page in the browser (`http://localhost:5600/`). Then, click on
the hamburger menu in the top-left corner, as shown in the following screenshot. After that, click on
the **Discover** option in the menu that appears:

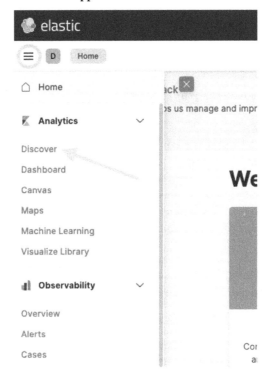

Figure 12.3 – Kibana hamburger menu

This should open the **Discover** page, as shown in the following screenshot. If this your first time opening the page, you must create an index pattern that will filter out the indexes available in Elasticsearch:

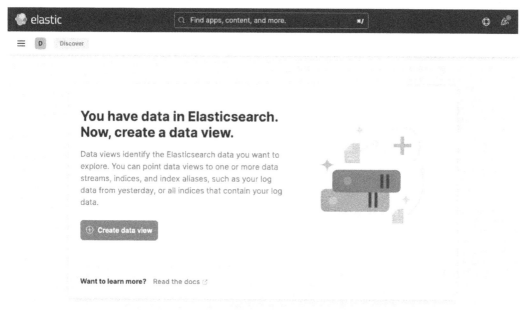

Figure 12.4 – Kibana's Discover page

Next, click on the **Create data view** button, which opens the following page to define the index pattern. Here, you should enter the index name (modern-api) given in the Logstash configuration in the ELK stack's Docker Compose file:

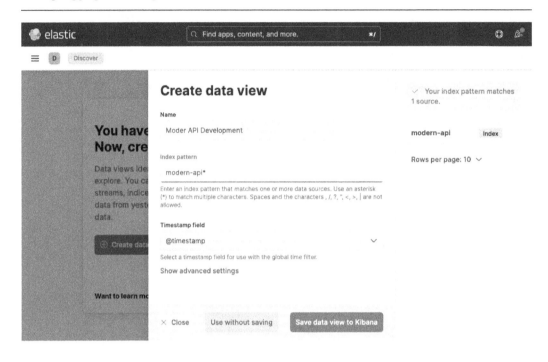

Figure 12.5 – Kibana's Create data view page

Here, you can also provide a name for the data view and an index pattern. You can keep the default value of @timestamp for **Timestamp field**.

Then, click on the **Save data view to Kibana** button. This action will create the data view and may show the following view (if you have called the client's REST APIs recently; otherwise, it will show no data view):

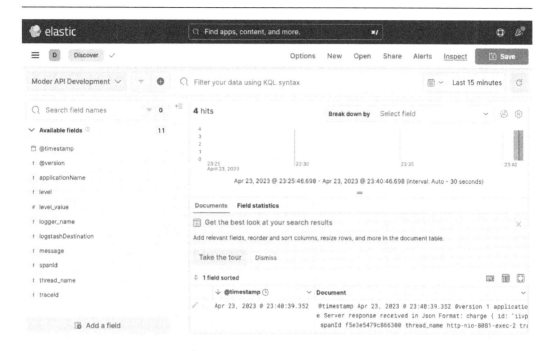

Figure 12.6 – Kibana's saved data view page

You can add the filter query to the **Search** textbox and the **Date/Duration** menu at the top right of the **Discover** page.

Query criteria can be input using the **Kibana Query Language** (**KQL**), which allows you to add different comparator and logical operators. For more information, refer to `https://www.elastic.co/guide/en/kibana/master/kuery-query.html`.

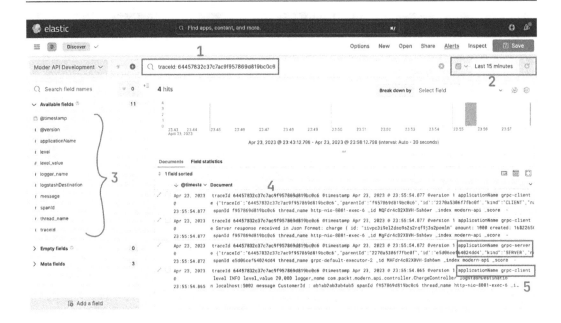

Figure 12.7 – Kibana's Discover page – filtering

As shown in the preceding figure, we have entered criteria (`traceId: 64457832c37c7ac9f957869d819bc0c6`) (*1*) and kept the **Duration** field as its default (the last 15 minutes) (*2*). The left-hand side also shows you how to select the Elasticsearch index and all the fields available in it (*3*).

Once you press the *Enter* key or click on the refresh button after entering the criteria, the search displays the available logs from all the services. The searched values are highlighted in yellow (*4*).

You can also observe that the searched trace ID shows logs from both server and client services (*5*).

The searched **Discovery** page also shows the graph that shows the number of calls made during a particular period. You can generate more logs and reveal any errors, and then you can use different criteria to filter the results and explore further.

You can also save the searches and perform more operations, such as customizing the dashboard. Please refer to `https://www.elastic.co/guide/en/kibana/master/index.html` for more information.

The ELK stack is good for log aggregation, filtering, and debugging using the trace ID and other fields. However, it can't check the performance of API calls – the time taken by the call. This is especially important when you have a microservice-based application.

This is where **Zipkin** (also known as **OpenZipkin**), along with Micrometer, comes in.

Distributed tracing with Zipkin and Micrometer

Spring Micrometer is a utility library that collects the metrics generated by the Spring Boot application. It provides vendor-neutral APIs that allow you to export the collected metrics to different systems, such as ELK. It collects different types of metrics. A few of them are the following:

- Metrics related to the JVM, CPU, and cache

- Latencies in Spring MVC, WebFlux, and the REST client

- Metrics related to Datasource and HikariCP

- Uptime and Tomcat usage

- Events logged to Logback

Zipkin, along with Micrometer, helps you not only to trace transactions across multiple service invocations but also to capture the response time taken by each service involved in the distributed transaction. Zipkin also shows this information using nice graphs. It helps you to locate the performance bottlenecks and drill down into the specific API call that creates the latency issue. You can find out the total time taken by the main API call as well as its internal API call time.

Services developed with Spring Boot facilitate their integration with Zipkin. You just need to make two code changes – the addition of the `zipkin-reporter-brave` dependency and the addition of the Zipkin endpoint property.

You can make these two changes to both the gRPC server and client, as shown next:

1. First, add the highlighted dependency to `build.gradle` (both the gRPC server and client projects):

    ```
    implementation 'net.logstash.logback:logstash-logback-
                    encoder:7.3'
    implementation 'io.micrometer:micrometer-tracing-
        bridge-brave'
    implementation 'io.zipkin.reporter2:zipkin-reporter-
        brave'

    implementation 'org.springframework.boot:
        spring-boot-starter-actuator'
    ```

2. Next, add the following properties to the `application.properties` file (for both the gRPC server and client):

    ```
    management.zipkin.tracing.endpoint=
                        http://localhost:9411/api/v2/spans
    management.tracing.sampling.probability=1.0
    ```

The Zipkin `tracing.endpoint` property points to the Zipkin API endpoint.

You are done with the changes required in the code for publishing the tracing information to Zipkin. Rebuild both the server and client services after making these changes.

Now, let's install and start Zipkin.

There are various ways to install and run Zipkin. Please refer to `https://zipkin.io/pages/quickstart` to find out about these options. You can add it in `docker-compose` too. However, for development purposes, we are going to fetch the latest release as a self-contained executable JAR from `https://search.maven.org/remote_content?g=io.zipkin&a=zipkin-server&v=LATEST&c=exec` and then start it using the following command (make sure to change the version in the JAR file based on the downloaded file):

```
$ java -jar zipkin-server-2.24.0-exec.jar
```

The previous command, when executed, will start Zipkin with an in-memory database. For production purposes, it is recommended to use a persistence store, such as Elasticsearch.

It should start with the default port `9411` at `http://127.0.0.1:9411/` if executed on `localhost`.

Once the Zipkin server and the ELK stack are up and running, you can start both the gRPC server and client services and execute the following command:

```
$ curl http://localhost:8081/charges
```

This command should print a log like the following log statement in the gRPC client service:

```
2023-04-24 11:35:10.313       INFO [grpc-client,64461c16391707ee95478f9
57f3ccb1d,95478f957f3ccb1d]       62484 --- [nio-8081-exec-2] c.p.m.api.
controller.ChargeController      : CustomerId : ab1ab2ab3ab4ab5
```

Keep the trace ID (`64461c16391707ee95478f957f3ccb1d` in the previous output) handy as you are going to use it in the Zipkin UI. Open the Zipkin home page by accessing `http://localhost:9411`. It will look as follows:

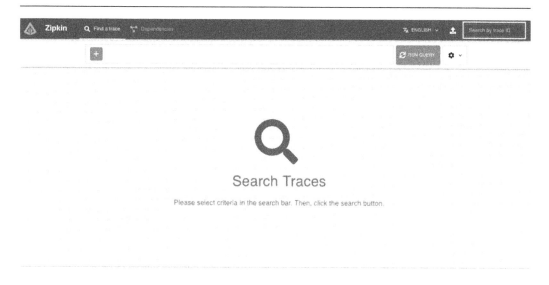

Figure 12.8 – Zipkin home page

You can observe that Zipkin also allows you to run queries. However, we'll make use of the trace ID. Paste the copied trace ID in the **Search by trace ID** textbox in the top-right corner (highlighted in green) and then press *Enter*:

Figure 12.9 – Zipkin search result page

In the preceding figure, the Zipkin trace ID shows complete API call information at the top (*1* and *2*) if the trace ID is available. In the left-hand side section, it shows all the corresponding API calls with a hierarchy, which shows the individual call times in a graphical way (*3*). These API call rows are selectable. Selected call details are displayed on the right-hand side (*4*).

The **Annotation** section on the right-hand side displays the individual call start and finish times, as shown in the following screenshot for the `grpc-server` call:

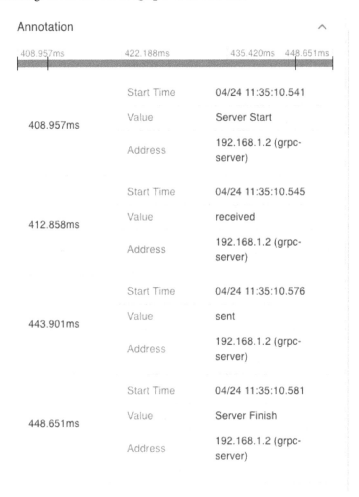

Figure 12.10 – Annotation details

Time tracking at a granular level for each distributed API call allows you to identify the latency issues and relative time tracking for performance tuning.

Summary

In this chapter, you learned how the trace/correlation ID is important and how it can be set up using Micrometer with Brave. You can use these generated IDs to find the relevant logs and API call durations. You integrated the Spring Boot services with the ELK stack and Zipkin.

You also implemented extra code and configurations, which are required for enabling distributed tracing for gRPC-based services.

You acquired log aggregation and distributed tracing skills using Micrometer, Brave, the ELK stack, and Zipkin.

In the next chapter, you are going to learn about the fundamentals of GraphQL APIs.

Questions

1. What is the difference between the trace ID and span ID?

2. Should you use a broker between services that generate the logs and the ELK stack? If yes, why?

3. How does Zipkin work?

Answers

1. Trace IDs and span IDs are created when the distributed transaction is initiated. A trace ID is generated for the main API call by the receiving service using Spring Cloud Sleuth. A trace ID is generated only once for each distributed call. Span IDs are generated by all the services participating in the distributed transaction. A trace ID is a correlation ID that will be common across the service for a call that requires a distributed transaction. Each service will have its own span ID for each of the API calls.

2. Yes, a broker such as Kafka, RabbitMQ, or Redis allows robust persistence of logs and removes the risk of losing log data in unavoidable circumstances. It also performs better and can handle sudden spikes of data.

3. A tracer such as Micrometer with Brave or Spring Cloud Sleuth (*which performs instrumentation*) does two jobs – records the time and metadata of the call being performed, and propagates the trace IDs to other services participating in the distributed transaction. Then, it pushes the tracing information to Zipkin using a *reporter* once the scan completes. The reporter uses *transport* such as HTTP and Kafka to publish the data in Zipkin.

The *collector* in Zipkin collects the data sent by the transporters from the running services and passes it to the storage layer. The storage persists the data. Persisted data is exposed by the Zipkin APIs. The Zipkin UI calls these APIs to show the information graphically:

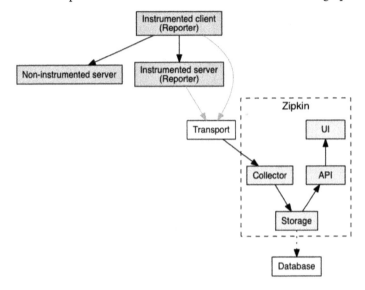

Further reading

- Elasticsearch documentation: `https://www.elastic.co/guide/en/elasticsearch/reference/current/index.html`

- Kibana documentation: `https://www.elastic.co/guide/en/kibana/master/index.html`

- Kibana Query Language: `https://www.elastic.co/guide/en/kibana/master/kuery-query.html`

- Logstash documentation: `https://www.elastic.co/guide/en/logstash/master/index.html`

- *Elasticsearch 8.x Cookbook – Fifth Edition*: `https://www.packtpub.com/product/elasticsearch-8x-cookbook-fifth-edition/9781801079815`

- Zipkin documentation: `https://zipkin.io/pages/quickstart`

Part 4 –
GraphQL

In this part, you will learn about GraphQL-based API development. After completing this section, you will know the in-depth fundamentals of GraphQL, be able to differentiate between REST, reactive, and gRPC APIs, in the context of GraphQL thoroughly, and understand when to use which API style. You will also learn how to design the GraphQL schema, which will be used to generate Java code. Finally, you will learn how to write data fetchers and loaders to resolve query fields to serve the GraphQL API requests.

This part contains the following chapters:

- *Chapter 13, Getting Started with GraphQL*
- *Chapter 14, GraphQL API Development and Testing*

13

Getting Started with GraphQL

In this chapter, you will learn about the fundamentals of **GraphQL**, including its **schema definition language** (**SDL**), queries, mutations, and subscriptions. The GraphQL API is popular in hand-held device-based apps such as mobile apps because it is fast and efficient in fetching the data and better than REST in certain cases. Therefore, it is important to learn about GraphQL. You will learn more about its comparison with REST in the *Comparing GraphQL with REST* section in this chapter. After completing this chapter, you will know the basics of GraphQL, including its semantics, schema design, and everything you need to develop a GraphQL-based API using Spring and Spring Boot.

We will cover the following topics in this chapter:

- Getting to know GraphQL
- Learning the fundamentals of GraphQL
- Designing a GraphQL schema
- Testing GraphQL queries and mutations
- Solving the N+1 problem

Technical requirements

This chapter covers the theory behind GraphQL and related concepts. It is advised to go through this chapter first to develop and test the GraphQL-based service code presented in the next chapter.

Getting to know GraphQL

You might have heard of or be aware of GraphQL, which has become more popular in the API space in past few years and is becoming the preferred way of implementing APIs for handheld devices and the web.

GraphQL is a declarative query and manipulation language and server-side runtime for APIs. GraphQL empowers the client to query exactly the data they want – no more, no less.

We'll discuss its brief history in the next subsection.

A brief history of GraphQL

In 2011, Facebook was facing challenges in terms of improving the performance of its website on mobile browsers. They started building their own mobile app with mobile-native technologies. However, APIs were not up to the mark because of hierarchical and recursive data. They wanted to optimize their network calls. Note that in those days, mobile network speed was in Kb/s in some parts of the world. Having a fast, high-quality mobile app was going to be the key to their success, since their consumers had started shifting to mobile devices.

In 2012, a few engineers on Facebook – Lee Byron, Dan Schafer, and Nick Schrock – teamed up to create GraphQL. Initially, it was used to design and develop Facebook's newsfeed feature, but later, it was used across its infrastructure, being used internally at Facebook exclusively until it was open sourced in 2015, when the GraphQL specification and its JavaScript implementation were made available to the public. Soon, other language implementations of the GraphQL specification started rolling out, including Java.

I think you would enjoy watching this GraphQL documentary at `https://www.youtube.com/watch?v=783ccP__No8`, which walks through GraphQL's journey from being an internal Facebook tool to its current success.

> **Did you know?**
> Netflix and Coursera were also working on a similar idea to build efficient and performant APIs. Coursera didn't take it forward, but Netflix open sourced Falcor.

Comparing GraphQL with REST

You developed APIs using REST in *Part 1: RESTful Web Services* of this book. In fact, an example e-commerce UI app also consumed REST APIs to implement its e-commerce functionality in the first part of this book. We are going to keep referring to REST in this chapter so that we can understand the necessary GraphQL concepts wherever they're applicable. This correlation should help you grasp the GraphQL concepts easily.

GraphQL is more powerful, flexible, and efficient than REST. Let's understand why.

Let's consider an example where a user logs in to an e-commerce UI app and automatically navigates to the product listing page. When this happens, the UI app consumes three different endpoints, as follows:

- The user endpoint, to fetch the user's information

- The product endpoint, to fetch the product list

- The cart endpoint, to fetch the cart items from the user's cart

So, basically, you must make three calls in REST to fetch the required information in a fixed structure (you can't change the fields that are sent in response) from the backend.

On the other hand, GraphQL can fetch a user's information, a user's cart data, and the product list in a single call. Thus, it reduces the number of network calls from three to one. GraphQL just exposes a single endpoint, unlike REST, where you must define an endpoint for each use case. You might say that you can write a new REST endpoint that does that. Yes, that may solve this specific use case, but it is not flexible; it won't allow for quick iteration.

Moreover, GraphQL lets you describe the fields you want to fetch from the backend in a request. The server response contains only those fields, which are sent as a part of the request– no more, no less.

For example, you may want to add user reviews to products. For this, you just need to add the reviews field to the GraphQL query. Similarly, you don't need to consume extra fields. You just add those fields that you need to the GraphQL query. On the other hand, REST's response contains predefined fields, regardless of whether you need certain fields in the response object or not. Then, you must filter the required fields at the client end. Therefore, we can say that GraphQL uses network bandwidth more efficiently by avoiding over-/under-fetching problems.

GraphQL APIs don't need constant changes as happens with REST, where you may need to change the API or add new APIs for a requirement change. This improves the development speed and iteration. You can easily add new fields or mark thsose that have been deprecated (fields not being used by a client anymore). Therefore, you can make the changes in the client without impacting the backend. In short, you can write evolving APIs without any versioning and breaking changes.

REST offers caching using built-in HTTP specifications. However, GraphQL does not follow the HTTP specifications; instead, it makes use of libraries such as Apollo/Relay for caching. However, REST is based on HTTP and does not follow any specification for implementation, which may lead to inconsistent implementations, as we discussed when comparing REST with gRPC. You can use the HTTP GET method to delete a resource.

GraphQL is superior to REST APIs in terms of its usage in mobile clients. The capabilities of GraphQL APIs are also defined using strong types. These types are part of the schema that contains the API definitions. These types are written in the schema using **SDL**.

GraphQL acts as a contract between the server and the client. You can correlate the GraphQL schema with the gRPC **interface definition language** (**IDL**) file and OpenAPI specification file.

We'll discuss the fundamentals of GraphQL in the next section.

Learning the fundamentals of GraphQL

GraphQL APIs contain three important **root types** – **query**, **mutation**, and **subscription**. These are all defined in the GraphQL schema using special SDL syntax.

GraphQL provides a single endpoint that returns the JSON response based on the request, which can be a query, a mutation, or a subscription.

First, let's understand queries.

Exploring the Query type

The Query type is used for reading operations that fetch information from the server. A single Query type can contain many queries. Let's write a query using SDL to retrieve the logged-in user, as shown in the following GraphQL schema:

```
type Query {
  me: LogginInUser
  # You can add other queries here
}
type LoggedInUser {
  id: ID
  accessToken: String
  refreshToken: String
  username: String
}
```

Here, you have done two things:

- You have defined the query root of the GraphQL interface, which contains the query you can run. It contains just a single query type, me, that returns an instance of the LoggedInUser type.

- You have specified the user-defined LoggedInUser object type, which contains four fields. These fields are followed by their types. In the preceding code, you used GraphQL's built-in *scalar types*, called ID and String, to define the types of the fields. We'll discuss these types later in this chapter when we discuss built-in scalar types in detail.

Once you have this schema implementation on the server and have fired the following GraphQL query, you will only get the fields you requested, along with their values, as a JSON object in response.

You can find the me query and its JSON response in the following code block:

```
# Request input
{
  me {
    id
    username
  }
}
#JSON response
{
  "data": {
    "me": {
      "id": "asdf90asdkqwe09kl",
      "username": "scott"
    }
```

```
    }
  }
```

Interestingly, here GraphQL's request input does not start with a query because `Query` is the default for the payload. This is called an **anonymous query**. The GraphQL handler on the server always assumes the payload to be a query unless you specify something specific, such as `Mutation`. However, if you want, you can also prefix the query request input with `query`, as shown here:

```
query {
  me {
    id
    username
  }
}
```

As you can see, this allows you to only query those fields that you need. Here, only the `id` and `username` fields were requested from the `LoggedInUser` type, and the server responded with only these two fields. The request payload is enclosed in curly braces { }. You can use # for commenting in the schema.

Now, you know how to define `Query` and `object` types in a GraphQL schema. You also learned how to form a GraphQL request payload according to its query type and the expected JSON response.

We'll learn about GraphQL mutations in the next subsection.

Exploring the Mutation type

The `Mutation` type is used in GraphQL requests for all the add, update, and delete operations that get performed on the server. A single `Mutation` type can contain many mutations. Let's define an `addItemInCart` mutation that adds a new item to the cart:

```
type Mutation {
  addItemInCart(productId: ID, qty: Int): [Item]
  # You can add other mutations here
}
type Item {
  id: ID!
  productId: ID
  qty: Int
}
```

Here, you have defined the `Mutation` type and a new object type called `Item`. The mutation is added and called `addItemInCart`. The `Query`, `Mutation`, and `Subscription` types can pass arguments. To define the necessary parameters, you can enclose the named arguments with ()

brackets; the arguments are divided by commas. The signature of `addItemInCart` contains two arguments and returns a list of cart items. A list is marked using `[]` brackets.

> **Optional and required arguments**
>
> Let's say you declare an argument with a default value, such as the following mutation:
>
> `pay(amount: Float, currency: String = "USD"): Payment`
>
> Here, `currency` is an optional argument. It contains the default value, whereas `amount` is a required field because it does not contain any default value.

Please note that `Int` is a built-in scalar type for signed 32-bit integers. Default values are null in GraphQL. If you want to force a non-nullable value for any field, then its type should be marked with an exclamation mark (`!`). Once it (`!`) has been applied to any field in the schema, the GraphQL server will always provide a value instead of a null for that field when it is placed in the request payload by the client. You can also declare a list with exclamation marks; for example, `items: [Item] !` and `items: [Item!] !`. Both declarations will provide zero or more items in a list. However, the latter would provide only a valid `Item` object (i.e., a non-nullable value).

Once you have this schema implementation on the server, you can use the following GraphQL query. You will get only the fields you requested, along with their values, as a JSON object:

```
# Request input
mutation {
  addItemInCart(productId: "qwer90asdkqwe09kl", qty: 2) {
    id
    productId
  }
}
```

You can see that this time, the GraphQL request input starts with the `mutation` keyword. If you don't start a mutation with the `mutation` keyword, then you might get an error with a message along the lines of **Field 'addItemInCart' doesn't exist on type 'Query'**. This is because the server treats the request payload as a query.

Here, you must add the required arguments to the `addItemInCart` mutation and then add the fields (`id` and `productId`) you want to retrieve in response. Once the request has been processed successfully, you will get a JSON output like the following:

```
#JSON response
{
  "data": {
    addItemInCart: [
      {
        "id": "zxcv90asdkqwe09kl",
```

```
        "productId": "qwer90asdkqwe09kl"
      }
    ]
  }
}
```

Here, the value of the `id` field is generated by the server. Similarly, you can write other mutations, such as delete and update, in the schema. Then, you can use the payload in the GraphQL request to process the mutation accordingly.

We'll explore the GraphQL `Subscription` type in the next subsection.

Exploring the Subscription type

The concept of subscriptions will be new to you if you are only familiar with REST. In the absence of GraphQL, you might use polling or WebSockets to implement similar functionality. There are many use cases where you will need the subscription feature, including the following:

- Live score updates or election results
- Batch processing updates

There are many such cases where you will need to immediately update events. GraphQL provides a subscription feature for this use case. In such cases, the client subscribes to the event by initiating and holding a steady connection. When the subscribed event occurs, the server pushes the resultant event data to the client. For example, let's say you want to know whenever there is a change in any item's inventory in an e-commerce app. Any change in the quantity of an item would trigger the event and the subscription would get a response with the updated quantity.

This data is sent as a stream through an initiated connection, rather than through a request/response kind of communication (which was used in the cases of query/mutation).

> **Recommended approach**
>
> It is recommended that a subscription should only be used when a small update occurs for a large object (such as batch processing), or there are live updates with low latency, such as a live score update. Otherwise, you should use polling (executing a query periodically at a specified interval).

Let's create a subscription in a schema, as follows:

```
type Subscription {
  orderShipped(customerID: ID!): Order
  # You can add other subscriptions here
}
```

```
# Order type contains order information and another object
# Shipping. Shipping contains id and estDeliveryDate and
# carrier fields
type Order {
  # other fields omitted for brevity
  shipping: Shipping
}
type Shipping {
  Id: ID!
  estDeliveryDate: String
  carrier: String
}
```

Here, we have defined an `orderShipped` subscription that accepts `customer ID` as an argument and returns `Order`. Clients subscribe to this event, and then whenever an order is shipped for the given `customerId`, the server will push the requested order details to the client using a stream.

You can use the following GraphQL request to subscribe to the GraphQL subscription:

```
# Request Input
subscription {
  orderShipped(customerID: "customer90asdkqwe09kl") {
    shipping {
      estDeliveryDate
      trackingId
    }
  }
}
# JSON Output
{
  "data": {
    "orderShipped": {
      "estDeliveryDate": "13-Aug-2022",
      "trackingId": "tracking90asdkqwe09kl"
    }
  }
}
```

The client will request a JSON response whenever any order belonging to a given customer is shipped. The server pushes these updates to all the clients who subscribed to this GraphQL subscription.

In this section, you learned how to declare the `Query`, `Mutation`, and `Subscription` types in a GraphQL schema.

You have defined scalar types and the user-defined object types in a schema. You also explored how to write a GraphQL request input for a query, mutation, or subscription.

Now, you know how to define the operation parameters in root types and pass arguments while sending GraphQL requests. Note that the non-nullable field in the schema can be marked by an exclamation mark (!). For arrays, or lists of objects, you must use square brackets ([ ]).

In the next section, we'll deep dive into GraphQL schema.

Designing a GraphQL schema

A schema is a GraphQL file that is written using DSL syntax. Primarily, it contains root types (query, mutation, and subscription), and the respective types that are used in root types, such as object types, scalar types, interfaces, union types, input types, and fragments.

First, let's discuss these types. You learned about root types (query, mutation, and subscription) and object types in the previous section. Let's now learn more about scalar types.

Understanding scalar types

Scalar types resolve concrete data. There are three kinds of scalar types – built-in scalar types, custom scalar types, and enumeration types. Let's discuss built-in scalar types first. GraphQL provides the following five kinds of built-in scalar types:

- `Int`: This stores integers, and is represented by a signed 32-bit integer.
- `Float`: This stores a signed, double-precision, floating-point value.
- `String`: This stores a sequence of UTF-8 characters.
- `Boolean`: This stores a Boolean value – true or false.
- `ID`: This is used to define the object identifier string. This can only be serialized as a string, and is not human-readable.

You can also define your own scalar types, which are known as custom scalar types. An example is the `Date` type, which can be defined like so:

```
scalar Date
```

You need to write an implementation that determines the serialization, deserialization, and validation of these custom scalar types. For example, the date can be treated as a Unix timestamp, or a string with a particular date format in a custom scalar `Date` type case.

Another special scalar type is the enumeration type (enum), which is used to define a particular set of allowed values. Let's define an order status enumeration, as shown here:

```
enum OrderStatus {
   CREATED
   CONFIRMED
   SHIPPED
   DELIVERED
   CANCELLED
}
```

Here, the OrderStatus enumeration type represents the order status at a given point in time.

We'll next examine GraphQL fragments in the following subsection before exploring other types.

Understanding fragments

You may encounter conflicting scenarios while querying on the client side. You may have two or more queries that return the same result (the same object or set of fields). To avoid this conflict, you can give the query result a name. This name is known as an **alias**.

Let's use an alias in the following query:

```
query HomeAndBillingAddress {
   home: getAddress(type: "home") {
      number
      residency
      street
      city
      pincode
   }
   billing: getAddress(type: "home") {
      number
      residency
      street
      city
      pincode
   }
}
```

Here, HomeAndBillingAddress is a named query that contains the getAddress query operation. getAddress is being used twice, which results in it returning the same set of fields. Therefore, the home and billing aliases are used to differentiate the result object.

The `getAddress` query may return the `Address` object. The `Address` object may have additional fields, such as `type`, `state`, `country`, and `contactNo`. So, when you have queries that may use the same set of fields, you can create a **fragment** and use it in queries. A fragment logically creates a subgroup of fields from the existing object in the GraphQL schema that can be reused at multiple places, as shown in the following code snippet.

Let's create a fragment and replace the common fields in the previous code block:

```
query HomeAndBillingAddress {
  home: getAddress(type: "home") {
    ...addressFragment
  }
  billing: getAddress(type: "home") {
    ...addressFragment
  }
}
fragment addressFragment on Address {
  number
  residency
  street
  city
  pincode
}
```

Here, the `addressFragment` fragment has been created and used in the query.

You can also create an **inline fragment** in the query. An inline fragment is a fragment that you can create on the fly in a GraphQL payload without declaring it explicitly. Such fragments are helpful when a response object contains the nested object and you just want a few fields of the nested object rather than all the object fields. Inline fragments can be used when a querying field returns an `Interface` or `Union` type. We will explore inline fragments in more detail later in the *Understanding interfaces* subsection under the *Designing a GraphQL schema* section.

We'll look at GraphQL interfaces in the next subsection.

Understanding interfaces

GraphQL interfaces are abstract. You may have a few fields that are common across multiple objects. You can create an `interface` type for such sets of fields. For example, a product may have some common attributes, such as ID, name, and description. The product can also have other attributes based on its type. For example, a book may have several pages, an author, and a publisher, while a bookcase may have material, width, height, and depth attributes.

Let's define these three objects (`Product`, `Book`, and `Bookcase`) using interfaces:

```
interface Product {
    id: ID!
    name: String!
    description: string
}
type Book implements Product {
    id: ID!
    name: String!
    description: string
    author: String!
    publisher: String
    noOfPages: Int
}
type Bookcase implements Product {
    id: ID!
    name: String!
    description: string
    material: [String!]!
    width: Int
    height: Int
    depth: Int
}
```

Here, an abstract type called `Product` has been created using the `interface` keyword. This interface can be implemented when we wish to create new object types – `Book` and `Bookcase`.

Now, you can simply write the following query that will return all the products (books and bookcases):

```
type query {
    allProducts: [Product]
}
```

Now, you can use the following query on the client side to retrieve all the products:

```
query getProducts {
    allProducts {
        id
        name
        description
    }
}
```

You might have noticed that the preceding code only contains attributes from the Product interface. If you want to retrieve attributes from Book and Bookcase, then you must use **inline fragments**, as shown here:

```
query getProducts {
  allProducts {
    id
    name
    description
    ... on Book {
      author
      publisher
    }
    ... on BookCase {
      material
      height
    }
  }
}
```

Here, an operation (...) is used to create inline fragments. This way, you can fetch the fields from the type that implements the interface.

We'll understand Union types in the next subsection.

Understanding Union types

Let's say there are two object types – Book and Author. Here, you want to write a GraphQL query that can return both books and authors. Note that the interface is not there; so, how can we combine both objects in the query result? In such cases, you can use a **Union type**, which is a combination of two or more objects.

Consider the following before creating a Union type:

- You don't need to have a common field.

- Union members should be of a concrete type. Therefore, you can't use union, interface, input, or scalar types.

Let's create a union type that can return any object included in the union type – books and bookcases – as shown in the following code block:

```
union SearchResult = Book | Author
type Book {
  id: ID!
  name: String!
```

```
        publisher: String
    }
    type Author {
        id: ID!
        name: String!
    }
    type Query {
        search(text: String): [SearchResult]
    }
```

Here, the union keyword is used to create a union type for the Book and Author objects. A pipe symbol (|) is used to separate the included objects. Finally, a query is defined that returns a collection of books or authors that contain the given text.

Now, let's write this query for the client, as shown here:

```
# Request Input
{
    search(text: "Malcolm Gladwell") {
        __typename
        ... on Book {
            name
            publisher
        }
        ... on Author {
            name
        }
    }
}
Response JSON
{
    "data": {
        "search": [
            {
                "__typename": "Book",
                "name": "Blink",
                "publisher": "Back Bay Books"
            },
            {
                "__typename": "Author",
                "name": " Malcolm Gladwell ",
            }
        ]
    }
}
```

As you can see, an inline fragment is used in the query. Another important point is the extra field, called __typename, which refers to the object it belongs to and helps you differentiate between different objects in the client.

We'll look at input types in the next subsection.

Understanding input types

So far, you have used scalar types as arguments. GraphQL also allows you to pass object types as arguments in mutations. The only difference is that you have to declare them with input instead of using the type keyword.

Let's create a mutation that accepts an input type as an argument:

```
type Mutation {
    addProduct(prodInput: ProductInput): Product
}
input ProductInput {
    name: String!
    description: String
    price: Float!
    # other fields…
}
type Product {
    # Product Input fields. Truncated for brevity.
}
```

Here, the addProduct mutation accepts ProductInput as an argument and returns Product.

Now, let's use the GraphQL request to add a product to the client, as shown here:

```
# Request Input
mutation AddProduct ($input: ProductInput) {
    addProduct(prodInput: $input) {
        name
    }
}
#---- Variable Section ----
{
    "input": {
        name: "Blink",
        description: "a book",
        "price": 10.00
    }
}
```

```
# JSON Output
{
  "data": {
    addProduct {
      "name": "Blink"
    }
  }
}
```

Here, you are running a mutation that uses an `input` variable. You might have observed that `Variable` is being used here to pass `ProductInput`. The named mutation is used for the variable. If variables, along with their types, are defined in the mutation, then they should be used in the mutation.

Variable values should be assigned in the variable section (or beforehand in the client). The value of a variable's input is assigned using a JSON object that should map to `ProductInput`.

We'll look at the tools we can use while designing a GraphQL schema in the next subsection.

Designing a schema with GraphQL tools

You can use the following tools for design and work with GraphQL, with each having its own offerings:

- **GraphiQL**: This is pronounced *graphical*. It is an official GraphQL Foundation project that provides the web-based GraphQL **IDE**. It makes use of **Language Server Protocol** (**LSP**), which uses the JSON-RPC-based protocol between the source code editor and the IDE. It is available at `https://github.com/graphql/graphiql`.

- **GraphQL Playground**: This is another popular GraphQL IDE that once provided better features than GraphiQL. However, GraphiQL now has feature parity with Playground. At the time of writing, GraphQL Playground is in maintenance mode. Check out `https://github.com/graphql/graphql-playground/issues/1366` for more details. It is available at `https://github.com/graphql/graphql-playground`.

- **GraphQL Faker**: This provides mock data for your GraphQL APIs. It is available at `https://github.com/APIs-guru/graphql-faker`.

- **GraphQL Editor**: This allows you to design your schema visually and then transform it into code. It is available at `https://github.com/graphql-editor/graphql-editor`.

- **GraphQL Voyager**: This converts your schema into interactive graphs, such as entity diagrams and all the relationships among these entities. It is available at `https://github.com/APIs-guru/graphql-voyager`.

In the next section, you'll test the knowledge that you have acquired throughout this chapter.

Testing GraphQL queries and mutations

Let's write queries and mutations in a real GraphQL schema to test the skills you have learned up to this point using GitHub's GraphQL API explorer. Let's perform the following steps:

1. First, go to `https://docs.github.com/en/graphql/overview/explorer`.

2. You might have to authorize it using your GitHub account, so that you can execute GraphQL queries.

3. GitHub Explorer is based on GraphiQL. It is divided into three vertical sections (from left to right in the gray area in *Figure 13.1*):

    *   The left-hand section is divided into two subsections – an upper section for writing queries and a bottom section for defining variables.

    *   The middle vertical section shows the response.

    *   Normally, the rightmost section is hidden. Click on the **Docs** link to display it. It shows the respective documentation and schema, along with the root types that you can explore.

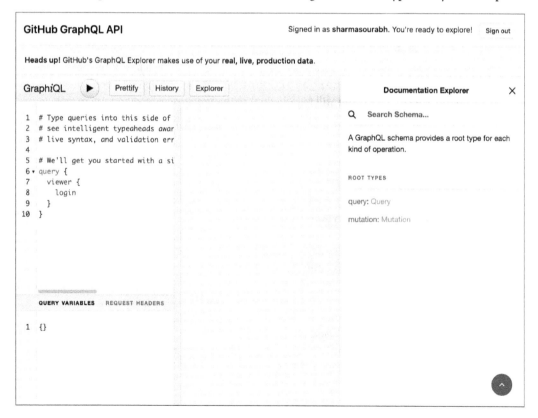

Figure 13.1 – GraphQL API Explorer

4. Let's fire the following query to find out the ID of the repository you wish to mark with a star:

```
{
  repository(
    name: "Modern-API-Development-with-Spring-6-and-Spring-
Boot-3"
    owner: "PacktPublishing"
  ) {
    id
    owner {
      id
      login
    }
    name
    description
    viewerHasStarred
    stargazerCount
  }
}
```

Here, you are querying the previous edition of this book's repository by providing two arguments – the repository's `name` and its `owner`. You are fetching a few of the fields. One of the most important ones is `stargazerCount` because we are going to perform an `addStar` mutation. This count will tell us whether the mutation was successful or not.

5. Click on the **Execute Query** button on the top bar or press *Ctrl + Enter* to execute the query. You should get the following output once this query executes successfully:

```
{
  "data": {
    "repository": {
      "id": "R_kgDOHzYNwg",
      "owner": {
        "id": "MDEyOk9yZ2FuaXphdGlvbjEwOTc0OTA2",
        "login": "PacktPublishing"
      },
      "name": "Modern-API-Development-with-Spring-6-
        and-Spring-Boot-3",
      "description": "Modern API Development with
      Spring 6 and Spring Boot 3, Published by Packt",
      "viewerHasStarred": false,
      "stargazerCount": 1
    }
  }
}
```

Here, you need to copy the value of `id` from the response, as it will be needed to mark this repository with a star.

6. Execute the following query to perform the `addStar` mutation:

```
mutation {
  addStar(input: {starrableId: "R_kgDOHzYNwg"}) {
    clientMutationId
  }
}
```

This performs the `addStar` mutation for the repository with the given ID.

7. Once the previous query has executed successfully, you must re-execute the query from *step 4* to find out about the change. If you get an access issue, then you can choose your own GitHub repository to perform these steps.

You can also explore other queries and mutations to deep dive into GraphQL.

Finally, let's understand the N+1 problem in GraphQL queries before we jump into the implementation in the next chapter.

Solving the N+1 problem

The N+1 problem is not new to Java developers. You might have encountered this problem while using Hibernate, which occurs if you don't optimize your queries or write entities properly.

Let's look at what the N+1 problem is.

What is the N+1 problem?

The **N+1 problem** normally occurs when associations are involved. There are one-to-many relationships between the customer and the order. One customer can have many orders. If you need to find all the customers and their orders, you can do the following:

1. First, find all the users. This find operation returns the list of user objects.

2. Then, find all the orders belonging to each user found in *step 1*. The `userId` field acts as the relation between the `Order` and `User` objects.

So, here, you fire two queries. If you further optimize the implementation, you can place a *join* between these two entities (`Order` and `User`) and receive all the records in a single query.

If this is so simple, then why does GraphQL encounter the N+1 problem? You need to understand the **resolver** function to answer this question.

If you refer to the database schema you created in *Chapter 4, Writing Business Logic for APIs*, you can see that the getUsersOrders query will lead to the following SQL statements being executed:

```
SELECT * FROM ecomm.user;
SELECT * FROM ecomm.orders WHERE customer_id in (1);
SELECT * FROM ecomm.orders WHERE customer_id in (2);
. . .
. . .
SELECT * FROM ecomm.orders WHERE customer_id in (n);
```

Here, to perform the getUsersOrders() operation, you execute a query on the user to fetch all the users. Then, you executes N queries on orders. This is why it is called the N+1 problem. This is not efficient because ideally you should execute a single query, or in the worst case, two queries.

GraphQL can only respond with the values of fields that have been requested in the query due to resolvers. Each field has its own resolver function in the GraphQL server implementation that fetches the data for its corresponding field. Let's assume we have the following schema:

```
type Mutation {
  getUsersOrders: [User]
}
type User {
  name: String
  orders: [Order]
}
type Order {
  id: Int
  status: Status
}
```

Here, we have a mutation that returns a collection of users. Each User may have a collection of orders. You could therefore use the following query in the client:

```
{
  getUsersOrders {
    name
    orders {
      id
      status
    }
  }
}
```

Let's understand how this query will be processed by the server.

In the server, each field will have its own resolver function that fetches the corresponding data. The first resolver will be for the user and will fetch all the users from the data store. Next, the resolver will fetch orders for each user. It will fetch the orders from the data store based on the given user ID. Therefore, the `orders` resolver will execute n times, where n is the number of users that have been fetched from the data store.

We'll learn how to resolve the N+1 problem in the next subsection.

How can we solve the N+1 problem?

The required solution will wait until all the orders have been loaded. Once all the user IDs have been retrieved, a database call should be made to fetch all the orders in a single data-store call. You can use a batch if the size of the database is huge. Then, the executer can resolve the individual order resolvers. However, this is easier said than done. GraphQL provides a library called **DataLoader** (`https://github.com/graphql/dataloader`) that does this job for you. This library mainly performs the batching and caching of queries.

Java provides a similar library called **java-dataloader** (`https://github.com/graphql-java/java-dataloader`) that can help you solve this problem. You can find out more about it at `https://www.graphql-java.com/documentation/batching`.

Summary

In this chapter, you learned about GraphQL, its advantages, and how it compares to REST. You learned how GraphQL solves over-fetching and under-fetching problems. You then learned about GraphQL's root types – queries, mutations, and subscriptions – and how different blocks can help you design the GraphQL schema. Finally, you understood how resolvers work, how they can lead to the N+1 problem, and the solution to this problem.

Now that you know about the fundamentals of GraphQL, you can start designing GraphQL schemas. You also learned about GraphQL's client-side queries and how to make use of aliases, fragments, and variables to resolve common problems.

In the next chapter, you will make use of the GraphQL skills you acquired in this chapter to implement GraphQL APIs.

Questions

1. Is GraphQL better than REST? If yes, then in what way?
2. When should you use fragments?
3. How can you use variables in a GraphQL query?

Answers

1. It depends on the use cases. However, GraphQL performs much better for mobile apps and web-based UI applications and is best suited for **service-to-service** (**s2s**) communications.

2. Fragments should be used while sending a request from the GraphQL client when the response contains an interface or union.

3. You can use a variable in a GraphQL query/mutation, as shown in the following code. This code used to modify the GraphQL request sent in *step 6* of the *Testing GraphQL queries and mutations* section:

```
mutation {
  addStar(input: {starrableId: $repoId }) {
    clientMutationId
  }
}
```

Here, you can see that the `$repoId` variable is used. You must declare that variable in the named mutation and it can then be used in the mutation's argument, as shown in the following code snippet:

```
{
  "repoId": "R_kgDOHzYNwg"
}
```

Further reading

- GraphQL specifications: `https://spec.graphql.org/`

- GraphQL documentation: `https://graphql.org/learn/`

- *Full-Stack Web Development with GraphQL and React*: `https://www.packtpub.com/product/full-stack-web-development-with-graphql-and-react-second-edition/9781801077880`

14

GraphQL API Development and Testing

In the previous chapter, we learned about the fundamental concepts of GraphQL. You are going to use that knowledge to develop and test GraphQL-based APIs in this chapter. You will implement GraphQL-based APIs for a sample application in this chapter. The GraphQL server implementation will be developed based on a **design-first** approach, the way you defined the OpenAPI specification in *Chapter 3, API Specifications and Implementation*, and designed the schema in *Chapter 11, gRPC API Development and Testing*.

After completing this chapter, you will have learned how to practically implement the GraphQL concepts learned about in the previous chapter and about the implementation of the GraphQL server using Java and Spring and its testing.

This chapter will cover the following main topics:

- Workflow and tooling for GraphQL
- Implementing the GraphQL server
- Documenting APIs
- Test automation

Technical requirements

The code for this chapter is available at

```
https://github.com/PacktPublishing/Modern-API-Development-with-
Spring-6-and-Spring-Boot-3/tree/dev/Chapter14
```

Workflow and tooling for GraphQL

As per the data graph (data structure) way of thinking in GraphQL, data is exposed using an API consisting of graphs of objects. These objects are connected using relations. GraphQL only exposes a single API endpoint. Clients query this endpoint, which uses a *single data graph*. On top of that, the data graph may resolve data from a single source, or multiple sources, by following the **OneGraph principle** of GraphQL. These sources could be a database, legacy system, or services that expose data using REST/gRPC/SOAP.

The GraphQL server can be implemented in the following two ways:

- **Standalone GraphQL service**: A standalone GraphQL service contains a single data graph. It could be a monolithic app or based on a microservice architecture that fetches the data from single or multiple sources (having no GraphQL API).

- **Federated GraphQL services**: It's very easy to query a single data graph for comprehensive data fetching. However, enterprise applications are made using multiple services; hence, you can't have a single data graph unless you build a monolithic system. If you don't build a monolithic system, then you will have multiple service-specific data graphs.

    This is where you make use of federated GraphQL services. A federated GraphQL service contains a **single distributed graph** exposed using a gateway. Clients call the gateway, which is an entry point into the system. The data graph is distributed among multiple services and each service can maintain its own development and release cycle independently. Having said that, federated GraphQL services still follow the OneGraph principle. Therefore, the client queries the single endpoint to fetch any part of the graph.

Let's assume that a sample e-commerce app is developed using GraphQL federated services. It has products, orders, shipping, inventory, customers, and other services that expose the domain-specific data graphs using the GraphQL API.

Let's look at a high-level diagram of GraphQL federated e-commerce services, as follows:

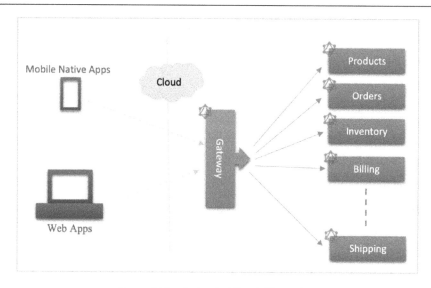

Figure 14.1 – Federated GraphQL services

Let's say the GraphQL client queries for a list of the most ordered products with the least inventory by calling the `Gateway` endpoint. This query may have fields from `Orders`, `Products`, and `Inventory`. Each service is responsible for resolving only the respective part of a data graph. `Orders` would resolve order-related data, `Products` would resolve product-related data, `Inventory` would resolve inventory-related data, and so on. `Gateway` then consolidates the graph data and sends it back to the clients.

The `graphql-java` library (`https://www.graphql-java.com`) provides the Java implementation of the GraphQL specification. Its source code is available at `https://github.com/graphql-java/graphql-java`.

Spring provides a Spring Boot Starter project for GraphQL based on `graphql-java`, available at `https://github.com/spring-projects/spring-graphql`. However, we are going to use the Spring-based Netflix **Domain Graph Service (DGS)** framework (`https://netflix.github.io/dgs`) because the Netflix DGS provides not only the GraphQL Spring Boot starter but also the full set of tools and libraries that you need to develop production-ready GraphQL services. It is built on top of Spring Boot and the `graphql-java` library.

Netflix open sourced the DGS framework after using it in production in February 2021. It is continuously being enhanced and supported by the community. Netflix uses the same open sourced DGS framework codebase in production environments, which gives the assurance of the code's quality and future maintenance. The OTT Disney+ platform was also built using the Netflix DGS framework (`https://webcache.googleusercontent.com/search?q=cache:ec4kC7jBjMQJ:https://help.apps.disneyplus.com/3rd-party-libs.html&cd=14&hl=en&ct=clnk&gl=in&client=firefox-b-d`).

It provides the following features:

- A Spring Boot starter and integration with Spring Security

- Full WebFlux support

- A Gradle plugin for code generation from a GraphQL schema

- Support for interfaces and union types, plus the provision of custom scalar types

- Support for GraphQL subscriptions using WebSocket and server-sent events

- Error handling

- Pluggable instrumentation and Micrometer integration

- GraphQL federated services with easy integration with GraphQL federation

- Dynamic schemas with hot reloading schemas

- Operation caching

- File upload

- GraphQL Java client

- GraphQL test framework

Let's write a GraphQL server using Netflix's DGS framework in the next section.

Implementing the GraphQL server

You are going to develop a standalone GraphQL server in this chapter. The knowledge you acquire while developing the standalone GraphQL server can be used to implement federated GraphQL services.

Let's create the Gradle project first in the next subsection.

Creating the gRPC server project

Either you can use the *Chapter 14* code by cloning the Git repository (`https://github.com/PacktPublishing/Modern-API-Development-with-Spring-6-and-Spring-Boot-3/tree/dev/Chapter14`) or you can start by creating a new Spring project from scratch using Spring Initializr for the server and client with the following options:

- **Project**: `Gradle - Groovy`

- **Language**: `Java`

- **Spring Boot**: `3.0.8`

  The preferred version is 3.0+; if not available, you can modify it later manually in the `build.gradle` file

- **Project Metadata**:

  - **Group**: `com.packt.modern.api`

  - **Artifact**: `chapter14`

  - **Name**: `chapter14`

  - **Description**: `Chapter 14 code of book Modern API Development with Spring and Spring Boot Ed 2`

  - **Package name**: `com.packt.modern.api`

  - **Packaging**: `JAR`

- **Java**: `17`

  You can change it to another version, such as 17/20/21, in the `build.gradle` file later, as shown in the following code block:

  ```
  // update following build.gradle file
  sourceCompatibility = JavaVersion.VERSION_17
  // or for Java 20
  // sourceCompatibility = JavaVersion.VERSION_20
  // or for Java 21
  // sourceCompatibility = JavaVersion.VERSION_20
  ```

- **Dependencies**: `org.springframework.boot:spring-boot-starter-web`

Then, you can click on the **GENERATE** button and download the project. The downloaded project will be used to create the GraphQL server.

Next, let's add the GraphQL DGS dependencies to the newly created project.

Adding the GraphQL DGS dependencies

Once the Gradle project is available, you can modify the `build.gradle` file to include the GDS dependencies and plugin as shown in the following code:

```
plugins {
    id 'org.springframework.boot' version '3.0.6'
    id 'io.spring.dependency-management' version '1.1.0'
    id 'java'
    id 'com.netflix.dgs.codegen' version '5.7.1'
}
// code truncated for brevity
def dgsVersion = '6.0.5'
dependencies {
    implementation platform("com.netflix.graphql.
        dgs:graphql-dgs-platform-dependencies:${
```

```
        dgsVersion}")
    implementation 'com.netflix.graphql.dgs:graphql-dgs-
        spring-boot-starter'
    implementation 'com.netflix.graphql.dgs:graphql-dgs-
        extended-scalars'
    implementation 'com.netflix.graphql.dgs:graphql-dgs-
        spring-boot-micrometer'
    runtimeOnly 'com.netflix.graphql.dgs:graphql-dgs-
        subscriptions-websockets-autoconfigure'
    implementation 'org.springframework.boot:spring-boot-
        starter-web'
    implementation 'org.springframework.boot:spring-boot-
        starter-actuator'
    testImplementation 'org.springframework.boot:
        spring-boot-starter-test'
    implementation 'net.datafaker:datafaker:1.9.0'
}
```

https://github.com/PacktPublishing/Modern-API-Development-with-Spring-6-and-Spring-Boot-3/tree/dev/Chapter14/build.gradle

Here, first, the DGS Codegen plugin is added, which will generate the code from the GraphQL schema file. Next, the following five dependencies are added:

- graphql-dgs-platform-dependencies: The DGS platform dependencies for the DGS bill of materials (BOM)

- graphql-dgs-spring-boot-starter: The DGS Spring Boot Starter library for DGS Spring support

- graphql-dgs-extended-scalars: The DGS extended scalars library for custom scalar types

- graphql-dgs-spring-boot-micrometer: The DGS library to provide integration with Micrometer to provide support for metrics and instrumentation along with Spring Actuator

- graphql-dgs-subscriptions-websockets-autoconfigure: To provide the autoconfiguration for GraphQL WebSocket support

Please note that the datafaker library is used here to generate the domain seed data.

Next, let's configure the DGS Codegen plugin in the same build.gradle file as shown in the next code block:

```
generateJava {
    generateClient = true
```

```
        packageName = "com.packt.modern.api.generated"
    }
```

You have configured the following two properties of DGS Codegen using the `generateJava` task, which uses the Gradle plugin `com.netflix.graphql.dgs.codegen.GenerateJavaTask` class:

- `generateClient`: This determines whether you would like to generate the client or not

- `packageName`: The Java package name of the generated Java classes

The DGS Codegen plugin picks GraphQL schema files from the `src/main/resources/schema` directory by default. However, you can modify it using the `schemaPaths` property, which accepts an array. You can add this property to the previous code of `generateTask` along with `packageName` and `generateClient` if you want to change the default schema location, as shown next:

```
schemaPaths = ["${projectDir}/src/main/resources/schema"]
```

You can also configure type mappings as you did for the `org.hidetake.swagger.generator` Gradle plugin while generating the Java code from OpenAPI specs in *step 4* of the *Converting OAS to Spring code* section in *Chapter 3*, *API Specifications and Implementation*. To add a custom type mapping, you can add the `typeMapping` property to the plugin task, as shown next:

```
typeMapping = ["GraphQLType": "mypackage.JavaType"]
```

This property accepts an array; you can add one or more type mappings here. You can refer to the plugin documentation at `https://netflix.github.io/dgs/generating-code-from-schema/` for more information.

Let's add the GraphQL schema next.

Adding the GraphQL schema

Netflix's DGS supports both the code-first and design-first approaches. However, we are going to use the design-first approach in this chapter as we have done throughout this book. Therefore, first, we'll design the schema using the GraphQL schema language and then use the generated code to implement the GraphQL APIs.

We are going to keep the domain objects minimal to reduce the complexity of business logic and keep the focus on the GraphQL server implementation. Therefore, you'll have just two domain objects – `Product` and `Tag`. The GraphQL schema allows the following operation using its endpoint as shown in the following schema definition:

```
type Query {
    products(filter: ProductCriteria): [Product]!
    product(id: ID!): Product
}
```

```
type Mutation {
    addTag(productId: ID!, tags: [TagInput!]!): Product
    addQuantity(productId: ID!, quantity: Int!): Product
}
type Subscription {
    quantityChanged: Product
}
```

https://github.com/PacktPublishing/Modern-API-Development-with-
Spring-6-and-Spring-Boot-3/tree/dev/Chapter14/src/main/resources/
schema/schema.graphqls

You need to add the schema.graphqls GraphQL schema file at the src/main/ resources/
schema location. You can have multiple schema files there to create the schema module-wise.

Here, the following root types have been exposed:

- Query: The product and product queries to fetch a product by its ID, and a collection of products matched by the given criteria.

- Mutation: The addTag mutation adds a tag to the product that matches the given ID. Another mutation, addQuantity, increases the product quantity. The addQuantity mutation can also be used as an event that triggers the subscription publication.

- Subscription: The quantityChanged subscription publishes the product where the quantity has been updated. The event quantity change is captured through the addQuantity mutation.

Let's add the object types and input types being used in these root types to schema. graphqls as shown in the next code block:

```
type Product {
    id: String
    name: String
    description: String
    imageUrl: String
    price: BigDecimal
    count: Int
    tags: [Tag]
}
input ProductCriteria {
    tags: [TagInput] = []
    name: String = ""
    page: Int = 1
    size: Int = 10
}
input TagInput {
```

```
        name: String
    }
    type Tag {
        id: String
        name: String
    }
```

These are straightforward object and input types. All fields of the `ProductCriteria` input type have been kept optional.

We have also used a `BigDecimal` custom scalar type. Therefore, we need to first declare it in the schema. We can do that by adding `BigDecimal` to the end of the schema file, as shown next:

```
scalar BigDecimal
```

Next, you also need to map it to `java.math.BigDecimal` in the code generator plugin. Let's add it to the `build.gradle` file as shown next (check the highlighted line):

```
generateJava {
    generateClient = true
    packageName = "com.packt.modern.api.generated"
    typeMapping = ["BigDecimal": "java.math.BigDecimal"]
}
```

After these changes, your project is ready to generate the GraphQL objects and client. You can run the following command from the project root directory to build the project:

```
$ ./gradlew clean build
```

This command will generate the Java classes in the `build/generated` directory.

Before you start implementing the GraphQL root types, let's discuss the custom scalar types in the next subsection.

Adding custom scalar types

You are going to use `BigDecimal` to capture the monetary values. This is a custom scalar type; therefore, you need to add this custom scalar to the code so that the DGS framework can pick it for serialization and deserialization. (You also need to add a mapping to the Gradle code generator plugin.)

There are two ways to add the custom scalar type – by implementing the `Coercing` interface and by making use of the `graphql-dgs-extended-scalars` library. We are going to use the latter because it has fewer lines and the actual implementation is provided by the Netflix DGS framework.

The first, crude way of adding the scalar type is to implement the `graphql.schema.Coercing` interface and annotate it with the `@DgsScalar` annotation. Here, you need to write the boiler plate code yourself.

Instead, we will opt for the second method, which involves using the scalar type provided by the DGS framework, which is well tested on production systems. The `graphql.schema.Coercing` interface is provided by the `graphql-java` library. The `DateTimeScalar` scalar type is implemented using `Coercing`, as shown in the following code:

```
@DgsScalar(name="DateTime")
public class DateTimeScalar
                implements Coercing<LocalDateTime, String> {
  @Override
  public String serialize(Object dataFetcherResult)
                throws CoercingSerializeException {
    if (dataFetcherResult instanceof LocalDateTime) {
      return ((LocalDateTime) dataFetcherResult)
          .format(DateTimeFormatter.ISO_DATE_TIME);
    } else {
      throw new CoercingSerializeException
          ("Invalid Dt Tm");
    }
  }

  @Override
  public LocalDateTime parseValue(Object input)
        throws CoercingParseValueException {
    return LocalDateTime.parse(input.toString(),
        DateTimeFormatter.ISO_DATE_TIME);
  }

  @Override
  public LocalDateTime parseLiteral(Object input)
        throws CoercingParseLiteralException {
    if (input instanceof StringValue) {
      return LocalDateTime.parse(((StringValue) input)
          .getValue(), DateTimeFormatter.ISO_DATE_TIME);
    }
    throw new CoercingParseLiteralException
        ("Invalid Dt Tm");
  }
}
```

Here, you have overridden the three methods – `serialize()`, `parseValue()`, and `parseLiteral()` – of the `Coercing` interface to implement serialization and parsing for the `DateTimeScalar` custom scalar type.

However, you are going to use the second method – the `graphql-dgs-extended-scalars` library – for registering new scalar types. This library has already been added in the `build.gradle` file. Let's make use of the `graphql-dgs-extended-scalars` library to register the `BigDecimalScaler` type.

Create a new Java file called `BigDecimalScaler.java` and add the following code to it:

```
@DgsComponent
public class BigDecimalScalar {

  @DgsRuntimeWiring
  public RuntimeWiring.Builder addScalar(
      RuntimeWiring.Builder builder) {
    return builder.scalar(GraphQLBigDecimal);
  }
}
```

```
https://github.com/PacktPublishing/Modern-API-Development-with-
Spring-6-and-Spring-Boot-3/blob/dev/Chapter14/src/main/java/com/
packt/modern/api/scalar/BigDecimalScalar.java
```

Here, you are making use of `DgsRuntimeWiring` to add the custom `GraphQLBigDecimal` scalar provided by the `graphql-dgs-extended-scalars` library. The `RuntimeWiring` class comprises data fetchers, type resolvers, and custom scalars that are needed to wire together a functional `GraphQLSchema` class. The `DgsRuntimeWiring` annotation marks the method as runtime wiring. So, you can perform the customization before the `RuntimeWiring` class gets executed. Basically, you are adding the `GraphQLBigDecimal` scalar type to `RuntimeWiring.Builder` for runtime wiring execution.

The `BigDecimalScalar` class is marked with the `@DgsComponent` annotation. The DGS framework is an annotation-based Spring Boot programming model. The DGS framework provides these types of annotations (such as `@DgsComponent`) for Spring Boot. A class marked with `@DgsComponent` is a DGS component as well as a regular Spring component.

Similarly, you have added the `DateTimeScalar` type. The code for the `DateTimeScalar` scalar type is available at `https://github.com/PacktPublishing/Modern-API-Development-with-Spring-6-and-Spring-Boot-3/tree/dev/Chapter14/src/main/java/com/packt/modern/api/scalar/DateTimeScalar.java`.

All the schema details and its documentation can be explored using the GraphQL documentation that is available in GraphiQL or similar tools. Let's find out how you find the auto-generated documentation next.

Documenting APIs

You can use GraphiQL or a playground tool that provides a graphical interface to explore the GraphQL schema and documentation.

On GraphiQL (`http://localhost:8080/graphiql`, which can be started by running the `jar` built out of this chapter code), you can open the document explorer by clicking on the book icon available in the top-left corner of the page. Once you click on it, it will display the documentation.

However, if you are looking for a static page, then you can use tools such as `graphdoc` (`https://github.com/2fd/graphdoc`) to generate the static documentation for the GraphQL APIs.

Next, let's start implementing GraphQL root types. First, you are going to implement GraphQL queries.

Implementing GraphQL queries

Both the queries we introduced in the schema in the previous section are straightforward. You pass a product ID to find a product identified by that ID – that's the product query for you. Next, you pass the optional product criteria to find the products based on the given criteria; otherwise, products are returned based on the default values of the fields of product criteria.

In REST, you implemented the controller class in the *Implementing the OAS code interfaces* section of *Chapter 3, API Specifications and Implementation*. You created a controller, passed the call to the service, and the service called the repository to fetch the data from the database. You are going to use the same design. However, you are going to use `ConcurrentHashMap` in place of the database to simplify the code. This can also be used in your automated tests.

Let's create a repository class for that, as shown in the next code block:

```
public interface Repository {
  Product getProduct(String id);
  List<Product> getProducts();
}
```

`https://github.com/PacktPublishing/Modern-API-Development-with-Spring-6-and-Spring-Boot-3/tree/dev/Chapter14/src/main/java/com/packt/modern/api/repository/Repository.java`

These are straightforward signatures for fetching the product and collection of products.

Let's implement the newly created repository interface using `ConcurrentHashMap`, as shown in the next code block:

```
@org.springframework.stereotype.Repository
public class InMemRepository implements Repository {
  private static final Map<String, Product>
```

```
productEntities = new ConcurrentHashMap<>();
    private static final Map<String, Tag> tagEntities =
        new ConcurrentHashMap<>();
    // rest of the code is truncated
```

https://github.com/PacktPublishing/Modern-API-Development-with-Spring-6-and-Spring-Boot-3/tree/dev/Chapter14/src/main/java/com/packt/modern/api/repository/InMemRepository.java

Here, you have created two instances of ConcurrentHashMap for storing the products and tags. Let's add the seed data to these maps using the constructor:

```
public InMemRepository() {
  Faker faker = new Faker();
  IntStream.range(0, faker.number()
      .numberBetween(20, 50)).forEach(number -> {
    String tag = faker.book().genre();
    tagEntities.putIfAbsent(tag,
      Tag.newBuilder().id(UUID.randomUUID().toString())
        .name(tag).build());
  });
  IntStream.range(0, faker.number().numberBetween(4, 20))
    .forEach(number -> {
      String id = String.format("a1s2d3f4-%d", number);
      String title = faker.book().title();
      List<Tag> tags = tagEntities.entrySet().stream()
        .filter(t -> t.getKey().startsWith(
          faker.book().genre().substring(0, 1)))
        .map(Entry::getValue).collect(toList());
    if (tags.isEmpty()) {
     tags.add(tagEntities.entrySet().stream()
       .findAny().get().getValue());
    }
    Product product = Product.newBuilder().id(id).
        name(title)
      .description(faker.lorem().sentence())
      .count(faker.number().numberBetween(10, 100))
      .price(BigDecimal.valueOf(faker.number()
        .randomDigitNotZero()))
      .imageUrl(String.format("/images/%s.jpeg",
        title.replace(" ", "")))
      .tags(tags).build();
    productEntities.put(id, product);
  });
  // rest of the code is truncated
```

This code first generates the tags and then stores them in the `tagEntities` map. The code also attaches the tags to new products before storing products in the `productEntities` map. This has been done for development purposes only. You should use the database in production applications.

Now, the `getProduct` and `getProducts` methods are straightforward, as shown in the next code block:

```
@Override
public Product getProduct(String id) {
  if (Strings.isBlank(id)) {
    throw new RuntimeException("Invalid Product ID.");
  }
  Product product = productEntities.get(id);
  if (Objects.isNull(product)) {
    throw new RuntimeException("Product not found.");
  }
  return product;
}
@Override
public List<Product> getProducts() {
  return productEntities.entrySet().stream()
    .map(e -> e.getValue()).collect(toList());
}
```

The `getProduct` method performs the basic validations and returns the product. The `getProducts` method simply returns the collection of products converted from the map.

Now, you can add the service and its implementation. Let's add the service interface as shown in the next code block:

```
public interface ProductService {
  Product getProduct(String id);
  List<Product> getProducts(ProductCriteria criteria);
  Product addQuantity(String productId, int qty);
  Publisher<Product> gerProductPublisher();
}
```

https://github.com/PacktPublishing/Modern-API-Development-with-Spring-6-and-Spring-Boot-3/tree/dev/Chapter14/src/main/java/com/packt/modern/api/services/ProductService.java

These service method implementations simply call the repository to fetch the data. Let's add the implementation shown in the next code block:

```
@Service
public class ProductServiceImpl implements ProductService {
```

```
    private final Repository repository;
    public ProductServiceImpl(Repository repository) {
      this.repository = repository;
    }
    @Override
    public Product getProduct(String id) {
      return repository.getProduct(id);
    }
    // continue …
```

https://github.com/PacktPublishing/Modern-API-Development-with-Spring-6-and-Spring-Boot-3/tree/dev/Chapter14/src/main/java/com/packt/modern/api/services/ProductServiceImpl.java

Here, the repository is injected using constructor injection.

Let's add the getProducts() method also, which performs filtering based on given filtering criteria, as shown in the next code block:

```
@Override
public List<Product> getProducts(ProductCriteria criteria) {
  List<Predicate<Product>> predicates = new ArrayList<>(2);
  if (!Objects.isNull(criteria)) {
    if (Strings.isNotBlank(criteria.getName())) {
      Predicate<Product> namePredicate =
        p -> p.getName().contains(criteria.getName());
      predicates.add(namePredicate);
    }
    if (!Objects.isNull(criteria.getTags()) &&
        !criteria.getTags().isEmpty()) {
      List<String> tags = criteria.getTags().stream()
        .map(ti -> ti.getName()).collect(toList());
      Predicate<Product> tagsPredicate =
        p -> p.getTags().stream().filter(
          t -> tags.contains(t.getName())).count() > 0;
      predicates.add(tagsPredicate);
    }
  }
  if (predicates.isEmpty()) {
    return repository.getProducts();
  }
  return repository.getProducts().stream()
    .filter(p -> predicates.stream().allMatch(
      pre -> pre.test(p))).collect(toList());
}
```

This method first checks whether criteria are given or not. If criteria are not given, then it calls the repository and returns all the products.

If criteria are given, then it creates the predicates list. These predicates are then used to filter out the matching products and return to the calling function.

Now comes the most critical piece of GraphQL query implementation: writing the data fetchers. First, let's write the data fetcher for the product query.

Writing fetchers for GraphQL queries

You are going to write data fetchers in this section. Data fetchers, as the name suggests, retrieve the information from the source that is from a persistent store such as a database or a third-party API/ document store. You will learn how to write data fetchers to retrieve a single field of data, a single object, and a collection of objects.

Writing the data fetcher for product

The data fetcher is a critical DSG component for serving GraphQL requests that fetches the data and the DSG internally resolves each of the fields. You mark them with the special @DgsComponent DGS annotation. These are types of Spring components that the DGS framework scans and uses for serving requests.

Let's create a new file called ProductDatafetcher.java in the datafetchers package to represent a DGS data fetcher component. It will have a data fetcher method for serving the product query. You can add the following code to it:

```
@DgsComponent
public class ProductDatafetcher {
  private final ProductService productService;
  public ProductDatafetcher(
      ProductService productService) {
    this.productService = productService;
  }
  @DgsData(parentType = DgsConstants.QUERY_TYPE,
          field = QUERY.Product)
  public Product getProduct(@InputArgument("id") String id) {
    if (Strings.isBlank(id)) {
      new RuntimeException("Invalid Product ID.");
    }
    return productService.getProduct(id);
  }
  // continue …
```

https://github.com/PacktPublishing/Modern-API-Development-with-Spring-6-and-Spring-Boot-3/tree/dev/Chapter14/src/main/java/com/packt/modern/api/datafetchers/ProductDatafetcher.java

Here, you create a product service bean injection using the constructor. This service bean helps you to find the product based on the given product ID.

Two other important DGS framework annotations have been used in the getProduct method. Let's understand what they do:

- @DgsData: This is a data fetcher annotation that marks the method as the data fetcher. The parentType property represents the type, and the field property represents the type's (parentType) field. Therefore, you can say that method will fetch the field of the given type.

  You have set Query as parentType. The field property is set as a product query. Therefore, this method works as an entry point for the GraphQL query product call. The @DsgData annotation properties are set using the DgsConstants constants class.

  DgsConstants is generated by the DGS Gradle plugin, which contains all the constant parts of the schema.

- @InputArgument: This annotation allows you to capture the arguments passed by the GraphQL requests. Here, the value of the id parameter is captured and assigned to the id string variable.

You can find the test cases related to this data fetcher method in the *Test automation* section.

Similarly, you can write the data fetcher method for the products query. Let's code it in the next subsection.

Writing the data fetcher for a collection of products

Let's create a new file called ProductsDatafetcher.java in the datafetchers package to represent a DGS data fetcher component. It will have a data fetcher method for serving the products query. You can add the following code to it:

```
@DgsComponent
public class ProductsDatafetcher {
  private ProductService service;
  public ProductsDatafetcher(ProductService service) {
    this.service = service;
  }
  @DgsData(
    parentType = DgsConstants.QUERY_TYPE,
    field = QUERY.Products
```

```
    )
    public List<Product> getProducts(@InputArgument("filter")
        ProductCriteria criteria) {
      return service.getProducts(criteria);
    }
  // continue …
```

https://github.com/PacktPublishing/Modern-API-Development-with-Spring-6-and-Spring-Boot-3/tree/dev/Chapter14/src/main/java/com/packt/modern/api/datafetchers/ProductsDatafetcher.java

This getProducts() method does not look different from the data fetcher method returned for getProduct() in the second-to-last code block. Here, the parentType and field properties of @DsgData indicate that this method will be used to fetch the collection of products for the products query (note that we are using the plural form here).

You are done with the GraphQL query implementation. You can now test your changes. You need to build the application before running the test. Let's build the application using the following command:

```
$ gradlew clean build
```

Once the build is done successfully, you can run the following command to run the application:

```
$ java -jar build/libs/chapter14-0.0.1-SNAPSHOT.jar
```

The application should be running on the default port 8080 if you have not made any changes to the port settings.

Now, you can open a browser window and open GraphiQL using the following URL: http://localhost:8080/graphiql (part of the DGS framework). Change the host/port accordingly if required.

You can use the following query to fetch the collection of products:

```
{
  products(
    filter: {name: "His Dark Materials",
        tags: [{name: "Fantasy"}, {name: "Legend"}]}
  ) {
    id
    name
    price
    description
    tags {
      id
      name
```

```
        }
    }
}
```

Once you run the preceding query, it will fetch the products matching the given criteria in the filter.

Figure 14.2 – GraphQL query execution in the GraphiQL tool

This will work great. However, what if you want to fetch the tags separately? You might have relations (such as orders with billing information) in objects that may be fetched from separate databases or services, or from two separate tables. In that case, you might want to add a field resolver using the data fetcher method.

Let's add a field resolver using the data fetcher method in the next subsection.

Writing the field resolver using the data fetcher method

So far, you don't have a separate data fetcher for fetching the tags. You fetch the products, and it also fetches the tags for you because we are using a concurrent map that stores both queries' data together. Therefore, first, you need to write a new data fetcher method for fetching the tags for a given product.

Let's add the `tags()` method to the `ProductsDatafetcher` class to fetch the tags, as shown in the next code block:

```
@DgsData(
    parentType = PRODUCT.TYPE_NAME,
    field = PRODUCT.Tags
)
public List<Tags> tags(String productId) {
   return tagService.fetch(productId);
}
```

https://github.com/PacktPublishing/Modern-API-Development-with-Spring-6-and-Spring-Boot-3/tree/dev/Chapter14/src/main/java/com/packt/modern/api/datafetchers/ProductsDatafetcher.java

Here, the `tags()` method has a different set of values for the `@DsgData` properties. The `parentType` property is not set to a root type like in earlier data fetcher methods, which were set to `Query`. Instead, it is set to an object type – `Product`. The `field` property is set to `tags`.

This method will be called for fetching the tags for each individual product because it is a field resolver for the `tags` field of the `Product` object. Therefore, if you have 20 products, this method will be called 20 times to fetch the tags for each of the 20 products. This is an *N+1* problem, which we learned about in the *Solving the N+1 problem* section in *Chapter 13, Getting Started with GraphQL*.

In the N+1 problem, extra database calls are made for fetching the data for relations. Therefore, given a collection of products, it may hit a database for fetching the tags for each product separately.

You know that you must use data loaders to avoid the N+1 problem. Data loaders cache all the IDs of products before fetching their corresponding tags in a single query.

Next, let's learn how to implement a data loader for fixing the N+1 problem in this case.

Writing a data loader for solving the N+1 problem

You are going to make use of the `DataFetchingEnvironment` class as an argument in the data fetcher methods. It is injected by the `graphql-java` library in the data fetcher methods to provide the execution context. This execution context contains information about the resolver, such as the object and its fields. You can also use them in special use cases such as loading the data loader classes.

Let's modify the `tags()` method in the `ProductsDatafetcher` class mentioned in the previous code block to fetch the tags without the N+1 problem, as shown in the next code block:

```
@DgsData(
   parentType = PRODUCT.TYPE_NAME,
   field = PRODUCT.Tags
)
```

```java
public CompletableFuture<List<Tags>> tags(
    DgsDataFetchingEnvironment env) {
  DataLoader<String, List<Tags>> tagsDataLoader =
      env.getDataLoader(TagsDataloaderWithContext.class);
  Product product = env.getSource();
  return tagsDataLoader.load(product.getId());
}
```

https://github.com/PacktPublishing/Modern-API-Development-with-Spring-6-and-Spring-Boot-3/tree/dev/Chapter14/src/main/java/com/packt/modern/api/datafetchers/ProductsDatafetcher.java

Here, the modified `tags()` data fetcher method performs the `fetch` method using a data loader and returns the collection of tags wrapped inside `CompletableFuture`. And it will be called only once even if the number of products is more than *1*.

What is CompletableFuture?

`CompletableFuture` is a Java concurrency class that represents the result of asynchronous computation, which is marked as completed explicitly. It can chain multiple dependent tasks asynchronously where the next task will be triggered when the current task's result is available.

You are using `DsgDataFetchingEnvironment` as an argument. It implements the `DataFetchingEnvironment` interface and provides ways to load the data loader class by both its class and name. Here, you are using the data loader class to load the data loader.

The `getSource()` method of `DsgDataFetchingEnvironment` returns the value from the `parentType` property of `@DsgData`. Therefore, `getSource()` returns `Product`.

This modified data fetcher method will fetch the tags for a given list of products in a single call. This method will fetch the tags for a list of products because the data loader class implements `MappedBatchLoader`, which performs the operation using batches.

The data loader class fetches the tags of the given product (by ID) using the data loader in batches. The magic lies in returning `CompletableFuture`. Therefore, though you are passing a single product ID as an argument, the data loader processes it in bunches. Let's implement this data loader class (`TagsDataloaderWithContext`) next to dig into it more.

You can create a data loader class in two ways – with context or without context. Data loaders without context implement `MappedBatchLoader`, which has the following method signature:

```java
CompletionStage<Map<K, V>> load(Set<K> keys);
```

On the other hand, data loaders with context implement the MappedBatchLoaderWithContext interface, which has the following method signature:

```
CompletionStage<Map<K, V>> load(Set<K> keys,
                    BatchLoaderEnvironment environment);
```

Both are the same as far as data loading is concerned. However, the data loader with context provides you with extra information (through BatchLoaderEnvironment) that can be used for various additional features, such as authentication, authorization, or passing the database details.

Create a new Java file called TagsDataloaderWithContext.java in the dataloaders package with the following code:

```
@DgsDataLoader(name = "tagsWithContext")
public class TagsDataloaderWithContext implements
        MappedBatchLoaderWithContext<String, List<Tag>> {
  private final TagService tagService;
  public TagsDataloaderWithContext(TagService tagService) {
    this.tagService = tagService;
  }
  @Override
  public CompletionStage<Map<String, List<Tag>>> load(
    Set<String> keys, BatchLoaderEnvironment environment) {
    return CompletableFuture.supplyAsync(() ->
        tagService.getTags(new ArrayList<>(keys)));
  }
}
```

https://github.com/PacktPublishing/Modern-API-Development-with-Spring-6-and-Spring-Boot-3/tree/dev/Chapter14/src/main/java/com/packt/modern/api/dataloaders/TagsDataloaderWithContext.java

Here, it implements the load() method from the MappedBatchLoaderWithContext interface. It contains the BatchLoaderEnvironment argument, which provides the environment context, which can contain user authentication and authorization information or database information. However, we are not using it because we don't have any additional information related to authentication, authorization, or the database to pass to the repository or underlying data access layer. If you do, you can make use of the environment argument.

You can also find the data loader without context at https://github.com/PacktPublishing/Modern-API-Development-with-Spring-6-and-Spring-Boot-3/tree/dev/Chapter14/src/main/java/com/packt/modern/api/dataloaders/TagDataloader.java. Its code is more or less like what we have written for the data loader with context. The only difference is we haven't used the context.

You can see that it makes use of the tag's service to fetch the tags. Then, it simply returns the completion stage by supplying the tags received from the tag service. This operation is performed in a batch by the data loader.

You can create a new tag service and its implementation as follows:

```
public interface TagService {
  Map<String, List<Tag>> getTags(List<String> productIds);
}
```

https://github.com/PacktPublishing/Modern-API-Development-with-Spring-6-and-Spring-Boot-3/tree/dev/Chapter14/src/main/java/com/packt/modern/api/services/TagService.java

This is the signature of the getTags method, which returns the map of product IDs with corresponding tags.

Let's implement this interface as shown in the next code block:

```
@Service
public class TagServiceImpl implements TagService {
  private final Repository repository;
  public TagServiceImpl(Repository repository) {
    this.repository = repository;
  }
  @Override
  public Map<String, List<Tag>> getTags(
      List<String> productIds) {
    return repository.getProductTagMappings(productIds);
  }
  @Override
  public Product addTags(
      String productId, List<TagInput> tags) {
    return repository.addTags(productId, tags);
  }
}
```

https://github.com/PacktPublishing/Modern-API-Development-with-Spring-6-and-Spring-Boot-3/tree/dev/Chapter14/src/main/java/com/packt/modern/api/services/TagServiceImpl.java

Here, the implemented method is straightforward. It passes the call to the repository, which fetches the tags based on the passed collection of product IDs.

You can add `getProductTagMappings` to the `src/main/java/com/packt/ modern/ api/repository/Repository.java` interface as shown in the next line:

```
Map<String, List<Tag>> getProductTagMappings(
    List<String> productIds);
```

Then, you can implement this method in the `src/main/java/com/packt/ modern/api/ repository/InMemRepository.java` class as shown in the next code block:

```
@Override
public Map<String, List<Tag>> getProductTagMappings(
    List<String> productIds) {
  return productEntities.entrySet().stream()
    .filter(e -> productIds.contains(e.getKey()))
    .collect(toMap(e -> e.getKey(),
      e -> e.getValue().getTags()));
}
```

Here, the code first creates the stream of the product map's entry set, then filters the products that match the product passed in this method. At the end, it converts filtered products to map the product ID with the Key and Tags values and then returns map.

Now, if you call the `product` GraphQL query, and even if products are fetched with a properly normalized database, it loads the product tags in batches without the *N+1* problem.

You are done with GraphQL query implementation and should be comfortable with implementing queries on your own.

Next, you are going to implement GraphQL mutations.

Implementing GraphQL mutations

As per the GraphQL schema, you are going to implement two mutations – `addTag` and `addQuantity`.

The `addTag` mutation takes `productId` and a collection of tags as arguments and returns the `Product` object. The `addQuantity` mutation takes `productId` and the quantity to add and returns `Product`.

Let's add this implementation to the existing `ProductDatafetcher` class as shown in the following code block:

```
// rest of the ProductDatafetcher class code
@DgsMutation(field = MUTATION.AddTag)
public Product addTags(
    @InputArgument("productId") String productId,
    @InputArgument(value = "tags", collectionType =
```

```
            TagInput.class) List<TagInput> tags) {
    return tagService.addTags(productId, tags);
  }


  @DgsMutation(field = MUTATION.AddQuantity)
  public Product addQuantity(
        @InputArgument("productId") String productId,
        @InputArgument(value = "quantity") int qty) {
    return productService.addQuantity(productId, qty);
  }
  // rest of the ProductDatafetcher class code
```

https://github.com/PacktPublishing/Modern-API-Development-with-Spring-6-and-Spring-Boot-3/tree/dev/Chapter14/src/main/java/com/packt/modern/api/datafetchers/ProductsDatafetcher.java

Here, these signatures follow the respective mutations written in the GraphQL schema. You are using another DGS framework @DgsMutation annotation, which is a type of @DgsData annotation that is marked on methods to denote them as a data fetcher method. The @DgsMutation annotation, by default, has the Mutation value set to the parentType property. You just must set the field property in this annotation. Both methods have their respective values set to the field property in the @DgsMutation annotation.

Notice that the @InputArgument annotation for tags uses another collectionType property that is used for setting the type of input. It is required when the input type is not scalar. If you don't use it, you'll get an error. Therefore, make sure to use the collectionType property whenever you have a non-scalar type input.

These methods use the tag and product services to perform the requested operations. So far, you have not added the tag service into the ProductDatafetcher class. Therefore, you need to add TagService first, as shown in the next code block:

```
  // rest of the ProductDatafetcher class code
  private final TagService tagService;

  public ProductDatafetcher(ProductService productService,
      TagService tagService) {
    this.productService = productService;
    this.tagService = tagService;
  }
  // rest of the ProductDatafetcher class code
```

Here, the TagService bean has been injected using the constructor.

Now, you need to implement the `addTag()` method in the `TagService` and `addQuantity` methods in `ProductService`. Both the interfaces and their implementations are straightforward and pass the call to the repository to perform the operations. You can find the complete source code `TagService` and `ProductService` classes in the GitHub code repository (https://github.com/PacktPublishing/Modern-API-Development-with-Spring-6-and-Spring-Boot-3/tree/dev/Chapter14/src/main/java/com/packt/modern/api/services).

Let's also add these two methods to the `Repository` interface as shown in the next code block:

```
// rest of the Repository class code
Product addTags(String productId, List<TagInput> tags);
Product addQuantity(String productId, int qty);
// rest of the Repository class code
```

These signatures in the `src/main/java/com/packt/modern/api/repository/Repository.java` interface also follow the respective mutations written in the GraphQL schema.

Let's implement the `addTags()` method first in the `src/main/java/com/packt/modern/api/repository/InMemRepository.java` class, as shown in the next code block:

```
@Override
public Product addTags(String productId, List<TagInput> tags) {
  if (Strings.isBlank(productId)) {
    throw new RuntimeException("Invalid Product ID.");
  }
  Product product = productEntities.get(productId);
  if (Objects.isNull(product)) {
    throw new RuntimeException("Product not found.");
  }
  if (tags != null && !tags.isEmpty()) {
    List<String> newTags = tags.stream().map(
        t -> t.getName()).collect(toList());
    List<String> existingTags = product.getTags().stream()
        .map(t -> t.getName())
        .collect(toList());
    newTags.stream().forEach(nt -> {
      if (!existingTags.contains(nt)) {
        product.getTags().add(Tag.newBuilder()
          .id(UUID.randomUUID().toString())
        .name(nt).build());
      }
    });
    productEntities.put(product.getId(), product);
  }
  return product;
}
```

```
https://github.com/PacktPublishing/Modern-API-Development-with-
Spring-6-and-Spring-Boot-3/tree/dev/Chapter14/src/main/java/com/
packt/modern/api/repository/InMemRepository.java
```

Here, you first perform the validation for the `productId` and `tags` arguments. If everything goes fine, then you add the tags to the product, update the concurrent map, and return the updated product.

You are done with the implementation of GraphQL mutations. You can now test your changes. You need to build the application before running the test. Let's build the application using the following command:

```
$ gradlew clean build
```

Once the build is done successfully, you can run the following command to run the application:

```
$ java -jar build/libs/chapter14-0.0.1-SNAPSHOT.jar
```

The application should be running on default port 8080 if you have not made any changes to the port settings.

Now, you can open a browser window and open *GraphiQL* using the following URL: `http://localhost:8080/graphiql` (part of the DGS framework). Change the host/port accordingly if required.

You can use the following GraphQL request to perform the `addTag` mutation:

```
mutation {
  addTag(productId: "a1s2d3f4-0",
         tags: [{name: "new Tags..."}]) {
    id
    name
    price
    description
    tags {
      id
      name
    }
  }
}
```

Here, you are adding the tags to the given `productId`; therefore, you pass `productId` and `tags` as arguments. You can use the following GraphQL request to perform the `addQuantity` mutation:

```
mutation {
  addQuantity(productId: "a1s2d3f4-0", quantity: 10) {
    id
    name
```

```
        description
        price
        count
        tags {
          id
          name
        }
      }
    }
  }
```

Here, you pass `productId` and `quantity` as arguments. You have learned how to implement GraphQL mutations in the GraphQL server. Let's implement GraphQL subscriptions in the next section.

Implementing and testing GraphQL subscriptions

Subscription is another GraphQL root type that sends the object to the subscriber (client) when a particular event occurs.

Let's assume an online shop offers a discount on products when the product's inventory reaches a certain level. You cannot track each product's quantity manually and then perform the computation and trigger the discount. To do things faster (or reduce manual intervention), this is where you can make use of a subscription.

Each change in the product's inventory (quantity) through the `addQuantity()` mutation should trigger the event and the subscriber should receive the updated product and hence the quantity. Then, the subscriber can place the logic and automate this process.

Let's write the subscription that will send the updated product object to the subscriber. You are going to use Reactive Streams and WebSocket to implement this functionality.

You need to enable CORS. Let's enable it by adding the following properties into the `application.properties` file:

```
management.endpoints.web.exposure.include=health,metrics
graphql.servlet.actuator-metrics=true
graphql.servlet.tracing-enabled=false
graphql.servlet.corsEnabled=true
```

https://github.com/PacktPublishing/Modern-API-Development-with-Spring-6-and-Spring-Boot-3/tree/dev/Chapter14/src/main/resources/application.properties

Here, you have also enabled the actuator metrics and tracing for GraphQL along with exposing the health and metrics actuator endpoints.

In build.gradle, you have graphql-dgs-subscriptions-websockets-autoconfigure to take care of the auto-configuration of WebSocket that is required for WebSocket-based GraphQL subscriptions.

You can add the following subscription data fetcher to the ProductDatafetcher class as shown in the following code:

```
// rest of the ProductDatafetcher class code @DgsSubscription(field =
SUBSCRIPTION.QuantityChanged)
public Publisher<Product> quantityChanged(
    @InputArgument("productId") String productId) {
  return productService.gerProductPublisher();
}
// rest of the ProductDatafetcher class code
```

https://github.com/PacktPublishing/Modern-API-Development-with-Spring-6-and-Spring-Boot-3/tree/dev/Chapter14/src/main/java/com/packt/modern/api/datafetchers/ProductDatafetcher.java

Here, you are using another DGS framework annotation, @DgsSubscription, which is a type of @DgsData annotation that is marked on a method to denote it as a data fetcher method. The @DgsSubscription annotation, by default, has the Subscription value set to the parentType property. You just must set the field property in this annotation. By setting field to quantityChanged, you are indicating to the DGS framework to use this method when the subscription request for quantityChanged is called.

The Subscription method returns the Publisher instance, which can send an unbound number of objects (in this case, Product instances) to multiple subscribers. Therefore, the client just needs to subscribe to the product publisher.

You need to add a new method to the ProductService interface and its implementation in the ProductServiceImpl class. The method signature in the ProductService interface and its implementation are straightforward. It passes the call to the repository to perform the operation. You can have a look at the source code in the book's GitHub code repository.

The actual work is being performed by the repository. Therefore, you need to make certain changes in the repository, as shown in the following steps:

1. First, add the following method signature to the repository interface:

    ```
    Publisher<Product> getProductPublisher();
    ```

2. Next, you have to implement the getProductPublisher() method in the InMemRepository class. This method returns the product publisher as shown in the following code:

    ```
    public Publisher<Product> getProductPublisher() {
    ```

```
        return productPublisher;
    }
```

3. Now, we need all the magic to be performed by Reactive Streams. First, let's declare the `FluxSink<Product>` and `ConnectableFlux<Product>` (which is returned by the repository) variables:

```
private FluxSink<Product> productsStream;
private ConnectableFlux<Product> productPublisher;
```

4. Now, we need to initialize these declared instances. Let's do so in the `InMemRepository` constructor, as shown in the following code:

```
Flux<Product> publisher = Flux.create(emitter -> {
    productsStream = emitter;
});
productPublisher = publisher.publish();
productPublisher.connect();
```

5. `Flux<Product>` is a product stream publisher that passes the baton to `productsStream` (FluxSink) to emit the next signals followed by `onError()` or `onComplete()` events. This means `productsStream` should emit the signal when the product quantity gets changed. When `Flux<Product>` calls the `publish()` method, it returns an instance of `connectableFlux`, which is assigned to `productPublisher` (the one that is returned by the subscription).

6. You are almost done with the setup. You just need to emit the signal (product) when the product gets changed. Let's add the following highlighted line to the `addQuantity()` method before it returns the product, as shown in the following code:

```
product.setCount(product.getCount() + qty);
productEntities.put(product.getId(), product);
productsStream.next(product);
return product;
```

You have completed the subscription `quantityChanged` implementation. You can test it next.

You need to build the application before running the test. Let's build the application using the following command:

```
$ gradlew clean build
```

Once the build is done successfully, you can run the following command to run the application:

```
$ java -jar build/libs/chapter14-0.0.1-SNAPSHOT.jar
```

The application should be running on the default port 8080 if you have not made any changes to the port settings.

Before testing the GraphQL subscription, you need to understand the GraphQL subscription protocol over WebSocket.

Understanding the WebSocket sub-protocol for GraphQL

You have implemented the GraphQL subscription over WebSocket in this chapter. In WebSocket-based subscription implementation, the network socket is the main communication channel between the GraphQL server and the client.

The graphql-dgs-subscriptions-websockets-autoconfigure dependency's current implementation (*version 6.0.5*) makes use of graphql-transport-ws sub-protocol specifications. In this sub-protocol, messages are represented using the JSON format, and over the network, these JSON messages are *stringified*. Both the server and client should conform to this message structure.

There are the following types of messages (code in Kotlin from the DGS framework):

```
object MessageType {
    const val CONNECTION_INIT = "connection_init"
    const val CONNECTION_ACK = "connection_ack"
    const val PING = "ping"
    const val PONG = "pong"
    const val SUBSCRIBE = "subscribe"
    const val NEXT = "next"
    const val ERROR = "error"
    const val COMPLETE = "complete"
}
```

You might have got the idea about the life cycle of a GraphQL subscription over WebSocket by looking at the message type. Let's understand the life cycle of a subscription in detail:

1. **Connection Initialization** (CONNECTION_INIT): The client initiates the communication by sending this type of message. The connection initialization message contains two fields – type ('connection_init') and payload. The payload field is an optional field. Its (ConnectionInitMessage) structure is represented as follows:

    ```
    {
      type: 'connection_init';
      payload: Map<String, Object>; // optional
    }
    ```

2. **Connection Acknowledgment** (`CONNECTION_ACK`): The server sends the connection acknowledgment in response to a successful connection initialization request. It means the server is ready for subscription. Its structure (`ConnectionAckMessage`) is represented as follows:

    ```
    {
      type: 'connection_ack';
      payload: Map<String, Any>; // optional
    }
    ```

3. **Subscribe** (`SUBSCRIBE`): The client now can send the `subscribe` request. If the client sends the `subscribe` request without getting a connection acknowledgment from the server, the client may get the error `4401: Unauthorized`.

    This request contains three fields – `id`, `type`, and `payload`. Here, each new subscription request should contain a unique `id`; otherwise, the server may throw `4409: Subscriber for <unique-operation-id> already exists`. The server keeps track of the `id`, until the subscription is active. The moment the subscription is complete, the client can reuse the `id`. The structure of this message type (`SubscribeMessage`) is as follows:

    ```
    {
      id: '<unique-id>';
      type: 'subscribe';
      payload: {
        operationName: ''; // optional operation name
        query: '';   // Mandatory GraphQL subscription
        query
        variables?: Map<String, Any>; // optional
        variables
        extensions?: Map<String, Any>; // optional
      };
    }
    ```

4. **Next** (`NEXT`): After a successful subscription operation, the client receives the messages of type `NEXT` from the server that contain the data related to the operation the client subscribes to. Data is part of the `payload` field. The server keeps sending these message types to the client until GraphQL subscription events occur. Once the operation gets completed, the server sends the complete message to the client. Its message type (`NextMessage`) is represented by the following:

    ```
    {
      id: '<unique-id>'; // one sent with subscribe
      type: 'next';
      payload: ExecutionResult;
    }
    ```

5. **Complete** (COMPLETE): `Complete` is a bi-directional message, which can be sent by both the server and client:

    - *Client to Server*: The client can send the complete message to the server when the client wants to stop listening to the message sent by the server. Since it's a duplex call, the client should ignore the messages that are en route when the client sends a complete request.

    - *Server to Client*: The server sends the complete message to the client when the requested operation is completed by the server. The server doesn't send the complete message when an error message is sent by the server for the subscription request by the client.

    The message type (`CompleteMessage`) is represented by the following structure:

    ```
    {
        id: '<unique-id>'; // one sent with subscribe
        type: 'complete';
    }
    ```

6. **Error** (ERROR): The server sends an error message when the server encounters any operation execution error. Its type (`ErrorMessage`) is represented by the following structure:

    ```
    {
        id: '<unique-id>';
        type: 'error';
        payload: GraphQLError[];
    }
    ```

7. `PING` and `PONG`: These are bi-directional message types and are sent by both the server and client. If the client sends a `ping` message, the server should immediately send a `pong` message and vice versa. These messages are useful for detecting networking problems and network latency. Both ping (`PingMessage`) and pong (`PongMessage`) contain the following structure:

    ```
    {
        type: String; // either 'ping' or 'pong'
        payload: Map<String, Object>; // optional
    }
    ```

Understanding the subscription life cycle will help you test the subscription thoroughly.

You can use any tool that supports GraphQL subscription testing. We'll test it using the Insomnia WebSocket request client – a bit of a crude way so you can understand the complete life cycle of the GraphQL subscription.

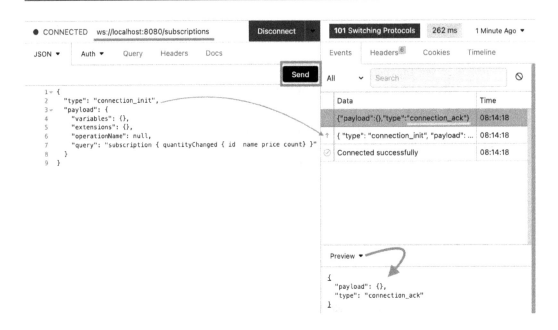

Figure 14.3 – GraphQL subscription connection_init call in the Insomnia client

Testing GraphQL subscriptions using Insomnia WebSocket

Let's perform the following steps to test the subscription manually:

1. First, add a new request using *WebSocket Request* by using the (+) drop-down menu available in the top-left corner.

2. Then add the following URL in the **URL** box:

    ```
    ws://localhost:8080/subscriptions
    ```

3. Then, add the following headers in the **Headers** tab:

    ```
    Connection: Upgrade
    Upgrade: websocket
    dnt: 1
    accept: */*
    accept-encoding: gzip, deflate, br
    host: localhost:8080
    origin: http://localhost:8080
    sec-fetch-dest: websocket
    sec-fetch-mode: websocket
    sec-fetch-site: same-origin
    Sec-WebSocket-Protocol: graphql-transport-ws
    ```

```
Sec-WebSocket-Version: 13
Sec-WebSocket-Key: 3dcYr9va5icM8VcKuCr/KA==
Sec-WebSocket-Extensions: permessage-deflate
```

Here, through the headers, you upgrade the connection to WebSocket; therefore, the server sends the `101 Switching Protocol` response. Also, you can see that you are using the `graphql-transport-ws` GraphQL sub-protocol.

4. Then, add the following payload in the **JSON** tab for connection initialization (see *Figure 14.3*):

```
{
   "type": "connection_init",
   "payload": {
      "variables": {},
   "extensions": {},
   "operationName": null,
   "query":"subscription { quantityChanged { id name price count}
}"}
}
```

5. Then, click on the **Send** button (don't click on the **Connect** button – if you do, then it needs to be followed by one more click on **Send**).

6. On a successful connection, you will receive the following acknowledgment message from the server. It means the server is ready to serve the subscription request (shown in *Figure 14.3*):

```
{
   "payload": {},
   "type": "connection_ack"
}
```

7. Then, use the following payload in the **JSON** tab:

```
{
   "id": "b",
   "type": "subscribe",
   "payload": {
      "variables": {},
   "extensions": {},
   "operationName": null,
      "operationName": null,
   "query":"subscription { quantityChanged { id name price count}
}"}
}
```

Here, you are adding a unique ID to the message. The type of message is set to `subscribe`. You can send a `subscribe` message because a connection acknowledgment is received by the client. The `query` field contains the GraphQL subscription query.

8. Then, again click on the **Send** button (don't click on the **Connect** button – if you do, then it needs to be followed by one more click on **Send**).

9. After clicking on the **Send** button, you need to fire the `addQuantity` mutation to trigger the publication of the event by using the following payload:

```
mutation {
    addQuantity(productId: "a1s2d3f4-0", quantity: 10) {
        id
        name
        price
        count
    }
}
```

10. After a successful mutation call, you can check the subscription output in the Insomnia client. You will find an incoming JSON message that will display the increased quantity, as shown in *Figure 14.4*.

11. You can repeat *steps 9 and 10* to get the (NEXT type) messages.

12. Once you are done, you can send the following JSON payload to complete the call as shown in *Figure 14.4*:

```
{
    "id": "b",
    "type": "complete"
}
```

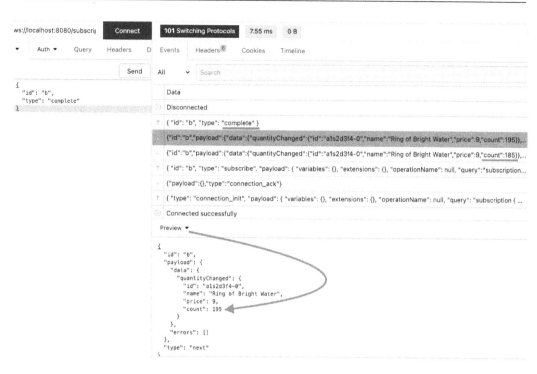

Figure 14.4 – GraphQL subscription's next and complete calls in the Insomnia client

This is the way you can implement and test the GraphQL subscription over WebSocket. You will automate the test for GraphQL subscription in the *Testing GraphQL subscriptions using automated test code* subsection in this chapter.

Next, you should know about the instrumentation that helps to implement the tracing, logging, and metrics collection. Let's discuss this in the next subsection.

Instrumenting the GraphQL APIs

The GraphQL Java library supports the instrumentation of the GraphQL APIs. This can be used to support metrics, tracing, and logging. The DGS framework also uses it. You just must mark the instrumentation class with the Spring `@Component` annotation.

The instrumentation bean can be implemented using the `graphql.execution.instrumentation.Instumentation` interface. Here, you have to write boilerplate code, which may increase the unit test automation code for you. Another way that is much easier is to extend the `SimpleInstrumentation` class, which does the simple implementation for you. However, you can override the methods for custom implementation.

Let's add instrumentation that will record the time taken by the data fetcher and complete GraphQL request processing. This metric may help you to fine-tune the performance and identify the fields that take more time to resolve.

Before adding the tracing, let's add the custom header in the response.

Adding a custom header

Let's create the DemoInstrumentation.java file in the instrumentation package and add the following code:

```
@Component
public class DemoInstrumentation
                extends SimpleInstrumentation {
  @NotNull
  @Override
  public CompletableFuture<ExecutionResult>
    instrumentExecutionResult(ExecutionResult exeResult,
            InstrumentationExecutionParameters params,
            InstrumentationState state) {
    HttpHeaders responseHeaders = new HttpHeaders();
    responseHeaders.add("myHeader", "hello");
    return super.instrumentExecutionResult(DgsExecutionResult
        .builder().executionResult(execResult)
        .headers(responseHeaders).build(),
      params,
      state
    );
  }
}
```

https://github.com/PacktPublishing/Modern-API-Development-with-Spring-6-and-Spring-Boot-3/tree/dev/Chapter14/src/main/java/com/packt/modern/api/instrumentation/DemoInstrumentation.java

Here, this class extends SimpleInstrumentation and is created as a Spring bean by marking it as @Component. The SimpleInstrumentation class allows you to instrument the execution result. Here, you can see that you have added the custom header in the response. Let's test it.

You can build and execute the project after adding the previous code and then execute the following mutation:

```
mutation {
  addQuantity(productId: "a1s2d3f4-0", quantity: 10) {
    id
```

```
        name
        price
        count
    }
}
```

You'll find the instrumented myHeader header and its value in the response headers.

Now, you can instrument the tracing information in your response by adding the following bean to your project:

```
@Configuration
public class InstrumentationConfig {
    @Bean
    @ConditionalOnProperty( prefix = "graphql.tracing",
            name = "enabled", matchIfMissing = true)
    public Instrumentation tracingInstrumentation(){
        return new TracingInstrumentation();
    }
}
```

https://github.com/PacktPublishing/Modern-API-Development-with-Spring-6-and-Spring-Boot-3/tree/dev/Chapter14/src/main/java/com/packt/modern/api/instrumentation/InstrumentationConfig.java

This configuration does the magic. You must remember that you need com.netflix.graphql.dgs:graphql-dgs-spring-boot-micrometer along with Spring Actuator dependencies in your build.gradle file to make it work.

The previous code adds the execution result metrics provided by the DGS framework to GraphQL API responses. This metric includes the tracing time and duration, validation time and duration, the resolver's information, and so on.

Once you have this instrumentation in place and execute any query or mutation, the result will include the extension fields consisting of the result metrics instrumented by the Instrumentation bean (GraphQL Tracing) created in the previous code.

Let's execute the following mutation in GraphiQL (http://localhost:8080/graphiql):

```
mutation {
    addQuantity(productId: "a1s2d3f4-0", quantity: 10) {
        id
        name
        price
        count
    }
}
```

The previous mutation will provide the following response with instrumented metrics:

```json
{
  "data": {
    "addQuantity": {
      "id": "a1s2d3f4-0",
      // output truncated for brevity
    }
  },
  "extensions": {
    "tracing": {
      "version": 1,
      "startTime": "2023-05-07T19:04:42.032422Z",
      "endTime": "2023-05-07T19:04:42.170516Z",
      "duration": 138103974,
      "parsing": {
        "startOffset": 11023640,
        "duration": 7465319
      },
      "validation": {
        "startOffset": 31688145,
        "duration": 20146090
      },
      "execution": {
        "resolvers": [
          {
            "path": [
              "addQuantity"
            ],
            "parentType": "Mutation",
            "returnType": "Product",
            "fieldName": "addQuantity",
            "startOffset": 92045595,
            "duration": 24507328
          },
          // output truncated for brevity
        ]
      }
    }
  }
}
```

Here, you can see that it not only returns `data` but also provides the instrumented metrics in the `extensions` field. Please note that you should keep this instrumentation enabled only for the development environment to fine-tune the GraphQL implementation and benchmarking, and keep it disabled for the production environment.

Let's find out more about the instrumentation metrics in the next subsection.

Integration with Micrometer

You have added `graphql-dgs-spring-boot-micrometer` as one of the dependencies in `build.gradle`. This library provides GraphQL metrics out of the box such as `gql.query`, `gql-resolver`, and so on.

You can expose the `metrics` endpoint by adding the following line in the `application.properties` file:

```
management.endpoints.web.exposure.include=health,metrics
```

You can fire the following endpoint to find out the available GraphQL metrics. Please note that it displays only triggered GraphQL metrics. For example, if there is no error while calling GraphQL APIs, it won't show `gql.error`:

```
http://localhost:8080/actuator/metrics
```

This endpoint displays the list of available metrics in your application, including GraphQL metrics.

The following four types of GraphQL metrics are provided by the DGS framework, which may help you to find out the code responsible for poor performance:

- `gql.query`: This captures the time taken by the GraphQL query or mutation.

- `gql.resolver`: This captures the time taken by each data fetcher invocation.

- `gql.error`: A single GraphQL request can have multiple errors. This metric captures the number of errors encountered during the GraphQL request execution. It will only be available when there are errors in execution.

- `gql.dataLoader`: This captures the time taken by the data loader invocation for the batch of queries.

The available GraphQL metrics from the actuator metrics endpoint output can be accessed using the following endpoint call:

```
http://localhost:8080/actuator/metrics/gql.query
```

It may provide output as shown here:

```
{
  "name": "gql.query",
  "baseUnit": "seconds",
  "measurements": [{
    "statistic": "COUNT",   "value": 4.0
  }, {
    "statistic": "TOTAL_TIME", "value": 1.403888175
  }, {
    "statistic": "MAX", "value": 0.0
  }],
  "availableTags": [{
    "tag": "gql.query.sig.hash",
    "values": ["10e750742768cb7c428699…",
               "a750f4b9bb5d40f2d23b01…"]
  }, {
    "tag": "gql.operation",
    "values": ["SUBSCRIPTION", "MUTATION"]
  }, {
    "tag": "gql.query.complexity", "values": ["10"]
  }, {
    "tag": "gql.operation.name", "values": ["anonymous"]
  }, {
    "tag": "outcome", "values": ["success", "failure"]
  }]
}
```

You can see that it provides the total elapsed time, a number of requests count, and the max time taken by the query/mutation. It also provides tags. These tags can be customized if required by implementing the following interfaces – `DgsContextualTagCustomizer` (to customize common tags such as application profile and version or deployment environment), `DgsExecutionTagCustomizer` (to customize the tags related to execution results), and `DgsFieldFetchTagCustomizer` (to customize the tags related to data fetchers).

You have learned how to instrument the GraphQL APIs in this section. Let's explore automating the testing of GraphQL code in the next section.

Test automation

The DGS framework provides classes and utilities that facilitate the automation of GraphQL API tests.

Create a new file called `ProductDatafetcherTest.java` inside the `datafetchers` package in the `test` directory and add the following code:

```
@SpringBootTest(classes = { DgsAutoConfiguration.class,
```

```
    ProductDatafetcher.class,BigDecimalScalar.class })
public class ProductDatafetcherTest {
  private final InMemRepository repo = new InMemRepository();
  private final int TEN = 10;
  @Autowired
  private DgsQueryExecutor dgsQueryExecutor;
  @MockBean
  private ProductService productService;
  @MockBean
  private TagService tagService;
  // continue …
```

https://github.com/PacktPublishing/Modern-API-Development-with-Spring-6-and-Spring-Boot-3/tree/dev/Chapter14/src/test/java/com/packt/modern/api/datafetchers/ProductDatafetcherTest.java

Here, you are using the @SpringBootTest annotation to auto-configure a Spring Boot based test. You are limiting the Spring context by providing specific classes such as DgsAutoConfiguration, ProductDatafetcher, and BigDecimalScalar. You should add only those classes here that are required to perform the test.

Then, you are auto-wiring the DgsQueryExecutor class, which provides the query execution capability to your test. After that, you add two Spring-injected mock beans for the Product and Tag services.

You are ready with the configuration and instances you need to run the tests.

Let's add the setup method that is required before running the tests. You can add the following method for this purpose in ProductDatafetcherTest.java:

```
@BeforeEach
public void beforeEach() {
 List<Tag> tags = new ArrayList<>();
 tags.add(Tag.newBuilder().id("tag1").name("Tag 1").build());
 Product product = Product.newBuilder().id("any")
       .name("mock title").description("mock description")
       .price(BigDecimal.valueOf(20.20)).count(100)
       .tags(tags).build();
 given(productService.getProduct
    ("any")).willReturn(product);
 tags.add(Tag.newBuilder().id("tag2")
       .name("addTags").build());
 product.setTags(tags);
 given(tagService.addTags("any",
```

```
      List.of(TagInput.newBuilder().name("addTags").build()))))
        .willAnswer(invocation -> product);
}
```

Here, you are using Mockito to stub the `productService.getProduct()` and `tagService.addTags()` calls.

You are done with the setup. Let's run our first test, which will fetch the JSON object after running the GraphQL `product` query next.

Testing GraphQL queries

Let's add the following code to `ProductDatafetcherTest.java` to test the `product` query:

```
@Test
@DisplayName("Verify JSON returned by the query 'product'")
public void product() {
    String name = dgsQueryExecutor.executeAndExtractJsonPath(
        "{product(id: \"any\"){ name }}",
            "data.product.name");
    assertThat(name).contains("mock title");
}
```

Here, the code is using the `DgsQueryExecutor` instance to execute the `product` query and extract the JSON property. Then, it validates the name extracted from the JSON and compares it with the value set in the `beforeEach()` method.

Next, you'll test the `product` query again, but this time, to test the exception.

You can add the following code to `ProductDatafetcherTest.java` to test the exception thrown by the `product` query:

```
@Test
@DisplayName("Verify exception to query product - invalid ID")
public void productWithException() {
    given(productService.getProduct("any"))
        .willThrow(new RuntimeException
            ("Invalid Product ID."));
    ExecutionResult res = dgsQueryExecutor.execute(
            "{ product (id: \"any\") { name }}");
    verify(productService, times(1)).getProduct("any");
    assertThat(res.getErrors()).isNotEmpty();
    assertThat(res.getErrors().get(0).getMessage()).isEqualTo(
            "java.lang.RuntimeException:
                Invalid Product ID.");
}
```

Here, the `productService` method is stubbed to throw the exception. When `DgsQueryExecutor` runs, the Spring-injected mock bean uses the stubbed method to throw the exception that is being asserted here.

Next, let's query `product` again, this time to explore `GraphQLQueryRequest`, which allows you to form the GraphQL query in a fluent way. The `GraphQLQueryRequest` construction takes two arguments – first, the instance of `GraphVQLQuery`, which can be a query/mutation or subscription, and second, the projection root type of `BaseProjectionNode`, which allows you to select the fields.

Let's add the following code to `ProductDatafetcherTest.java` to test the `product` query using `GraphQLQueryRequest`:

```
@Test
@DisplayName("Verify JSON using GraphQLQueryRequest")
void productsWithQueryApi() {
  GraphQLQueryRequest gqlRequest = new GraphQLQueryRequest(
    ProductGraphQLQuery.newRequest().id("any").build(),
    new ProductProjectionRoot().id().name());
  String name = dgsQueryExecutor.executeAndExtractJsonPath(
      gqlRequest.serialize(), "data.product.name");
  assertThat(name).contains("mock title");
}
```

Here, the `ProductGraphQLQuery` class is part of the auto-generated code by the DGS GraphQL Gradle plugin.

One thing we have not yet tested in previous tests is verifying the subfields in the `tags` field of `product`.

Let's verify it in the next test case. Add the following code in `ProductDatafetcherTest.java` to verify the tags:

```
@Test
@DisplayName("Verify Tags returned by the query 'product'")
void productsWithTags() {
  GraphQLQueryRequest gqlRequest = new GraphQLQueryRequest(
      ProductGraphQLQuery.newRequest().id("any").build(),
      new ProductProjectionRoot().id().name().tags()
      .id().name());
  Product product = dgsQueryExecutor
    .executeAndExtractJsonPathAsObject(gqlRequest.serialize(),
        "data.product", new TypeRef<>() {});
  assertThat(product.getId()).isEqualTo("any");
  assertThat(product.getName()).isEqualTo("mock title");
  assertThat(product.getTags().size()).isEqualTo(2);
  assertThat(product.getTags().get(0).getName())
```

```
        .isEqualTo("Tag 1");
    }
```

Here, you can see that you have to use a third argument (`TypeRef`) in the `executeAndExtractJsonPathAsObject()` method if you want to query the subfields. If you don't use it, you will get an error.

You are done with GraphQL query testing. Let's move on to testing the mutations in the next subsection.

Testing GraphQL mutations

Testing a GraphQL mutation is no different from testing GraphQL queries.

Let's test the `addTag` mutation to `ProductDatafetcherTest.java` as shown in the following code:

```
@Test
@DisplayName("Verify the mutation 'addTags'")
void addTagsMutation() {
  GraphQLQueryRequest gqlRequest = new GraphQLQueryRequest(
      AddTagGraphQLQuery.newRequest().productId("any")
          .tags(List.of(TagInput.newBuilder()
          .name("addTags").build())).build(),
        new AddTagProjectionRoot().name().count());
  ExecutionResult exeResult = dgsQueryExecutor.execute(
      gqlRequest.serialize());
  assertThat(exeResult.getErrors()).isEmpty();
  verify(tagService).addTags("any", List.of(
      TagInput.newBuilder().name("addTags").build()));
}
```

Here, the `AddTagGraphQLQuery` class is part of the code auto-generated by the DGS GraphQL Gradle plugin. You fire the request and then validate the results based on the existing configuration and setup.

Similarly, you can test the `addQuantity` mutation. Only the arguments and assertions will change; the core logic and classes will remain the same.

You can add the test to `ProductDatafetcherTest.java` as shown in the next code block to test the `addQuantity` mutation:

```
@Test
@DisplayName("Verify the mutation 'addQuantity'")
void addQuantityMutation() {
  given(productService.addQuantity("a1s2d3f4-1", TEN))
      .willReturn(repo.addQuantity("a1s2d3f4-1", TEN));
  GraphQLQueryRequest gqlRequest = new GraphQLQueryRequest(
```

```
        AddQuantityGraphQLQuery.newRequest()
        .productId("a1s2d3f4-1").quantity(TEN).build(),
          new AddQuantityProjectionRoot().name().count());
   ExecutionResult exeResult = dgsQueryExecutor
                              .execute(gqlRequest.serialize());
   assertThat(executionResult.getErrors()).isEmpty();
   Object obj = executionResult.getData();
   assertThat(obj).isNotNull();
   Map<String, Object> data = (Map)((Map
       )exeResult.getData()).get(MUTATION.AddQuantity);
   org.hamcrest.MatcherAssert.assertThat(
       (Integer) data.get("count"), greaterThan(TEN));
 }
```

You are done with GraphQL mutation testing. Let's move on to testing subscriptions in the next subsection.

Testing GraphQL subscriptions using automated test code

Testing a subscription needs extra effort and care, as you can see in the following code, which performs the test for the `quantityChanged` subscription. It uses the existing `addQuantity` mutation to trigger the subscription publisher that sends a `product` object on each call. You capture the product of the first call and store the value of the `count` field. Then, use it to perform the assertion as shown in the following code:

```
@Test
@DisplayName("Verify the subscription 'quantityChanged'")
void reviewSubscription() {
  given(productService.gerProductPublisher())
      .willReturn(repo.getProductPublisher());
  ExecutionResult exeResult = dgsQueryExecutor.execute(
      "subscription {quantityChanged
         {id name price count}}");
  Publisher<ExecutionResult> pub = exeResult.getData();
  List<Product> products = new CopyOnWriteArrayList<>();
  pub.subscribe(new Subscriber<>() {
    @Override
    public void onSubscribe(Subscription s) {s.request(2);}
    @Override
    public void onNext(ExecutionResult result) {
        if (result.getErrors().size() > 0) {
          System.out.println(result.getErrors());
        }
        Map<String, Object> data = result.getData();
        products.add(
```

```
                new ObjectMapper().convertValue(data.get(
                SUBSCRIPTION.QuantityChanged), Product.class));
        }
        @Override
        public void onError(Throwable t) {}
        @Override
        public void onComplete() {}
    });
    addQuantityMutation();
    Integer count = products.get(0).getCount();
    addQuantityMutation();
    assertThat(products.get(0).getId())
        .isEqualTo(products.get(1).getId());
    assertThat(products.get(1).getCount())
        .isEqualTo(count + TEN);
    }
}
```

Here, the core logic lies in the subscription that is done by calling the `publisher.subscribe()` method (check the highlighted line). You know that the GraphQL `quantityChanged` subscription returns the publisher. This publisher is received from the `data` field of the execution result.

The publisher subscribes to the stream by passing a `Subscriber` object, which is created on the fly. The subscriber's `onNext()` method is used to receive the product sent by the GraphQL server. These objects are pushed into the list. Then, you use this list to perform the assertion.

Summary

In this chapter, you learned about the different ways of implementing the GraphQL server, including federated GraphQL services. You have also explored the complete standalone GraphQL server implementation, which performs the following operations:

- Writing the GraphQL schema

- Implementing the GraphQL query APIs

- Implementing the GraphQL mutation APIs

- Implementing the GraphQL subscription APIs

- Writing the data loaders to solve the N+1 problem

- Adding custom scalar types

- Adding the GraphQL API's instrumentation

- Writing the GraphQL API's test automation using Netflix's DGS framework

You learned about GraphQL API implementation using Spring and Spring Boot skills that will help you implement GraphQL APIs for your work assignments and personal projects.

Questions

1. Why should you prefer frameworks such as Netflix's DGS in place of the `graphql-java` library to implement GraphQL APIs?

2. What are federated GraphQL services?

Answers

1. You should prefer a framework such as Netflix DGS in place of the `graphql-java` library to implement GraphQL APIs because it bootstraps the development and avoids writing boilerplate code.

   Apart from the ease of development, the framework uses `graphql-java` internally; therefore, it keeps itself in sync with the GraphQL specification's Java implementation. It also supports developing federated GraphQL services.

   It also provides plugins, the Java client, and testing utilities that help you to automate the development. The Netflix DGS framework is well tested and has been used by Netflix in production for quite some time.

2. A federated GraphQL service contains a single distributed graph exposed using a gateway. Clients call the gateway, which is an entry point to the system. A data graph will be distributed among multiple services and each service can maintain its own development and release cycle independently. Having said that, federated GraphQL services still follow the OneGraph principle. Therefore, the client would query a single endpoint for fetching any part of the graph.

Further reading

- GraphQL Java implementation: `https://www.graphql-java.com/` and `https://github.com/graphql-java/graphql-java`

- Netflix DGS documentation: `https://netflix.github.io/dgs/getting-started/`

- *Full-Stack Web Development with GraphQL and React*: `https://www.packtpub.com/product/full-stack-web-development-with-graphql-and-react-second-edition/9781801077880`

Index

Packtpub.com

Subscribe to our online digital library for full access to over 7,000 books and videos, as well as industry leading tools to help you plan your personal development and advance your career. For more information, please visit our website.

Why subscribe?

- Spend less time learning and more time coding with practical eBooks and Videos from over 4,000 industry professionals

- Improve your learning with Skill Plans built especially for you

- Get a free eBook or video every month

- Fully searchable for easy access to vital information

- Copy and paste, print, and bookmark content

Did you know that Packt offers eBook versions of every book published, with PDF and ePub files available? You can upgrade to the eBook version at packtpub.com and as a print book customer, you are entitled to a discount on the eBook copy. Get in touch with us at customercare@packtpub.com for more details.

At www.packtpub.com, you can also read a collection of free technical articles, sign up for a range of free newsletters, and receive exclusive discounts and offers on Packt books and eBooks.

Other Books You May Enjoy

If you enjoyed this book, you may be interested in these other books by Packt:

Full Stack Development with Spring Boot and React - Third Edition

Juha Hinkula

ISBN: 978-1-80181-678-6

- Make fast and RESTful web services powered by Spring Data REST.
- Create and manage databases using ORM, JPA, Hibernate, and more.
- Explore the use of unit tests and JWTs with Spring Security.
- Employ React Hooks, props, states, and more to create your frontend.
- Discover a wide array of advanced React and third-party components.
- Build high-performance applications complete with CRUD functionality.

Learning Spring Boot 3.0 - Third Edition

Greg L. Turnquist

ISBN: 978-1-80323-330-7

- Create powerful, production-grade web applications with minimal fuss.

- Support multiple environments with one artifact and add production-grade support with features.

- Find out how to tweak your Java apps through different properties.

- Enhance the security model of your apps.

- Make use of enhancing features such as native deployment and reactive programming in Spring Boot.

- Build anything from lightweight unit tests to fully running embedded web container integration tests.

Packt is searching for authors like you

If you're interested in becoming an author for Packt, please visit `authors.packtpub.com` and apply today. We have worked with thousands of developers and tech professionals, just like you, to help them share their insight with the global tech community. You can make a general application, apply for a specific hot topic that we are recruiting an author for, or submit your own idea.

Share Your Thoughts

Now you've finished *Modern API Development with Spring 6 and Spring Boot 3*, we'd love to hear your thoughts! Scan the QR code below to go straight to the Amazon review page for this book and share your feedback or leave a review on the site that you purchased it from.

https://packt.link/r/1-804-61327-4

Your review is important to us and the tech community and will help us make sure we're delivering excellent quality content.

Download a free PDF copy of this book

Thanks for purchasing this book!

Do you like to read on the go but are unable to carry your print books everywhere? Is your eBook purchase not compatible with the device of your choice?

Don't worry, now with every Packt book you get a DRM-free PDF version of that book at no cost.

Read anywhere, any place, on any device. Search, copy, and paste code from your favorite technical books directly into your application.

The perks don't stop there, you can get exclusive access to discounts, newsletters, and great free content in your inbox daily

Follow these simple steps to get the benefits:

1. Scan the QR code or visit the link below

https://packt.link/free-ebook/9781804613276

2. Submit your proof of purchase

3. That's it! We'll send your free PDF and other benefits to your email directly